# Multiple Comparisons

This book is dedicated to my family,
which did nothing, except everything
to make it possible for me to write this book.

# Multiple Comparisons
## Theory and methods

Jason C. Hsu

*Department of Statistics*
*The Ohio State University, USA*

CRC Press
Taylor & Francis Group
Boca Raton London New York

CRC Press is an imprint of the
Taylor & Francis Group, an **informa** business

A CHAPMAN & HALL BOOK

First published 1996 by Chapman & Hall/CRC
First edition 1996

First published 1999 by CRC Press
Taylor & Francis Group
6000 Broken Sound Parkway NW, Suite 300
Boca Raton, FL 33487-2742

First issued in paperback 2022

ISBN 13: 978-1-03-247802-9 (pbk)
ISBN 13: 978-0-412-98281-1 (hbk)

DOI: 10.1201/b15074

**Visit the Taylor & Francis Web site at
http://www.taylorandfrancis.com**

**and the CRC Press Web site at
http://www.crcpress.com**

## Library of Congress Cataloging-in-Publication Data

Catalog record is available from the Library of Congress

Library of Congress Card Number 95-72199

# Contents

# Preface

This is a book on multiple comparisons, the comparison of two or more treatments. According to one survey, the only statistical method applied more frequently than multiple comparison methods is the $F$-test (Mead and Pike 1975). However, whereas there are a large number of books dealing with the $F$-test in analysis of variance, there are few comprehensive treatments of multiple comparisons. The book by the late Ruppert Miller (1981) is highly accessible, but does not cover recent developments. The book by Hochberg and Tamhane (1987), while a fabulous resource for research statisticians, seems formidable for the typical practitioner. The recent book by Westfall and Young (1993) capably shows how modern computers enable one to adjust the $p$-values of tests of hypotheses for multiplicity, but it is often desirable to go beyond stating $p$-values and infer the direction and the magnitude of the differences among the treatments being compared. I thus undertook the task of writing an accessible multiple comparisons book, with an emphasis on proper application of the latest methods for confident directions inference and confidence intervals inference, empowered by modern computers.

If multiple comparison methods rank second in frequency of *use*, they perhaps rank first in frequency of *abuse*. Abuses have continued more or less unabated due to, I believe, a lack of both clear guidance toward proper application of multiple comparison methods, and a strong refutation of abuses. To remedy this situation, a classification system in this book guides the reader to the multiple comparison method appropriate for each problem. Methods for all-pairwise multiple comparisons, multiple comparisons with a control, and multiple comparisons with the best are discussed. In addition to chapters on what to do, there is also a chapter on what not to do, exposing various misconceptions associated with multiple comparisons. Applications are illustrated with real data examples. Sample size computation and graphical representations are also discussed.

The historical nature of abuses in multiple comparisons makes my task difficult. Editors of many scientific journals apparently are not yet aware of the fact that reported findings in some articles they publish do not possess the strength of evidence usually demanded for publication. That is, they have not seen through certain meaningless nomenclature to realize that the multiple comparison analyses performed by some authors may have error rates well beyond any acceptable level. From the experimenters' point of view, having more multiple comparison methods at their disposal increases their chance of finding 'significance,' a prerequisite to publication. Therefore, many statistical computer packages support invalid multiple comparison methods because of 'market demand.' As statistical computer packages with large user bases are sometimes considered 'self-validating,' there is almost a vicious circle.

Nevertheless, merely citing abuses as amusement for other statisticians is too cynical. Highlighting the abuses and providing useful solutions seem more productive.

The writing of this book follows the credo 'everything should be made as simple as possible, but not simpler,' which I have seen attributed to both Albert Einstein and Chinese fortune cookies.

In my opinion, the first and most important step in a statistical analysis is the formulation of the question or questions to be answered. Thus, the organization of the book is by type and strength of desired inference. It is suggested that the desired inference dictates the proper multiple comparison method, which then takes on different forms depending on model and distributional assumptions appropriate for the data. It is hoped that when a real-life multiple comparisons problem does not fit exactly into one of my 'pigeon-holes,' the classification scheme in the book may still lead one to an *approximate* solution to the *right* problem. Whether a ready-made solution or one conjured up following the theoretical development in the book, I believe it will be better than an *exact* solution to the *wrong* problem, the latter accounting for a significant portion of the abuses.

The chapters are arranged in the order I hope they will be read. Chapter 1 introduces the reader to simultaneous statistical inference in the ANOVA setting, deferring until later issues unrelated to complications caused by dependencies among point estimators of multiple comparison parameters. Chapter 2 is a road map to proper multiple comparisons analysis by type and strength of inference desired. Chapter 3 describes the simplest type of multiple comparisons, namely, multiple comparisons with a control. Multiple comparisons with the best, discussed in Chapter 4, are intimately related to multiple comparisons with a control. Therefore, Chapter 4 should only be read after reading Chapter 3. But Chapters 3 and 4 are not crucial to the understanding of Chapter 5 on all-pairwise comparisons.

Chapter 6 is a rogues' gallery of common abuses in multiple comparisons. Chapter 7 discusses multiple comparisons in the general linear model. Although it contains methods that have not been as time-tested as those in the previous chapters, I feel it important to provide my best suggestions at this time given the prominence of the general linear model in practice.

Appendices A and B provide some general probabilistic and statistical tools useful in multiple comparisons. Appendix C discusses the sample size computation aspect of designing multiple comparison experiments. Computer implementation of multiple comparison methods, particularly those in JMP, MINITAB and the SAS system, are illustrated throughout the chapters. Software developed by the author for multiple comparisons described in the book can be accessed via the World Wide Web (WWW), as indicated in Appendix D.

This work was supported in part by Grant no. R01 311287, awarded by the National Cancer Institute, and by National Science Foundation Grant DDM-8922721. Some computing resources were provided by the Ohio Supercomputer Center.

Many people influenced the writing of this book, which started out as a set of notes prepared for an American Statistical Association tutorial that Barbara Mann suggested I give. I learned more than statistics from my mentor Shanti S. Gupta. My joint work with Roger Berger, Jane Chang, Don Edwards, W. C. Kim, Barry Nelson, Mario Peruggia, Steve Ruberg, W. C. Soong, Gunnar Stefansson, and Dan Voss had a direct impact on the book. Special thanks go to Eve Bofinger, Charles Dunnett, Helmut Finner, Tony Hayter, George McCabe and Julie Shaffer, who reviewed various portions of this book and made excellent comments and suggestions. Many others, particularly Tim Costigan, Angela Dean, Bekka Denning, Sue Leurgans, and Hong Long Wang, also influenced the book. Dr Hans Frick and Dr Volker Rahlfs provided illuminating references on relevant European statistical guidelines. Dr Richard Hantz of Monsanto and Dr Larry Fisher of Eli Lilly made valuable suggestions improving the presentation of analyses of the bovine growth hormone toxicity studies. Mark Berliner gave me an endless supply of issues of *Science*, from which several interesting data sets were culled.

It was a joy to work with Jay Aubuchon of Minitab, Inc. and John Sall of SAS Institute, who had the courage to implement, in MINITAB and JMP, respectively, only the multiple comparison methods considered most useful in this book. My continuing dialog with Randy Tobias and Russ Wolfinger of SAS Institute has also been very fruitful.

Data in Table 2.4 is reprinted by permission of *The New England Journal of Medicine*. Data in Table 2.5 is reprinted by permission of Dr Greg Guyer, AAAS and the Monsanto Agricultural Group. Data in Table 3.2 is reprinted by permission of Dr Greg Guyer and AAAS. Data in Table 4.2 is reprinted by permission of Professor George P. McCabe and the Association for Computing Machinery, Inc. Data in Table 5.1 is reprinted by permission of the Entolomological Society of America. Data in Table 5.3 and 7.8 is reprinted by permission of Professor Harold W. Stevenson and AAAS. Data in Table 7.1 and 7.2 is reprinted by permission of the American Statistical Association. Data in Table 7.5 is reprinted by permission of Professor Stanley P. Azen and Academic Press.

Permission to reproduce BMDP output was granted by BMDP Statistical

Software, Inc. Permission to reproduce MINITAB output was granted by
Minitab, Inc., 3081 Enterprise Drive, State College, PA 16801-3008, USA.

I am indebted to Dr. Tetsuhisa Miwa for meticulous help with correcting
the typographic errors of the first printing.

Jason C. Hsu
Thanksgiving, 1995
Dublin, Ohio

# CHAPTER 1

# Introduction to simultaneous statistical inference

This chapter discusses simultaneous statistical inference, that is, inference on several parameters at once, in a setting simpler than multiple comparisons inference, which is simultaneous inference on certain functions of the *differences* of the treatment effects. This chapter serves the dual purpose of giving some useful statistical methods, and introducing the reader to issues in simultaneous inference that are unrelated to complications caused by multiple comparisons, namely, dependencies among point estimators of parameters of interest.

## 1.1 The one-way model

Let $\mu_1, \mu_2, \ldots, \mu_k$ be the mean responses under $k$ treatments. These *treatments* may be medical treatments, computer systems, or inventory policies. Suppose under the $i$th treatment, a random sample $Y_{i1}, Y_{i2}, \ldots, Y_{in_i}$ is taken, where between the treatments the random samples are independent. Then under the usual normality and equality of variance assumptions, we have the one-way model

$$Y_{ia} = \mu_i + \epsilon_{ia}, \quad i = 1, \ldots, k, \quad a = 1, \ldots, n_i, \tag{1.1}$$

where $\epsilon_{11}, \ldots, \epsilon_{kn_k}$ are independent and identically distributed (i.i.d.) normal with mean 0 and variance $\sigma^2$ unknown. We use the notation

$$\hat{\mu}_i \quad = \bar{Y}_i \quad = \sum_{a=1}^{n_i} Y_{ia}/n_i,$$

$$\hat{\sigma}^2 \quad = MSE \quad = \sum_{i=1}^{k}\sum_{a=1}^{n_i}(Y_{ia} - \bar{Y}_i)^2 / \sum_{i=1}^{k}(n_i - 1)$$

for the sample means and the pooled sample variance. It is a standard result that $\hat{\mu}_1, \ldots, \hat{\mu}_k$ are independent normally distributed random variables with means $\mu_1, \ldots, \mu_k$ and variances $\sigma^2/n_1, \ldots, \sigma^2/n_k$, and they are independent of $\hat{\sigma}$. If we let $\nu = \sum_{i=1}^{k}(n_i - 1)$, then $\nu\hat{\sigma}^2/\sigma^2$ has a $\chi^2$ distribution with $\nu$ degrees of freedom.

One type of simultaneous statistical inference is inference on the treatment means $\mu_1, \mu_2, \ldots, \mu_k$ themselves. The following simple example illustrates one situation in which such inference is of primary interest. Some additional applications, in time series analysis for example, are cited in Alexander (1993).

### 1.1.1 Example: treatment × age combination

Suppose it is desired to compare a new treatment against a standard treatment, and there is reason to suspect that which treatment is better may depend on the age of the patient. An experiment is conducted, dividing 60 patients into five age groups, with 12 patients in each group further divided into six pairs matched on various characteristics. Within each pair, the flip of a fair coin decides who gets the standard treatment and who gets the new treatment. The (hypothetical) data obtained from this experiment is given in Table 1.1, with the scores being survival time in months.

Let $Y_{ia}$ denote the difference between the score of the patient receiving the standard treatment and the score of the patient receiving the new treatment for the $a$th pair of patients in the $i$th age group. ($Y_{11} = -15, Y_{12} = 26$, etc.) Given appropriate experimental conditions, the differences $Y_{ia}$ can be assumed to be independent since they correspond to different pairs of patients. Figure 1.1 is a scatterplot of the difference $Y_{ia}$ versus the age group index $i$. The differences $Y_{ia}$ are independent since they correspond to dif-

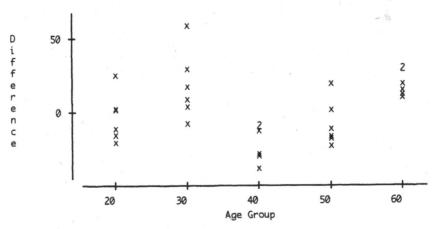

Figure 1.1 *Scatterplot of difference vs. age*

ferent pairs of patients, and one might entertain the model (1.1).

For this experiment, of primary interest are the mean differences between the standard treatment and the new treatment for the five age groups, that is, the parameters of interest are $\mu_1, \ldots, \mu_5$. In order to have confidence that all age groups receive their proper treatments, with possibly different treatments for different age groups, clearly the inferences on $\mu_1, \ldots, \mu_5$ must be

Table 1.1 *Response scores under standard and new treatments*

| Patient pair | Age index | Age group | Standard treatment | New treatment | Difference |
|---|---|---|---|---|---|
| 1 | 1 | | 57 | 72 | −15 |
| 2 | 1 | | 53 | 27 | 26 |
| 3 | 1 | 20–29 | 28 | 26 | 2 |
| 4 | 1 | | 60 | 71 | −11 |
| 5 | 1 | | 40 | 60 | −20 |
| 6 | 1 | | 48 | 45 | 3 |
| 7 | 2 | | 70 | 52 | 18 |
| 8 | 2 | | 85 | 26 | 59 |
| 9 | 2 | 30–39 | 50 | 46 | 4 |
| 10 | 2 | | 61 | 52 | 9 |
| 11 | 2 | | 83 | 53 | 30 |
| 12 | 2 | | 51 | 58 | −7 |
| 13 | 3 | | 55 | 83 | −28 |
| 14 | 3 | | 36 | 65 | −29 |
| 15 | 3 | 40–49 | 31 | 40 | −9 |
| 16 | 3 | | 28 | 66 | −38 |
| 17 | 3 | | 41 | 50 | −9 |
| 18 | 3 | | 32 | 44 | −12 |
| 19 | 4 | | 18 | 40 | −22 |
| 20 | 4 | | 39 | 55 | −16 |
| 21 | 4 | 50–59 | 53 | 70 | −17 |
| 22 | 4 | | 44 | 55 | −11 |
| 23 | 4 | | 63 | 61 | 2 |
| 24 | 4 | | 80 | 60 | 20 |
| 25 | 5 | | 76 | 60 | 16 |
| 26 | 5 | | 67 | 37 | 30 |
| 27 | 5 | 60–69 | 75 | 45 | 30 |
| 28 | 5 | | 78 | 58 | 20 |
| 29 | 5 | | 67 | 54 | 13 |
| 30 | 5 | | 80 | 69 | 11 |

correct *simultaneously* with a high probability. Though we will not go into the details of this computation, it turns out that for this example, if each two-sided confidence interval for $\mu_i$ has coverage probability 95%, then the probability that they all cover their respective parameters is only 78.3%. To guarantee that the inferences are correct simultaneously with a probability of $1 - \alpha$, one must adjust for multiplicity, ensuring each individual inference on $\mu_1, \ldots, \mu_5$ is correct with a probability somewhat higher than

$1 - \alpha$. Otherwise, as has been noted in the scientific literature (see Nowak 1994), the researcher may be tempted to dredge through subgroups in the hope of showing that at least some types of patients benefit from the new treatment. We first discuss inference on each $\mu_i$, then present methods of combining individual inferences such that the inferences are correct simultaneously with a probability of at least $1 - \alpha$.

*Remark.* There are situations in which it is not necessary for all the inferences to be simultaneously correct in order to guarantee that a correct decision is made, even if the data structure is exactly the same as in this example. See Exercise 1 of Chapter 2 for an example of such a situation.

To decide which treatment is appropriate for the $i$th age group, one might test for

$$H_{i0} : \mu_i = 0 \text{ vs. } H_{ia} : \mu_i \neq 0.$$

When $H_{i0}$ is rejected, then depending on the direction of the inequality indicated by the sample, one can conclude either $\mu_i < 0$ or $\mu_i > 0$. However, a confidence interval for $\mu_i$ is more informative. If the confidence interval for $\mu_i$ is to the left of 0, then $\mu_i < 0$, while if the confidence interval for $\mu_i$ is to the right of 0, then $\mu_i > 0$. But a confidence interval gives additional information on the magnitude of the difference, which is not conveyed by the associated $p$-value (which is the probability of observing a test statistic at least as extreme as the one observed, assuming the null hypothesis is true). For example, two data sets may give rise to the two confidence intervals $\mu_i \in 1 \pm 0.2$ months and $\mu_i \in 10 \pm 2$ months, which convey very different information about $\mu_i$, yet the same $p$-value associated with $H_{i0}$ (see Exercise 5).

When $H_{i0}$ is accepted, examining the associated $p$-value cannot differentiate between the possibility that $\mu_i$ may not be close to 0 but the sample size is too small to let the truth manifest itself, and the possibility that $\mu_i$ is indeed close to 0 (see Exercise 5). Again, a confidence interval for $\mu_i$ is more informative. The former possibility is indicated by a wide confidence interval covering 0 (e.g. $\mu_i \in 10 \pm 20$ months), in which case additional sampling is desirable, while the later possibility is indicated by a narrow confidence interval covering 0 (e.g. $\mu_i \in 0.1 \pm 0.2$ months), in which case one infers either treatment is appropriate for the $i$th age group.

With this motivating example in mind, we proceed to discuss several general methods of obtaining simultaneous confidence intervals for $\mu_1, \ldots, \mu_k$.

## 1.2 Modeling

Assuming the validity of the model (1.1), one might modify the one-sample, two-sided $t$-test to test the null hypothesis

$$H_{i0} : \mu_i = 0 \text{ vs. } H_{ia} : \mu_i \neq 0$$

so that one rejects $H_{i0}$ if

$$T_i = \frac{|\hat{\mu}_i|}{\hat{\sigma}/\sqrt{6}} > q,$$

or one might modify the two-sided $t$ confidence interval

$$\mu_i \in (\hat{\mu}_i - q\hat{\sigma}/\sqrt{6}, \hat{\mu}_i + q\hat{\sigma}/\sqrt{6}),$$

with $q$ to be determined so that the inferences are correct simultaneously with a probability of $1 - \alpha$ or higher. However, prior to executing any statistical inference, one should verify the appropriateness of the model on which inference is based. In this respect, simultaneous statistical inference is no different from any other statistical inference.

Many statistics books, particularly those on regression, describe diagnostic techniques for detecting departures from the assumed model. Some books in fact deal entirely with regression diagnostics, e.g., Atkinson (1987), Belsley, Kuh and Welsch (1980) and Cook and Weisberg (1982). Since these diagnostics techniques are most often discussed in the regression context, they are usually implemented in the general linear model (GLM) module of a statistical computer package (e.g., on the `Fit Model` platform of JMP, or under the `GLM` command of MINITAB). But they may be absent from a more specialized module for analyzing one-way data, if such a specialized module exists (e.g., the `Fit Y by X` platform of JMP, or the `ONEWAY` command of MINITAB). This does not present any real problem, as the one-way model is a special case of the general linear model.

For the one-way model, simple diagnostic techniques center around the examination of the residuals

$$e_{ia} = Y_{ia} - \bar{Y}_i, \quad i = 1, \ldots, k, \quad a = 1, \ldots, n_i,$$

principally by checking whether their behavior conforms to the expected behavior of the unobserved errors $\epsilon_{ia}$, which they estimate. However, the residuals $e_{ia}$ do not mimic the behavior of the unobserved errors $\epsilon_{ia}$ faithfully. In particular, the variances of $e_{ia}$ are not equal if the sample sizes are unequal:

$$Var(e_{ia}) = (1 - 1/n_i)\sigma^2.$$

Thus, more sophisticated diagnostic techniques may examine what Cook and Weisberg (1982) call *internally Studentized residuals* (which can be saved by the `Studentized Residuals` option on the `Fit Model` platform of JMP, or the `SRESIDS` subcommand under the `GLM` command of MINITAB)

$$e_{ia}^{in} = \frac{e_{ia}}{\hat{\sigma}\sqrt{1 - 1/n_i}},$$

which, under the assumed model, have equal variances. Note, however, the numerator $e_{ia}$ enters into the computation of the denominator $\hat{\sigma}$. This might make the detection of an outlier $Y_{ia}$ difficult. Thus, it is often recommended that diagnostic techniques be applied to what Cook and Weisberg (1982) call *externally Studentized residuals* (which Atkinson, 1987, calls *deletion residuals* and which can be saved by the TRESIDS subcommand under the GLM command of MINITAB)

$$e_{ia}^{ex} = \frac{e_{ia}}{\hat{\sigma}_{(ia)}\sqrt{1 - 1/(n_i - 1)}},$$

where $\hat{\sigma}_{(ia)}$ is the estimate of $\sigma$ computed with $Y_{ia}$ deleted from the data. This makes the numerator and denominator of $e_{ia}$ independent. Consequently, under the assumed model, each $e_{ia}^{ex}$ individually has a $t$ distribution with $N - k - 1$ degrees of freedom, where

$$N = \sum_{i=1}^{k} n_i.$$

Departure from the normality assumption might be revealed by comparing a normal probability plot of $e_{ia}^{ex}$ against a straight line. Figure 1.2 shows a normal probability plot of the externally Studentized residuals from the treatment × age combination data, with a robust regression line (obtained by the rank regression command RREG in MINITAB) superimposed. The mild non-normality indicated is due to the rather large 3.20 externally Studentized residual from observation 8, with a corresponding normal score value of 2.04. As one form of departure from normality is contamination of the normal distribution by another distribution with larger dispersion, resulting in one or more outliers, we will later examine whether observation 8 might be an outlier. A formal test for normality, such as the Shapiro–Wilk

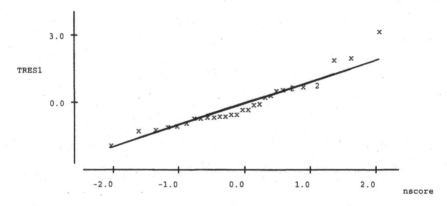

Figure 1.2 *Normal probability plot of externally Studentized residuals*

test, can be performed. For the treatment × age combination data, the

$p$-value of the Shapiro–Wilk test performed on the internally Studentized residuals $e_{ia}^{in}$ is 0.0970, indicating mild evidence of non-normality.

A scatterplot of either the response $Y_{ia}$ versus $i$ or of the externally Studentized residuals $e_{ia}^{ex}$ versus $i$ might reveal departures from the equality of variances assumption. Figure 1.1 does not indicate clearly unequal variances. A formal test for equality of variances, such as Bartlett's test (which assumes normality), can be performed. For the treatment × age combination data, the $p$-value of Bartlett's test applied to the residuals $e_{ia}$ is 0.2947, indicating insufficient evidence against unequal variances.

Singular outliers can be detected by comparing the absolute values of the externally Studentized residuals $e_{ia}^{ex}$ with $t_{\alpha^*/2N, N-k-1}$, the $1 - \alpha^*/2N$ quantile of a $t$ distribution with $N - k - 1$ degrees of freedom, where $\alpha^*$ is the maximum tolerable probability of detecting an outlier when there is none. For the treatment × age combination data, $t_{0.05/60,24} = 3.54$, so at $\alpha^* = 0.05$, there is insufficient evidence that observation 8 is an outlier.

We thus (cautiously) proceed with analyzing the treatment × age combination data assuming the model (1.1).

## 1.3 Simultaneous confidence intervals

One can ask the question: is it appropriate to present the $k$ confidence intervals for $\mu_i, i = 1, \ldots, k$, each at the $100(1 - \alpha)\%$ confidence level:

$$(\hat{\mu}_i - t_{\alpha/2,\nu}\hat{\sigma}/\sqrt{n_i}, \hat{\mu}_i + t_{\alpha/2,\nu}\hat{\sigma}/\sqrt{n_i}), \ i = 1, \ldots, k?$$

The top half of Figure 1.3 displays individual 95% confidence intervals for $\mu_1, \ldots, \mu_5$ from the treatment × age combination data (tips of the diamonds correspond to limits of the confidence intervals in the statistical package JMP for the Macintosh). The answer is 'no,' because the probability that all $k$ confidence intervals cover what they are supposed to cover is surely less than $1 - \alpha$ for $k > 1$. (As mentioned earlier, it is 78.3% for the treatment × age combination example.) If one does not adjust for 'multiplicity,' then the rate of making incorrect consequent decisions may be unacceptably high. We describe several methods of adjusting for multiplicity, to obtain simultaneous confidence intervals and to test simultaneous hypotheses.

### 1.3.1 The Studentized maximum modulus method

The Studentized maximum modulus (SMM) method, which gets its name because it is based on the Studentized maximum modulus statistic

$$\max_{1 \leq i \leq k} \frac{|\hat{\mu}_i - \mu_i|}{\hat{\sigma}/\sqrt{n_i}}, \tag{1.2}$$

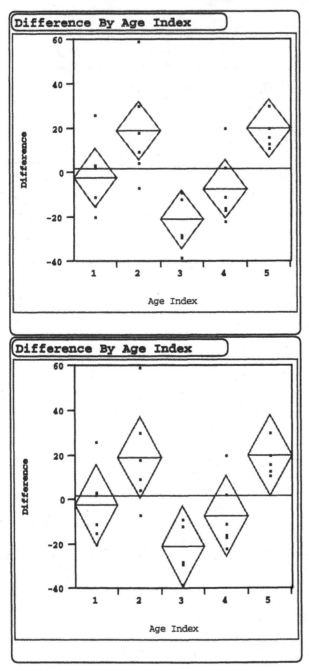

Figure 1.3 *Individual vs. simultaneous (product inequality) 95% confidence intervals*

provides exact $100(1 - \alpha)\%$ simultaneous confidence intervals for $\mu_i$ in this situation:

$$\mu_i \in \hat{\mu}_i \pm |m|_{\alpha,k,\nu}\hat{\sigma}/\sqrt{n_i} \text{ for } i = 1,\ldots,k, \tag{1.3}$$

where $|m|_{\alpha,k,\nu}$ is the $1 - \alpha$ quantile of the Studentized maximum modulus statistic (1.2), and is computed as the solution of the equation

$$\int_0^\infty [\Phi(|m|_{\alpha,k,\nu}s) - \Phi(-|m|_{\alpha,k,\nu}s)]^k \gamma_\nu(s)ds = 1 - \alpha. \tag{1.4}$$

In (1.4), $\Phi$ is the standard normal distribution function and $\gamma_\nu$ is the density of $\hat{\sigma}/\sigma$.

**Theorem 1.3.1**

$$P\{\mu_i \in \hat{\mu}_i \pm |m|_{\alpha,k,\nu}\hat{\sigma}/\sqrt{n_i} \text{ for } i = 1,\ldots,k\} = 1 - \alpha.$$

*Proof.* To prove that the coverage probability is $1 - \alpha$, note that conditional on $\hat{\sigma}/\sigma$, $\sqrt{n_i}(\hat{\mu}_i - \mu_i)/\sigma, i = 1,\ldots,k$, are i.i.d. standard normal random variables. Thus,

$$P\{\mu_i \in \hat{\mu}_i \pm |m|_{\alpha,k,\nu}\hat{\sigma}/\sqrt{n_i} \text{ for } i = 1,\ldots,k\}$$
$$= E_{\hat{\sigma}/\sigma}[P\{|\sqrt{n_i}(\hat{\mu}_i - \mu_i)/\sigma| \leq |m|_{\alpha,k,\nu}(\hat{\sigma}/\sigma) \text{ for } i = 1,\ldots,k\} \mid \hat{\sigma}/\sigma]$$
$$= \int_0^\infty [\Phi(|m|_{\alpha,k,\nu}s) - \Phi(-|m|_{\alpha,k,\nu}s)]^k \gamma(s)ds$$
$$= 1 - \alpha. \quad \square$$

The qmcc computer program described in Appendix D, with inputs the $k \times 1$ vector $\boldsymbol{\lambda} = (0,\ldots,0)$, $\nu$ and $\alpha$, computes $|m|_{\alpha,k,\nu}$. Alternatively, tables of $|m|_{\alpha,k,\nu}$ are available in Hochberg and Tamhane (1987), for example.

*1.3.2 Example: treatment × age combination (continued)*

For the treatment × age combination data,

$$\begin{aligned}
\hat{\mu}_1 &= -2.50, \\
\hat{\mu}_2 &= +18.83, \\
\hat{\mu}_3 &= -20.83, \\
\hat{\mu}_4 &= -7.33, \\
\hat{\mu}_5 &= +20.00,
\end{aligned}$$

and $\hat{\sigma} = 16.0854$. For $\alpha = 0.05$, $k = 5$ and $\nu = 25$ degrees of freedom, $t_{\alpha/2,\nu} = 2.0595$ and $|m|_{\alpha,k,\nu} = 2.766$. The 95% simultaneous confidence intervals as well as 95% individual confidence intervals for the treatment × age combination data are as displayed in Table 1.2. The wider SMM confidence intervals can be viewed as the individual confidence intervals with exact adjustment for multiplicity.

Table 1.2 *SMM and individual confidence intervals for mean difference*

| Age group | Simultaneous confidence intervals | Individual confidence intervals |
|---|---|---|
| 20–29 | −2.50 ±18.16 | −2.50 ±13.52 |
| 30–39 | 18.83 ±18.16 | 18.83 ±13.52 |
| 40–49 | −20.83 ±18.16 | −20.83 ±13.52 |
| 50–59 | −7.33 ±18.16 | −7.33 ±13.52 |
| 60–69 | 20.00 ±18.16 | 20.00 ±13.52 |

If you do not have access to computer software which computes $|m|_{\alpha,k,\nu}$ (e.g., qmcc), but have access to software which computes univariate $t$ quantiles, then the following two methods based on probabilistic inequalities are viable conservative alternatives to the exact SMM method.

### 1.3.3 A product inequality method

The random variables

$$|T_i| = \frac{|\hat{\mu}_i - \mu_i|}{\hat{\sigma}/\sqrt{n_i}}, \ i = 1, \ldots, k, \tag{1.5}$$

are almost independent, the only dependency being a result of their having the common divisor $\hat{\sigma}$. You might wonder what would happen if one pretended that they were independent, in which case the appropriate simultaneous confidence intervals for $\mu_1, \ldots, \mu_k$ would be as in (1.3), except the quantile $|m|_{\alpha,k,\nu}$ is replaced by $t_{[1-(1-\alpha)^{1/k}]/2,\nu}$, i.e., what if we infer

$$\mu_i \in \hat{\mu}_i \pm t_{[1-(1-\alpha)^{1/k}]/2,\nu}\hat{\sigma}/\sqrt{n_i} \text{ for } i = 1, \ldots, k, \tag{1.6}$$

where $t_{[1-(1-\alpha)^{1/k}]/2,\nu}$ is the $1 - [1 - (1 - \alpha)^{1/k}]/2$ quantile of the $t$ distribution with $\nu$ degrees of freedom? Qualitatively, would the simultaneous confidence intervals be conservative or liberal? Quantitatively, how much would $t_{[1-(1-\alpha)^{1/k}]/2,\nu}$ differ from $|m|_{\alpha,k,\nu}$? Since $\hat{\sigma}/\sigma$ approaches the constant 1 as $\nu$ approaches infinity, when $\nu = \infty$ (read when $\nu$ is *large*), $|T_1|, \ldots, |T_k|$ are indeed independent (read *almost* independent), in which case $t_{[1-(1-\alpha)^{1/k}]/2,\nu}$ equals (read *is close to*) $|m|_{\alpha,k,\nu}$, both equaling $z_{[1-(1-\alpha)^{1/k}]/2}$, where $z_\gamma$ denotes the $1 - \gamma$ quantile of the standard normal distribution. But how large a $\nu$ is large? We answer the first question first.

**Theorem 1.3.2** *The inference obtained by pretending* $|T_1|, \ldots, |T_k|$ *are independent is conservative, that is,*

$$P\{\mu_i \in \hat{\mu}_i \pm t_{[1-(1-\alpha)^{1/k}]/2,\nu}\hat{\sigma}/\sqrt{n_i} \text{ for } i = 1, \ldots, k\} \geq 1 - \alpha.$$

*Proof.* Let

$$E_i = \{|\hat{\mu}_i - \mu_i| < t_{[1-(1-\alpha)^{1/k}]/2,\nu}\hat{\sigma}/\sqrt{n_i}\}.$$

Then $P(E_i|\hat{\sigma}), i = 1, \ldots, k$, are all increasing in $\hat{\sigma}$. Therefore, by Corollary A.1.1 in Appendix A,

$$P\{\mu_i \in \hat{\mu}_i \pm t_{[1-(1-\alpha)^{1/k}]/2,\nu}\hat{\sigma}/\sqrt{n_i} \text{ for } i = 1, \ldots, k\}$$

$$= P(\bigcap_{i=1}^{k} E_i)$$

$$= E_{\hat{\sigma}}[\prod_{i=1}^{k} P(E_i \mid \hat{\sigma})]$$

$$\geq \prod_{i=1}^{k} E_{\hat{\sigma}}[P(E_i|\hat{\sigma})]$$

$$= \prod_{i=1}^{k} P(E_i)$$

$$= 1 - \alpha. \quad \square$$

We call this method obtained by pretending $|T_1|, \ldots, |T_k|$ are independent the *product inequality* method, since it is based on the product inequality given by Corollary A.1.1.

To answer the second question, we note that $\hat{\sigma}/\sigma$ approaches the constant 1 rather fast as the degrees of freedom $\nu$ increases. For example, with degrees of freedom $\nu = 25$, $\hat{\sigma}/\sigma$ lies in the interval $(0.764, 1.227)$ 90 percent of the time. Figure 1.4 shows the density of $\hat{\sigma}/\sigma$ for 25 degrees of freedom. (Exercise 2 is designed to give you a feeling of how fast the density of $\hat{\sigma}/\sigma$ becomes more concentrated as the degrees of freedom $\nu$ increases.) Therefore, if the $MSE$ degrees of freedom $\nu$ is at least moderate, the difference between $t_{[1-(1-\alpha)^{1/k}]/2,\nu}$ and $|m|_{\alpha,k,\nu}$ will be small and the product inequality method described above can be expected to be only slightly conservative.

Note that the usual tables for $t$ distribution quantiles are not useful for the execution of this product inequality method, since $[1 - (1 - \alpha)^{1/k}]/2$ typically will not correspond to the usual $\alpha$ of $0.10, 0.05$, or $0.01$. However, one can obtain $t_{[1-(1-\alpha)^{1/k}]/2,\nu}$ using the inverse $t$ distribution function included in most statistical packages. In fact, the Fit Y by X platform of the statistical package JMP for the Macintosh has the capability of plotting side-by-side confidence intervals for $\mu_i$ where the non-coverage probability $\alpha$ of each *individual* confidence interval can be optionally specified by the user. Therefore, a JMP user can conveniently execute the product inequality method by specifying the individual non-coverage probability $\alpha$ to equal $1 - (1 - \alpha)^{1/k}$.

Later on, it will be seen that a product inequality method is often available in multiple comparisons as well. That is, in multiple comparisons, it is often the case that conservative simultaneous inference is obtained if one

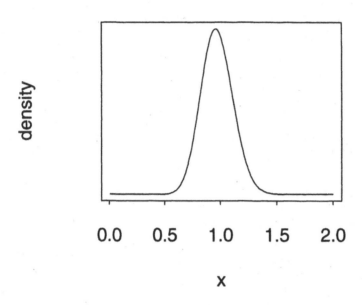

Figure 1.4 *Density of $x = \hat{\sigma}/\sigma$ with degrees of freedom $\nu = 25$*

pretends the individual inferences were independent. However, in contrast to the situation in this section, product inequality-based multiple comparison methods typically are quite inaccurate relative to exact multiple comparison methods. That is because the statistics involved in multiple comparisons typically have two sources of dependency, one due to relationship among the parameters of interest (e.g., $\mu_1 - \mu_3$ and $\mu_2 - \mu_3$ have $\mu_3$ in common), the other due to Studentization by a common $\hat{\sigma}$. Discussion in this section demonstrates that the second source of dependence, an issue orthogonal to multiple comparisons, is not that significant unless the degrees of freedom $\nu$ is small. However, as will be seen, the effect of the first source of dependency is typically quite significant in multiple comparisons.

### 1.3.4 Example: treatment × age combination (continued)

For the treatment × age combination data, with $\alpha = 0.05$ and $\nu = 25$, $t_{[1-(1-\alpha)^{1/k}]/2,\nu} = 2.779$, which is only slightly bigger than $|m|_{\alpha,k,\nu} = 2.766$.

The bottom half of Figure 1.3 displays the product inequality simultaneous 95% confidence intervals for the treatment × age combination data (which are substantially longer than the *individual* 95% confidence intervals for the same data displayed in the top half of Figure 1.3).

### 1.3.5 The Bonferroni inequality method

The familiar Bonferroni inequality states, for any events $E_1, \ldots, E_p$,

$$P\left(\bigcup_{m=1}^{p} E_m^c\right) \leq \sum_{m=1}^{p} P(E_m^c).$$

Applying the Bonferroni inequality to

$$E_i = \{\mu_i \in \hat{\mu}_i \pm q\hat{\sigma}/\sqrt{n_i}\},$$

it is easy to see that

$$\mu_i \in \hat{\mu}_i \pm t_{\alpha/2k,\nu}\hat{\sigma}/\sqrt{n_i} \text{ for } i = 1, \ldots, k \tag{1.7}$$

forms a set of conservative $100(1 - \alpha)\%$ simultaneous confidence intervals for $\mu_1, \ldots, \mu_k$.

One might wonder whether the Bonferroni inequality confidence intervals (1.7) are more or less conservative than the product inequality confidence intervals (1.6). The answer is the Bonferroni inequality confidence intervals are always more conservative, because

$$(1 - \alpha)^{1/k} < 1 - \alpha/k$$

for all $\alpha > 0$ and $k > 1$.

One can then ask the question 'How much more?' The answer is 'Not much,' because the difference between $(1-\alpha)^{1/k}$ and $1-\alpha/k$ is small for the $\alpha$ typically considered. Figure 1.5 compares $t_{[1-(1-\alpha)^{1/k}]/2,\nu}$ with $t_{\alpha/2k,\nu}$ for $k = 5$ and $\nu = 25$ degrees of freedom. This closeness between the product inequality method and the Bonferroni inequality method is a mathematical fact, unrelated to the correlation structure of the statistics involved. Thus, the two inequality methods are almost equally good in the setting of this chapter, and almost equally bad in multiple comparisons.

### 1.3.6 Example: treatment × age combination (continued)

For the treatment × age combination data, with $\alpha = 0.05$ and $\nu = 25$, the critical value for the Bonferroni method is $t_{\alpha/2k,\nu} = 2.787$, which is only slightly bigger than the critical value $t_{[1-(1-\alpha)^{1/k}]/2,\nu} = 2.779$ for the product inequality method.

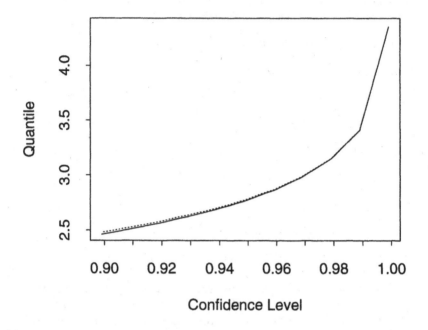

Figure 1.5 *Product inequality quantile and Bonferroni inequality quantile with* $k = 5$ *and degrees of freedom* $\nu = 25$

### 1.3.7 Scheffé's method

Since

$$\frac{\hat{\mu}_i - \mu_i}{\sigma/\sqrt{n_i}}, \quad i = 1, \ldots, k,$$

are i.i.d. standard normal random variables, the statistic

$$\frac{\sum_{i=1}^{k} \left(\sqrt{n_i}\hat{\mu}_i - \sqrt{n_i}\mu_i\right)^2}{\sigma^2}$$

has a $\chi^2$ distribution with $k$ degrees of freedom. It is independent of $\nu\hat{\sigma}^2/\sigma^2$, which has a $\chi^2$ distribution with $\nu$ degrees of freedom. Therefore, the statistic

$$\frac{\sum_{i=1}^{k} \left(\sqrt{n_i}\hat{\mu}_i - \sqrt{n_i}\mu_i\right)^2 / k}{\hat{\sigma}^2}$$

has an $F$ distribution with $k$ numerator and $\nu$ denominator degrees of freedom. Thus,

$$P\left\{\sum_{i=1}^{k}\left(\sqrt{n_i}\hat{\mu}_i - \sqrt{n_i}\mu_i\right)^2 \leq k\hat{\sigma}^2 F_{\alpha,k,\nu}\right\} = 1 - \alpha,$$

where $F_{\alpha,k,\nu}$ is the $1-\alpha$ quantile of an $F$ distribution with $k$ and $\nu$ degrees of freedom. This probability statement can be pivoted to give the following simultaneous confidence intervals for all linear combinations of $\mu_1, \ldots, \mu_k$:

$$\sum_{i=1}^{k} l_i \mu_i \in \sum_{i=1}^{k} l_i \hat{\mu}_i \pm \sqrt{kF_{\alpha,k,\nu}}\hat{\sigma}(\sum_{i=1}^{k} l_i^2/n_i)^{1/2} \text{ for all } l = (l_1, \ldots, l_k) \in \Re^k$$

which we shall attribute to Scheffé, who pioneered the pivoting of $F$ statistics to give simultaneous confidence intervals.

**Theorem 1.3.3**

$$P\left\{\sum_{i=1}^{k} l_i \mu_i \in \sum_{i=1}^{k} l_i \hat{\mu}_i \pm \sqrt{kF_{\alpha,k,\nu}}\hat{\sigma}(\sum_{i=1}^{k} l_i^2/n_i)^{1/2}\right.$$

$$\left. \text{for all } l = (l_1, \ldots, l_k) \in \Re^k \right\} = 1 - \alpha.$$

*Proof.* Letting $z_i = \sqrt{n_i}\mu_i$, we see that $z = (z_1, \ldots, z_k)$ satisfying

$$\sum_{i=1}^{k}\left(z_i - \sqrt{n_i}\hat{\mu}_i\right)^2 \leq kF_{\alpha,k,\nu}\hat{\sigma}^2$$

constitutes the interior of a $k$-dimensional sphere centered at the point $\left(\sqrt{n_1}\hat{\mu}_1, \ldots, \sqrt{n_k}\hat{\mu}_k\right)$ with radius $\sqrt{kF_{\alpha,k,\nu}}\hat{\sigma}$. Therefore, by applying Lemma B.1.1 in Appendix B to

$$a = \left(\frac{l_1}{\sqrt{n_1}}, \ldots, \frac{l_k}{\sqrt{n_k}}\right),$$

we obtain

$$E = \left\{\sum_{i=1}^{k}\left(\sqrt{n_i}\hat{\mu}_i - \sqrt{n_i}\mu_i\right)^2 \leq k\hat{\sigma}^2 F_{\alpha,k,\nu}\right\}$$

$$= \left\{\sum_{i=1}^{k} \frac{l_i}{\sqrt{n_i}}\sqrt{n_i}\mu_i \in \sum_{i=1}^{k} \frac{l_i}{\sqrt{n_i}}\sqrt{n_i}\hat{\mu}_i \pm \sqrt{kF_{\alpha,k,\nu}}\hat{\sigma}(\sum_{i=1}^{k} l_i^2/n_i)^{1/2}\right.$$

$$\left. \text{for all } l = (l_1, \ldots, l_k) \in \Re^k \right\}$$

$$= \left\{\sum_{i=1}^{k} l_i \mu_i \in \sum_{i=1}^{k} l_i \hat{\mu}_i \pm \sqrt{kF_{\alpha,k,\nu}}\hat{\sigma}(\sum_{i=1}^{k} l_i^2/n_i)^{1/2}\right.$$

$$\left. \text{for all } l = (l_1, \ldots, l_k) \in \Re^k \right\}.$$

The result follows from the fact that $P(E) = 1 - \alpha$.    □

One can obviously deduce simultaneous confidence intervals on $\mu_1, \ldots, \mu_k$ from the Scheffé confidence set by applying Theorem 1.3.3 to $l = 1_i$, the $k$-dimensional unit vector with 1 in the $i$th coordinate and 0 in all other coordinates, for $i = 1, \ldots, k$, resulting in a rectangular confidence set for $\mu_1, \ldots, \mu_k$. Since this rectangle circumscribes the ellipsoidal Scheffé confidence set, its coverage probability is higher than $1 - \alpha$.

**Corollary 1.3.1**

$$P\{\mu_i \in \hat{\mu}_i \pm \sqrt{kF_{\alpha,k,\nu}}\,\hat{\sigma}/\sqrt{n_i},\ i = 1, \ldots, k\} > 1 - \alpha.$$

*1.3.8 Example: treatment × age combination (continued)*

For the treatment × age combination data, with $\alpha = 0.05$, $k = 5$ and $MSE$ degrees of freedom $\nu = 25$, $\sqrt{kF_{\alpha,k,\nu}} = 3.608$. Table 1.3 displays 95% simultaneous confidence intervals based on Scheffé's method for the treatment × age combination data.

Table 1.3 *Scheffé's confidence intervals for mean difference*

| Age group | Simultaneous confidence intervals |
|:---------:|:---------------------------------:|
| 20–29 | $-2.50 \pm 23.69$ |
| 30–39 | $18.83 \pm 23.69$ |
| 40–49 | $-20.83 \pm 23.69$ |
| 50–59 | $-7.33 \pm 23.69$ |
| 60–69 | $20.00 \pm 23.69$ |

Note how wide the intervals based on Scheffé's method are, as compared to the intervals based on the Studentized maximum modulus method. One way to explain this, for a spherical normal distribution in $k$ dimensions, is that the probability content of the sphere inscribed inside the cube (see Figure 1.6) with 95% probability content *decreases* rapidly as $k$ increases. (Exercise 4 asks you to verify this.)

**1.4 Simultaneous testing**

Given any set of $100(1-\alpha)\%$ simultaneous confidence intervals for $\mu_1, \ldots, \mu_k$, one can obtain a level-$\alpha$ test for the hypotheses

$$H_{i0} : \mu_i = \mu_{i0} \text{ vs. } H_{ia} : \mu_i \neq \mu_{i0},\ i = 1, \ldots, k,$$

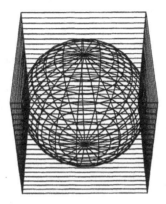

Figure 1.6 *Sphere inscribed inside a cube with 95% probability content*

by rejecting $H_{i0}$ if and only if the confidence interval for $\mu_i$ does not contain $\mu_{i0}$. For example, the Studentized maximum modulus method would lead one to reject $H_{i0} : \mu_i = \mu_{i0}$ if and only if

$$|T_i| = \frac{|\hat{\mu}_i - \mu_{i0}|}{\hat{\sigma}/\sqrt{n_i}} > |m|_{\alpha, k, \nu}.$$

Given a set of simultaneous confidence intervals for $\mu_i, i = 1, \ldots, k$, there is little reason to present the inference in the form of the less informative tests of simultaneous hypotheses. The acceptance of the null hypothesis $H_0$ does not imply $\mu_i$ is close to $\mu_{i0}$, for the acceptance may be due to the noise in the data being large relative to the sample size of $n_i$, resulting in a large $\hat{\sigma}/\sqrt{n_i}$. These two cases can be distinguished by examining whether the confidence interval for $\mu_i$ is a narrow or a wide interval bracketing $\mu_{i0}$. Likewise, the rejection of the null hypothesis $H_{i0}$ with $\hat{\mu}_i > 0$ (say) does not imply $\mu_i$ is much larger than $\mu_{i0}$, for the rejection may be due to the sample size $n_i$ being large relative to the noise in the data, resulting in a small $\hat{\sigma}/\sqrt{n_i}$. These two cases can be distinguished by examining whether the lower end point confidence interval for $\mu_i$ is much larger than $\mu_{i0}$. (See Exercise 5.)

However, if one is only interested in directional inference, either $\mu_i > \mu_{i0}$ or $\mu_i < \mu_{i0}$, but not the magnitude of the difference $\mu_i - \mu_{i0}$, then the stepdown testing procedure, due to Shaffer (1980) and Holm (1979b), gives sharper inference than deduction from simultaneous confidence intervals.

The idea of a stepdown method is to decide, at step 1, whether there is sufficient evidence to infer that the treatment that appears to be most significantly different from its hypothesized value is indeed different. If the answer is 'no,' then proceed no further. If the answer is 'yes,' then at step 2, one decides whether the treatment that appears to be the second most significantly different from its hypothesized value is indeed different, and so on. For simplicity, we describe the Shaffer–Holm method for the balanced one-way model, with $n_1 = \cdots = n_k = n$.

### 1.4.1 The Shaffer–Holm stepwise testing method

Let

$$T_i = \frac{\hat{\mu}_i - \mu_{i0}}{\hat{\sigma}/\sqrt{n}}, \ i = 1, \ldots, k,$$

and let $[1], [2], \ldots, [k]$ be the random indices such that

$$|T_{[1]}| \leq \cdots \leq |T_{[k]}|.$$

(Since the $T_i$'s are continuous random variables, ties occur with probability zero.) In other words, $[i]$ is the *anti-rank* of $|T_i|$ among $|T_1|, \ldots, |T_k|$. For example, suppose $k = 3$ and

$$|T_2| < |T_3| < |T_1|,$$

then

$$[1] = 2, \ [2] = 3, \ [3] = 1.$$

In the following, $|m|_{\alpha,h,\nu}$ denotes the solution to (1.4) with $k = h$. Note that $|m|_{\alpha,1,\nu} < |m|_{\alpha,2,\nu} < \cdots < |m|_{\alpha,k,\nu}$ and $|m|_{\alpha,1,\nu} = t_{\alpha/2,\nu}$, the $1 - \alpha/2$ quantile of the univariate $t$ distribution with $\nu$ degrees of freedom. Consider the following stepwise method.

$$\boxed{\text{Step 1}}$$

If  $|T_{[k]}| \leq |m|_{\alpha,k,\nu}$
    then           stop

    else           reject $H_{[k]0}$ and infer $\mu_{[k]} \gtrless \mu_{[k]0}$ as $T_{[k]} \gtrless \mu_{i0}$;
                      go to step 2.

$$\boxed{\text{Step 2}}$$

If   $|T_{[k-1]}| \le |m|_{\alpha,k-1,\nu}$
   then                                    stop
   else                                    reject $H_{[k-1]0}$ and infer $\mu_{[k-1]} \overset{>}{<} \mu_{i0}$ as $T_{[k-1]} \overset{>}{<} \mu_{[k-1]0}$;
                                           go to step 3.

$$\vdots$$

$$\boxed{\text{Step } k}$$

If   $|T_{[1]}| \le t_{\alpha/2,\nu}$
   then                            stop
   else                            reject $H_{[1]0}$ and infer $\mu_{[1]} \overset{>}{<} \mu_{[1]0}$ as $T_{[1]} \overset{>}{<} \mu_{i0}$.

It is relatively easy to prove that this stepdown method controls the supremum of the probability of rejecting any true hypothesis. Let $I$ denote the set of indices of hypotheses that are true, that is, $I = \{i : \mu_i = \mu_{i0}\}$, and let $|I|$ denote the number of elements in $I$.

**Theorem 1.4.1**

$$\sup_\mu P_\mu\{\text{Reject } H_{i0} \text{ for some } i, i \in I\} \le \alpha. \tag{1.8}$$

*Proof.* If $I = \emptyset$, then there is no true hypothesis to be rejected incorrectly, and the theorem trivially holds. Thus assume $I \ne \emptyset$. Consider the event

$$E = \{|T_i| \le |m|_{\alpha,|I|,\nu} \text{ for all } i \in I\}.$$

Suppose $M$ is the integer such that

$$\max_{i \in I} |T_i| = |T_{[M]}|,$$

then $M \ge |I|$. Therefore, when $E$ occurs,

$$|T_{[M]}| \le |m|_{\alpha,|I|,\nu} \le |m|_{\alpha,M,\nu},$$

and none of the hypotheses $H_{i0}, i \in I$ is rejected. The result follows from the fact that $P(E) = 1 - \alpha$.   $\square$

Note, however, the guarantee (1.8) is often insufficient in practice, because a false null hypothesis can be rejected for the wrong reason. Suppose $\mu_i > 0$ and the false null hypothesis $H_{i0} : \mu_i = 0$ is rejected, but the inference is $\mu_i < 0$; then an error, called a *Type III* error, has been made. The consequence of such an error is potentially serious. In the treatment $\times$ age combination example, a Type III error causes the inferior treatment to be given. If we consider a *directional* error has been made if either we reject $H_{i0}$ and infer $\mu_i < 0$ when $\mu_i \ge 0$ or we reject $H_{i0}$ and infer $\mu_i > 0$ when $\mu_i \le 0$ for some $i, i = 1, \ldots, k$, that is, if either a Type I or a Type III error is made, then it is important to ascertain whether the stepdown method controls the directional error rate. Shaffer (1980) and Holm (1979b) proved that it does. (Technically, Shaffer's proof requires independence among the test statistics and applies here only when $\nu = \infty$. On the other hand, her proof applies to a variety of distributions.)

**Theorem 1.4.2 (Shaffer 1980; Holm 1979b)** *For the stepwise method,*

$$sup_\mu P_\mu\{directional\ error\} \leq \alpha.$$

*Proof.* While a formal proof is beyond the scope of this book, the idea of Holm's proof is that the directional error rate is equal to $\alpha$ at

$$\mu_1 = \cdots = \mu_m = 0, |\mu_{m+1}| = \cdots = |\mu_k| = \infty,$$

less than $\alpha$ at

$$\mu_1 = \cdots = \mu_{m-1} = 0, \mu_m = 0+, |\mu_{m+1}| = \cdots = |\mu_k| = \infty,$$

monotone as $\mu_m$ increases from 0+ to $\infty$, and one can proceed by induction on $m$. $\square$

There remains the question whether a confidence set for $\mu_1, \ldots, \mu_k$ exists which corresponds to this stepwise procedure, in terms of inference available on $\mu_1, \ldots, \mu_k$ being non-zero. The answer is 'yes' when $k = 2$ and $\nu = \infty$, for which Hayter and Hsu (1994) constructed such a confidence set using Lemma B.3.1. But even for this simple special case, the confidence set is rather complicated. So a confidence set for the general case, if it exists, is perhaps even more complicated.

## 1.5 Unequal variances

Suppose the model (1.1) holds but that the variances under the $k$ treatments are unequal:

$$Y_{ia} = \mu_i + \epsilon_{ia}, \quad i = 1, \ldots, k, \quad a = 1, \ldots, n_i, \tag{1.9}$$

where the $\epsilon_{ia}$'s are independent and, for each $i$, $\epsilon_{ia}, a = 1, \ldots, n_i$, are normally distributed with mean 0 and variance $\sigma_i^2$. Since the variances are unequal, we estimate $\sigma_i^2$ individually by

$$\hat{\sigma}_i^2 = \sum_{a=1}^{n_i}(Y_{ia} - \hat{\mu}_i)^2/(n_i - 1).$$

Individual confidence intervals for $\mu_i$ based on $(\hat{\mu}_i, \hat{\sigma}_i)$ are now independent, so adjusting for multiplicity is trivial in this case. By the product rule governing the joint probability of independent events,

$$\mu_i \in \hat{\mu}_i \pm t_{0.5-(1-\alpha)^{1/k}/2, n_i-1}\hat{\sigma}_i/\sqrt{n_i} \text{ for } i = 1, \ldots, k \tag{1.10}$$

is a $100(1 - \alpha)\%$ confidence statement for $\mu_1, \ldots, \mu_k$.

The confidence intervals (1.10) are of course applicable when the variances are equal too, so it might appear safer always to give these confidence intervals instead of the confidence intervals in (1.3). However, by not pooling the $\sigma^2$ estimates as in the construction of the confidence intervals in (1.3), a disadvantage of the confidence intervals (1.10) is that the error degrees of freedom decreases from $\sum(n_i - 1)$ to $n_1 - 1, \ldots, n_k - 1$, resulting in wider confidence intervals in general, particularly when $n_1 - 1, \ldots, n_k - 1$ are small.

## 1.6 Nonparametric methods

Suppose the model (1.9) holds but that the error distribution, though symmetric about zero, may not be normal:

$$Y_{ia} = \mu_i + \epsilon_{ia}, \quad i = 1, \ldots, k, \quad a = 1, \ldots, n_i, \qquad (1.11)$$

where the $\epsilon_{ia}$'s are independent and, for each $i$, $\epsilon_{ia}, a = 1, \ldots, n_i$, have distribution $H_i$, where $H_i$ is absolutely continuous with $H_i(x) = 1 - H_i(-x)$ (so that $\mu_i$ remains the mean and the median of the response under the $i$th treatment). Individual level $100(1-\alpha)\%$ nonparametric confidence intervals for $\mu_1, \ldots, \mu_k$ (based on the sign statistic or the signed rank statistic, for example) can be readily obtained (cf. Chapter 3 of Hollander and Wolfe 1973). Since these individual confidence intervals are statistically independent of each other, if one sets the confidence level of each individual confidence interval at $(1 - \alpha)^{1/k}$, then jointly they form $100(1 - \alpha)\%$ simultaneous confidence intervals for $\mu_1, \ldots, \mu_k$.

## 1.7 Deduced inference versus direct inference

Scheffé's method presupposes confidence intervals for many linear combinations of $\mu_1, \ldots, \mu_k$ are of interest, while the other methods presented in this chapter presuppose inferences on $\mu_1, \ldots, \mu_k$ are of primary interest. But clearly one can deduce confidence bounds on all linear combinations of $\mu_1, \ldots, \mu_k$ from a set of confidence intervals for $\mu_1, \ldots, \mu_k$. For example,

$$
\begin{aligned}
1 &< \mu_1 &< 2 \\
3 &< \mu_2 &< 4
\end{aligned}
$$

implies

$$
\begin{aligned}
1 &< \mu_2 - \mu_1 &< 3 \\
2 &< (\mu_2 + \mu_1)/2 &< 3.
\end{aligned}
$$

That is, any confidence set for $\mu_1, \ldots, \mu_k$ is actually a confidence set for all linear combinations

$$\sum_{i=1}^{k} l_i \mu_i, \quad -\infty < l_1, \ldots, l_k < +\infty$$

of $\mu_1, \ldots, \mu_k$. For a general method of deducing simultaneous confidence intervals on linear combinations of $\mu_1, \ldots, \mu_k$ from the simultaneous confidence intervals on $\mu_1, \ldots, \mu_k$ given by the Studentized maximum modulus method, see Richmond (1982).

One may wonder, if in addition to $\mu_1, \ldots, \mu_k$ some other linear combinations are of interest, how good are such deduced confidence intervals compared to Scheffé's. For instance, in the treatment × age combination experiment, besides the means differences $\mu_1, \ldots, \mu_5$ for each age group, one might be interested in the *average* difference

$$\bar{\mu} = \frac{\mu_1 + \cdots + \mu_k}{k}$$

between the two treatments across the five age groups.

At the 95% confidence level, Scheffé's method gives the confidence interval

$$\bar{\mu} \in 1.634 \pm 10.603,$$

while the confidence interval deduced from the Studentized maximum modulus method is

$$\bar{\mu} \in 1.634 \pm 18.16,$$

which is much wider.

This illustrates the fact that, while it does not pay to presuppose all possible parameters are of primary interest, secondary inference deduced from the primary inference for which a method is intended typically is weak. There is no free lunch, so to speak. One should carefully decide which parameters are of primary interest, and then employ a method providing the sharpest inference on these parameters.

## 1.8 Exercises

1. *Testing main effects in a two-way model.* Consider a randomized complete block model

$$Y_{ab} = \mu + \alpha_a + \beta_b + \epsilon_{ab}, \quad a = 1, \ldots A, \quad b = 1, \ldots, B,$$

where $\epsilon_{11}, \ldots, \epsilon_{AB}$ are i.i.d. $N(0, \sigma^2)$. The usual size-$\alpha$ $F$-test for

$$H_\alpha : \alpha_1 = \cdots = \alpha_A$$

rejects if

$$\frac{MSA}{MSE} > F_{\alpha, A-1, (A-1)(B-1)},$$

while the usual size-$\alpha$ $F$-test for

$$H_\beta : \beta_1 = \cdots = \beta_B$$

rejects if

$$\frac{MSB}{MSE} > F_{\alpha, B-1, (A-1)(B-1)}.$$

Prove that, if both $H_\alpha$ and $H_\beta$ are true and the two tests are performed simultaneously, then the probability of at least one Type I error is at most $1 - (1 - \alpha)^2$.

2. Plot the density of $\hat\sigma/\sigma$ for increasing degrees of freedom $\nu$.

3. For the treatment × age combination data, execute the product inequality based method using your favorite statistical package.

4. Suppose $Z_1, \ldots, Z_k$ are i.i.d. $N(0, 1)$.

   (a) Find an expression for the value $z$ such that the probability content of the cube $\{|Z_i| \le z$ for $i = 1, \ldots, k\}$ equals $1 - \alpha$.

   (b) Find an expression for the probability content of the sphere $\sum_{i=1}^{k} Z_i^2 \le z^2$ inscribed inside the cube.

   (c) For $\alpha = 0.05$, compute and plot the probability in (b), as a function of $k$, for a sufficiently large number of $k$.

5. *Non-informativeness of p-values on effect size.* A panel of $2n$ expert cats are invited to taste two brands of cat food, using a two-period crossover design, say. Let $\delta$ denotes the mean difference between the two brands, measured on the standard Yummy scale for cat-foodivores. Let $D$ and $\hat\sigma_D$ be respectively estimates of $\delta$ and the standard deviation of $D$ such that, under the usual model,

$$\frac{D - \delta}{\hat\sigma_D}$$

has a $t$ distribution with $2n - 2$ degrees of freedom. Assume a difference of 10 on the Yummy scale is considered practically significant.

   (a) Construct two data sets, in terms of $n$ and the statistics $D, \hat\sigma_D$, having the same large $p$-values for testing $H_0 : \delta = 0$ versus $H_a : \delta \ne 0$, but where only one of the two data sets indicates $\delta$ is close to zero, that is, the two brands are not practically significantly different.

   (b) Construct two data sets, again in terms of $n$ and the statistics $D$, $\hat\sigma_D$, having the same small $p$-values for testing $H_0 : \delta = 0$ versus $H_a : \delta \ne 0$, but where only one of the two data sets indicates $\delta$ is far from zero, that is, the two brands are practically significantly different.

# CHAPTER 2

# Classification of multiple comparison methods

In contrast to inference on the treatment means themselves, as described in Chapter 1, the subject of multiple comparisons is concerned with the *comparison* of the treatment means $\mu_1, \mu_2, \ldots, \mu_k$. Thus, the parameters of interest in multiple comparison methods are functions of *contrasts* of $\mu_1, \mu_2, \ldots, \mu_k$. A contrast of the $\mu_i$'s is a linear combination $\sum_{i=1}^{k} c_i \mu_i$ with $\sum_{i=1}^{k} c_i = 0$. In this chapter multiple comparison methods will be classified primarily according to the type of inference each is designed to give, and secondarily according to the strength of the inference of which each is capable. We first describe various types of multiple comparisons.

## 2.1 Types of multiple comparisons inference

The following types of multiple comparisons will be discussed in detail in this book:

1. **ACC** All-Contrast Comparisons

2. **MCA** All-Pairwise Comparisons

3. **MCB** Multiple Comparisons with the Best

4. **MCC** Multiple Comparisons with the Control

We can obtain a better understanding of these different types of multiple comparisons by considering the parameters of primary interest associated with each comparison type. The parameters of primary interest are the parameters upon which we obtain direct inference.

1. For all-contrast comparisons (ACC), all contrasts

$$\sum_{i=1}^{k} c_i \mu_i$$

with

$$\sum_{i=1}^{k} c_i = 0$$

are assumed to be of primary interest.

2. For all-pairwise comparisons (MCA), the parameters of primary interest are

$$\mu_i - \mu_j, \text{ for all } i < j,$$

the $k(k-1)/2$ pairwise differences of treatment means.

3. Multiple comparisons with the best (MCB) compares each treatment with the best of the other treatments. The parameters of interest for MCB are the differences between each treatment and the best among the other treatments. More precisely, suppose a larger treatment effect implies a better treatment; then the parameters of primary interest are

$$\mu_i - \max_{j \neq i} \mu_j, \text{ for } i = 1, \ldots, k,$$

for if

$$\mu_i - \max_{j \neq i} \mu_j < 0,$$

then treatment $i$ is not the best, as there is another treatment better than it, while if

$$\mu_i - \max_{j \neq i} \mu_j > 0,$$

then treatment $i$ is the best treatment, for it is better than every other treatment. On the other hand, if a small treatment effect implies a better treatment, then the parameters of primary interest in MCB are

$$\mu_i - \min_{j \neq i} \mu_j, \text{ for } i = 1, \ldots, k.$$

4. For multiple comparisons with a control (MCC), suppose treatment $k$ is the control. Then the parameters of primary interest are

$$\mu_i - \mu_k, \text{ for } i = 1, \ldots, k - 1,$$

the difference between the mean of each new treatment and the mean of the control.

These four types of multiple comparisons have been developed to the point where correct computer implementations of the most useful methods to analyze data from the one-way model (balanced or unbalanced) are readily accessible, and correct computer implementations to analyze data from the general linear model (GLM) can reasonably be anticipated in the near future.

There are, of course, situations where the set of parameters of primary interest do not correspond to any of those already discussed. For example, the parameters of primary interest may be

$$\mu_i - \bar{\mu} = \mu_i - \frac{\sum_{j=1}^{k} \mu_j}{k}, \text{ for } i = 1, \ldots, k,$$

the difference between the mean of each treatment and the mean of all treatment means. This type of multiple comparisons can be called multiple comparisons with the mean (MCM). See Exercise 2. These other types of multiple comparisons are not discussed extensively in this book. In general, if the parameters of interest are not sufficiently close to those that have been studied in the literature, then some approximation technique may be necessary to effect the probabilistic computations.

### 2.1.1 Direct inference versus deduced inference

One might wonder why there is a need for multiple comparisons with the best or multiple comparisons with the a control, since one can deduce MCB and MCC from all-pairwise comparisons. For example, if all-pairwise comparisons declares treatment 1 to be better than treatment 2, then regardless of how treatment 2 compares with the other treatments, it is not the best, which is a MCB-type inference. Further, $k - 1$ of the $k(k - 1)/2$ all-pairwise comparisons are precisely the treatments versus control comparisons of MCC.

This line of questioning can in fact be carried further: Why not deduce all pairwise comparisons from all-contrast comparisons, since pairwise differences are special cases of contrasts? As a matter of fact, one might question why we should not deduce multiple comparisons from inference on the means themselves. For example, one can deduce the confidence interval $\mu_1 - \mu_2 \in (-3, +1)$ from the confidence intervals $\mu_1 \in (1, 3)$ and $\mu_2 \in (2, 4)$.

There are two reasons for considering specific types of multiple comparisons. The first reason is: direct inference is sharper than deduced inference. All else being equal, it is better to employ a method designed for the desired inference. For example, as illustrated in Section 1.7, inference on the means $\mu_i$ deduced from Scheffé's method, designed for inference on all linear combinations of the means, is much weaker than inference given by the Studentized maximum modulus method, designed specifically for inference on the means.

The second reason is that, as Tukey (1992) states, the human mind often finds it difficult to comprehend the result of a large number of simultaneous comparisons. The number of all-pairwise comparisons, $k(k-1)/2$, increases rapidly as the number of treatments $k$ increases. For example, with $k = 10$ the number of all-pairwise comparisons is already 45. The increased difficulty in comprehension is much less severe as $k$ increases with multiple comparisons with the best or multiple comparisons with a control, with $k$ and $k - 1$ parameters of primary interest, respectively. Tukey (1992) in fact suggests multiple comparisons with the mean, which he calls MAXAD (maximum absolute deviation) and attributes to Halperin *et al.* (1955), as one way of maintaining comprehension as $k$ increases.

## 2.2 Strength of multiple comparisons inference

The purpose of a typical comparative experiment is to help the experimenter reach a decision concerning the treatments being compared. In employing a statistical method, the experimenter's natural question is: 'How often will a mistake be made if the decision is based on the inference given by this statistical method?' A multiple comparison method makes one or more assertions if the data so warrants; it makes no assertion at all if the data is inconclusive. If the user chooses the multiple comparison method providing the inference most relevant to the decision of interest, then in most situations any incorrect assertion given by the multiple comparison method may lead to an incorrect decision. (For an example of a situation in which some incorrect assertions may not lead to an incorrect decision, see Exercise 1.) We thus define the *error rate* of a multiple comparison method as follows.

**Definition 2.2.1** *The* error rate *of a multiple comparison method is the supremum of the probability of making at least one incorrect assertion:*

$$\text{error rate} = \sup_{\mu} P_{\mu}\{ \text{ at least one incorrect assertion } \}.$$

This error rate is what Tukey calls the *experimentwise* error rate (Braun and Tukey 1983). Controlling such an error rate is termed *controlling the familywise error rate in the strong sense* by Hochberg and Tamhane (1987). Note, however, some authors use the term *experimentwise error rate* (e.g., Steel and Torrie 1980, p. 182; Hollander and Wolfe 1973, p. 448) to mean the probability that a method makes an incorrect assertion when $\mu_1 = \cdots = \mu_k$. It agrees with Definition 2.2.1 when the supremum of the probability of making an incorrect assertion occurs at $\mu_1 = \cdots = \mu_k$. But, as will be seen, there are popular multiple comparison methods for which this supremum does not occur at $\mu_1 = \cdots = \mu_k$. There are also popular multiple comparison methods for which the location of the supremum is not known. Given that $\mu_1 = \cdots = \mu_k$ is unlikely to be true, controlling the error rate only at this configuration, termed *controlling the familywise error rate in the weak sense* by Hochberg and Tamhane (1987), offers little protection against making an incorrect decision. Given a multiple comparison method which controls the so-called experimentwise error rate, the user needs to check whether it controls the probability of an incorrect assertion over all parameter configurations. For this reason, we avoid the term 'experimentwise error rate' in this book.

Multiple comparisons methods can be classified, in decreasing order of strength of inference, as follows. In our discussion, the symbol $\theta$ denotes a generic parameter of primary interest and, in the following, $p$ denotes the generic number of parameters of primary interest. For example, in all-pairwise comparisons, $\theta$ would be $\mu_i - \mu_j$ and $p$ would be $k(k-1)/2$. In multiple comparisons with the best, $\theta$ would be $\mu_i - \max_{j\neq i} \mu_j$ and $p$ would be $k$. In multiple comparisons with a control, $\theta$ would be $\mu_i - \mu_k$ and $p$ would be $k - 1$.

### 2.2.1 Confidence intervals methods

A *confidence intervals* multiple comparisons method with confidence level $1 - \alpha$ asserts

$$\theta_i \in I_i \text{ for } i = 1, \ldots, p,$$

while guaranteeing that the simultaneous coverage probability of the intervals is at least $100(1 - \alpha)\%$. (The intervals $I_i$ may be open or closed, and be of either finite or infinite length.)

For example, a confidence intervals multiple comparisons with the best method may infer

$$
\begin{aligned}
0 &\leq \mu_1 - \max_{j\neq 1} \mu_j &\leq 0.1 \\
-0.2 &\leq \mu_2 - \max_{j\neq 2} \mu_j &\leq 0 \\
-0.1 &\leq \mu_3 - \max_{j\neq 3} \mu_j &\leq 0
\end{aligned}
$$

at the 95% confidence level.

### 2.2.2 Confident directions methods

A *confident directions* multiple comparisons method with confidence level $1 - \alpha$ asserts, for each $i$, the inequality

$$\theta_i > 0$$

or

$$\theta_i < 0$$

if data so warrants (nothing is asserted about $\theta_i$ if data is inconclusive), while guaranteeing that the probability that all assertions it does make are correct is at least $1 - \alpha$.

For example, a confident directions multiple comparisons with a control method may assert

$$\mu_1 - \mu_3 > 0$$

and

$$\mu_2 - \mu_3 < 0,$$

at the 95% confidence level.

Confidence intervals inference implies confident directions inference, because if the confidence interval for $\mu_i - \mu_j$ is entirely to the right (left) of zero, then the inference $\mu_i > (<) \mu_j$ is implied.

### 2.2.3 Confident inequalities methods

A *confident inequalities* multiple comparisons method with confidence level $1 - \alpha$ asserts, for each $i$, the inequality

$$\theta_i \neq 0$$

if data so warrants (nothing is asserted about $\theta_i$ if data is inconclusive), while guaranteeing that the probability that all the assertions it does make are correct is at least $1 - \alpha$.

For example, a confident inequalities all-pairwise comparisons method may infer

$$\mu_1 \neq \mu_2$$

and

$$\mu_3 \neq \mu_4.$$

Confident inequalities methods are often formulated as tests for

$$H_{i,j} : \mu_i = \mu_j.$$

The rejection of $H_{i,j}$ corresponds to the assertion $\mu_i \neq \mu_j$, while the acceptance of $H_{i,j}$, a non-assertion, can equivalently be interpreted as the assertion $\mu_i - \mu_j \in (-\infty, \infty)$.

Confident directions inference implies confident inequalities inference, because if $\mu_i > \mu_j$ then $\mu_i \neq \mu_j$.

### 2.2.4 Tests of homogeneity

A level-$\alpha$ *test of homogeneity* asserts

$$\theta_i \neq 0 \text{ for some } i$$

if data so warrants, which corresponds to the *rejection* of the null hypothesis of homogeneity

$$H_0 : \mu_1 = \cdots = \mu_k; \tag{2.1}$$

it asserts nothing otherwise, which corresponds to the *acceptance* of the null hypothesis of homogeneity. A level-$\alpha$ test of homogeneity guarantees that the probability an incorrect assertion, rejecting $H_0$ when it is true, is no more than $\alpha$.

For example, an $F$-test may reject the homogeneity null hypothesis $H_0$ at the 5% level, without indicating which pair(s) of treatments are different, or what contrasts are non-zero.

Confident inequalities inference implies test of homogeneity inference, because the assertion

$$\mu_2 \neq \mu_3,$$

for example, implies

$$\mu_i \neq \mu_j \text{ for some } i \neq j.$$

## 2.2.5 *Individual comparison methods*

An *individual comparison* method with confidence level $1 - \alpha$ does not guarantee the *simultaneous* correctness of its assertions. Instead, it guarantees that *each* assertion it makes is correct with a probability of at least $1 - \alpha$.

Some statisticians do not believe multiplicity needs to be adjusted for, questioning why inference on the difference between two treatments should be affected by data on other treatments. But reporting the per comparison error rate alone is often insufficient. For example, with a per comparison of $\alpha$, the probability that all the assertions are correct simultaneously can be as high as $1 - \alpha$, when the assertions correspond to one and the same event, or as low as $(1 - \alpha)^p$, when the assertions are statistically independent, or even lower, when the random variables associated with the assertions are negatively correlated. For instance, suppose 10 comparisons are made at the per comparison error rate of 5%. Then the probability that all 10 comparisons are correct simultaneously can be close to 95%, when the 10 comparisons are highly positively correlated, or close to 60%, when the 10 comparisons are nearly independent. If decisions are made based on joint assertions, it would be impossible to accurately predict the rate of making an incorrect decision from per comparison error rates alone.

Suppose assertions $a_1, \ldots, a_p$ are to be made simultaneously. (If the formulation is in terms of hypotheses testing, then we interpret the acceptance of a null hypothesis as an assertion that the parameter can be any value in the parameter space, which is always correct.) Let $B_i$ equal 0 if assertion $a_i$ is correct, equal 1 if it is incorrect. Ideally, one would perhaps like to report the joint distribution of $B_1, \ldots, B_p$. From this distribution, the user can deduce the probability that all assertions are correct, the probability that all but one of the assertions are correct, the probability that a particular set of assertions are correct, etc. But not only can the computations of these probabilities be difficult (due to possible dependence on unknown parameters), the presentation of these probabilities may be challenging as well (due to high dimensionality). So often it is only feasible to report simple summaries of this distribution. (See Exercise 1 for an exception.) For example, the marginal probabilities of $B_i = 1$ are the per comparison error rates. The error rate as defined by Definition 2.2.1 ( = 1 − probability that all assertions are correct) is a particularly useful summary of this distribution, because it is an upper bound on the probability of making an incorrect decision based on the joint assertions.

The specific situations with which this book is concerned do call for multiplicity adjustments. For instance, suppose ten observations are taken from each of ten treatments under the one-way model (1.1). Suppose each pair of treatments are compared using a 5% two-sided $t$-test (standardized by the pooled estimate of $\sigma^2$), and any treatment thus judged to be inferior to some other treatment is asserted to not be the best (assuming a larger treatment effect is better, say). Then, even though the per comparison error rate is only 5%, the probability that the true best treatment is eliminated by mistake can exceed 13%. Multiple comparison with the best, discussed in Chapter 4, considers an error to have been made if the true best treatment is eliminated, and properly adjusts for multiplicity such that the probability of this error is no more than 5%.

Table 2.1 shows how our definition of error rate specializes to the more traditional definitions of error rate, as a function of the strength of inference. For confidence intervals methods, the error rate we have defined becomes the supremum of the *non-coverage probability* of the simultaneous confidence intervals. For confidence directions methods, the error rate we have defined can be called the *maximum directional* error rate. It is the supremum of the probability of making either a Type I error (asserting $\theta_i < 0$ or $\theta_i > 0$ when $\theta_i = 0$) or a Type III error (asserting $\theta_i > 0$ when $\theta_i < 0$ or asserting $\theta_i < 0$ when $\theta_i > 0$). For confident inequalities methods, the error rate we have defined can be called the *maximum Type I* error rate. For tests of homogeneity, the error rate as we have defined it becomes the familiar Type I error rate. For individual comparisons methods, the error rate as we have defined it becomes the so-called *per comparison Type I* error rate.

Table 2.1 *Error rates corresponding to different strengths of inference*

| Inference | Name for error rate |
|---|---|
| $\theta \in I$ | $\sup_\mu P\mu\{$ non-coverage $\}$ |
| $\theta > 0$ or $\theta < 0$ | maximum directional (Type I + Type III) |
| $\theta \neq 0$ | maximum Type I |
| not $\mu_1 = \cdots = \mu_k$ | Type I |
| individual comparisons | per comparison |

In anticipation of detailed discussions in subsequent chapters of this book, Table 2.2 classifies multiple comparison methods according to the type of inference each gives directly, and the strength of inference of which each is capable with a guaranteed confidence level of $1 - \alpha$. (Again, in this table, $\theta$ denotes a generic parameter of primary interest.)

Since the only all-contrast comparisons method discussed in this book is Scheffé's simultaneous confidence intervals, ACC type inference is not listed in Table 2.2.

Table 2.2  *Classification of multiple comparison methods*

|  | MCA $\mu_i - \mu_j$ | MCB $\mu_i - \max_{j \neq i} \mu_j$ | MCC $\mu_i - \mu_k$ |
|---|---|---|---|
| Confidence intervals $\theta \in I$ | Tukey Bofinger Hayter Steel/Dwass | Hsu Edwards–Hsu | Dunnett Bofinger/ Stefansson– Kim–Hsu Steel |
| Confident directions $\theta > 0$ or $\theta < 0$ | | | Naik/ Marcus– Peritz– Gabriel Dunnett– Tamhane (one-sided) |
| Confident inequalities $\theta \neq 0$ | Ryan Einot–Gabriel Welch Peritz Finner/Royen | | Dunnett– Tamhane (two-sided) |
| Test of homogeneity not $\mu_1 = \cdots = \mu_k$ | Newman–Keuls Protected LSD Dunn | | |
| Individual comparisons | Duncan Unprotected LSD | | |

There are no methods called Šidák's, Bonferroni's, or Hunter–Worsley's in Table 2.2, because Šidák's inequality, Bonferroni's inequality, and the Hunter–Worsley inequality are probabilistic inequalities useful in obtaining easy to compute multiple comparison methods of various types, as illustrated throughout the book. They are not multiple comparison methods *per se*.

We say a level $1 - \alpha$ confidence set corresponds to a level $1 - \alpha$ confident directions method if, for every data set, the confident directions method asserts

$$\theta > 0 \text{ or } \theta < 0$$

if and only if the confidence interval for $\theta$ given by the confidence set method is to the right or left of 0 respectively, for every parameter $\theta$ of primary interest. (It is understood that we are not looking for confidence intervals with end points that can only take on the values $-\infty, 0$, or $\infty$; giving the confidence interval

$$\theta \in (0, \infty) \text{ or } \theta \in (-\infty, 0)$$

whenever the level $1 - \alpha$ confident directions method asserts

$$\theta > 0 \text{ or } \theta < 0$$

does not count.) For a long time it was thought that no stepwise method has an exact corresponding confidence sets. However, Bofinger (1987) and Stefansson, Kim and Hsu (1988) showed (see Chapter 3) that the one-sided MCC method of Naik (1975) and Marcus, Peritz and Gabriel (1976), which was originally proposed as a confident directions method, has a corresponding confidence set. The one-sided stepup MCC method of Dunnett and Tamhane (1992), which is believed to be a confident directions method, should have an associated confidence set. The stepwise MCA methods of Ryan (1960), Einot and Gabriel (1975) and Welsch (1977) and the closed methods of Peritz (1970), Finner (1987) and Royen (1989) have not been mathematically proven to be confident directions methods, but they may be, and may have corresponding confidence sets. The same comment applies to the two-sided stepup MCC method of Dunnett and Tamhane (1992). Methods such as Newman–Keuls multiple range tests (Newman 1939; Keuls 1952), Duncan's (1955) multiple range tests and Fisher's (1935) protected and unprotected least significant differences (LSD), cannot have valid corresponding confidence sets since they are known to *not* be confident inequalities methods. (Recall stronger inference implies weaker inference.)

## 2.3 Inferential tasks of multiple comparison methods

To understand what inferential tasks multiple comparison methods must perform, let us consider the following two examples.

### 2.3.1 Example: effect of smoking on pulmonary health

In a retrospective study on how smoking affects pulmonary health, White and Froeb (1980) studied subjects who had been evaluated during a 'physical fitness profile.' Among the subjects, 2208 were disqualified because of their history of disease. The remaining 3002 subjects were then assigned, based on their smoking habits, to one of six groups as shown in Table 2.3.

The investigators randomly selected 200 female subjects from each group, except for the group of non-inhaling smokers, and recorded their pulmonary functions. Due to the small number of non-inhaling (cigar and pipe) smokers in the study, the sample from that group was limited to 50 female

Table 2.3 *Six groups of smokers*

| Group label | Definition |
|---|---|
| NS | non-smokers |
| PS | passive smokers |
| NI | non-inhaling (cigar and pipe) smokers |
| LS | light smokers |
|    | (1–10 cigarettes per day for at least the last 20 years) |
| MS | moderate smokers |
|    | (11–39 cigarettes per day for at least the last 20 years) |
| HS | heavy smokers |
|    | (≥ 40 cigarettes per day for at least the last 20 years) |

subjects. Summary statistics for female forced vital capacity (FVC) are given in Table 2.4.

Table 2.4 *FVC data for smoking and non-smoking female subjects*

| Group label | Group number | Sample size | Mean FVC | Std. dev. FVC |
|---|---|---|---|---|
| NS | 1 | 200 | 3.35 | 0.63 |
| PS | 2 | 200 | 3.23 | 0.46 |
| NI | 3 | 50  | 3.19 | 0.52 |
| LS | 4 | 200 | 3.15 | 0.39 |
| MS | 5 | 200 | 2.80 | 0.38 |
| HS | 6 | 200 | 2.55 | 0.38 |

To determine *how much* smoking affects an individual's pulmonary health relative to non-smokers, one can examine how far away from 0 the confidence intervals for the differences in mean FVC between each group of smokers and the group of non-smokers are. On the other hand, to make a policy decision on whether to ban smoking in public places, it may only be necessary to assess the existence of a *significant directional difference* between passive smokers and non-smokers, in the direction that passive smokers have lower average FVC than non-smokers.

### 2.3.2 Example: Bovine growth hormone toxicity studies

The use of bovine growth hormone is a controversial issue (Consumer Union 1992; Horowitz and Thompson 1993). Writing for the Federal Food and Drug Administration (FDA), Juskevich and Guyer (1990) reported on a number of experiments conducted which did not indicate bovine growth hormones would be harmful if present in milk consumed by humans. A subset of the data from one experiment included in that article gave weight changes in rbGH (recombinant bovine growth hormone) treated rats. The treatments included a negative control diet or placebo treatment (labeled level 1), a positive control treatment which was rbGH by injection (labeled level 2), and four different doses of bovine growth hormones given orally (labeled levels 3–6). Changes in weights of male rats after 85 days are given in Table 2.5.

Table 2.5 *Body weight changes (in grams) of male rats*

| Level | Method | Dosage (mg/kg per day) | Sample size | Mean weight change | Std. dev. weight change |
|-------|--------|------------------------|-------------|--------------------|-------------------------|
| 1 | oral | 0 | 30 | 324 | 39.2 |
| 2 | injection | 1.0 | 30 | 432 | 60.3 |
| 3 | oral | 0.1 | 30 | 327 | 39.1 |
| 4 | oral | 0.5 | 30 | 318 | 53.0 |
| 5 | oral | 5 | 30 | 325 | 46.3 |
| 6 | oral | 50 | 30 | 328 | 43.0 |

For this experiment, when comparing the positive control and the negative control, the desired inference is *significant directional difference*, in the direction that the positive control accelerates weight gain. The positive control was included in the experiment to verify that the measurement process was capable of detecting known differences. Thus a declaration of significant directional difference suffices; a confidence interval for the difference is not needed.

On the other hand, when comparing the four levels of orally fed bovine growth hormone with the negative control, the desired inference is *practical equivalence*: weight gains in rats given any growth hormone are close to weight gains of rats given the placebo. Such a declaration can be made if the confidence intervals for the weight gain differences between rats given growth hormone and rats given the negative control turn out to be tight around 0.

Thus, as illustrated by the two examples above, the desired inference in multiple comparisons may be either *significant directional difference* or *practical equivalence*. Within each of these two inferences, one may simply desire to make a 0-1 (yes-no) decision about the difference between the two treatments, or else construct a confidence interval which indicates, in addition, the magnitude of the difference. Table 2.6 summarizes the possible inferences. In Table 2.6, $\theta$ again denotes a generic parameter of primary interest.

Table 2.6 *Possible multiple comparison inferences*

|                     | Directional difference | Practical equivalence |
|---------------------|:----------------------:|:---------------------:|
| Confidence interval | $\theta \in I$ where $0 \notin I$ | $\theta \in I$ where $0 \in I$ (narrow) |
| 0-1 decision        | $\theta > 0$ or $\theta < 0$ | $-\delta_1 < \theta < \delta_2$ with $\delta_1, \delta_2 > 0$ prespecified |

Displaying $p$-values associated with tests of equality facilitates 0-1 *significant difference* inference, eliminating the need for the user to prespecify an error rate $\alpha$. However, such $p$-values do not enable one to make 0-1 *practical equivalence* inference, or confidence interval inference of either the *significant directional difference* type or the *practical equivalence* type. This is because a small $p$-value associated with a test of equality may indicate either a truly large difference between the treatments, or simply a very large sample size relative to noise. Likewise, a large $p$-value associated with a test of equality may indicate either a truly small difference between the treatments, or simply too small a sample size relative to noise. It is interesting to note that, recognizing the inadequacy of $p$-values, the *British Medical Journal* has had an editorial policy which requires the presentation of confidence intervals whenever possible (Langman 1986; Gardner and Altman 1986), with 'confidence intervals' being an item on its statistical check list (Gardner, Machin, and Campbell 1986).

## 2.4 Choosing a multiple comparison method

The proper choice of a multiple comparison method should depend primarily on the type of inference desired, and secondarily on the strength of inference that must be achieved.

At a given strength of inference, the smaller the number of parameters of primary interest, the smaller the critical value associated with the multiple comparison method. For all-contrast comparison methods, the number of parameters of primary interest is infinite, so these methods require the largest critical values. All-pairwise comparisons methods require the next largest critical values, as the number of parameters of primary interest in MCA is $k(k-1)/2$, compared to $k$ and $k-1$ for multiple comparisons with the best and multiple comparisons with a control, which have the smallest critical values. This can be seen, for example, in Figure 2.1, which compares the critical values of the four types of confidence intervals methods for

$k = 10$, $\alpha = 0.05$, and $\nu = 40$. Thus, a logical first step is to choose the type of multiple comparison with the smallest number of parameters of primary interest such that inference on these parameters is sufficient for the purpose of the analysis.

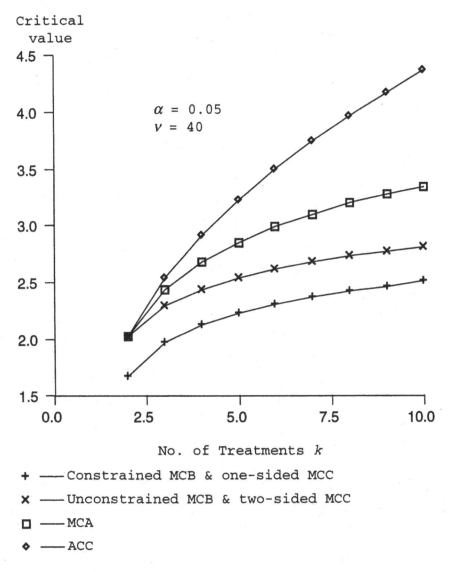

Figure 2.1 *Critical values of four types of confidence intervals method*

Within each type of inference, the critical value(s) employed by a method giving weaker assertions will be smaller than the critical value(s) employed by a method giving stronger assertions. Figure 2.2 compares the critical values employed by various MCA methods, as a function of the number of treatments being compared. Note that, in this figure, the *total* number of treatments being compared is fixed at $k = 10$. Whereas single-step methods such as Tukey's method and the unprotected LSD use a single critical value for all the comparisons, stepwise methods such as the Ryan/Einot–Gabriel/Welsch method, the Newman–Keuls multiple range test, and Duncan's multiple range test, use different critical values for comparing subsets of treatments of different size. Thus, having chosen the appropriate *type* of multiple comparison, a logical second step is to choose the *strength* of multiple comparisons inference to be the lowest while still sufficient for the purpose of the analysis.

Confident inequalities methods and tests of homogeneity generally do not give useful inference, because it is often known that the treatment effects are not exactly the same. (In a classic paper, Beecher, 1955, showed that even placebos have powerful effects.) As Tukey (1991) states:

> Statisticians classically asked the wrong question – and were willing to answer with a lie, one that was often a downright lie. They asked 'Are the effects of A and B different?' and they were willing to answer 'no.'
> ... asking 'Are the effects different?' is foolish.
> What we should be answering first is 'Can we tell the direction in which the effects of A differ from the effects of B?'

Informed decisions can hardly be made based on assertions of mere inequalities. Instead, as illustrated in the previous section, they must be based on assertions of which treatments are better, which ones are worse, and, possibly, by how much, as provided by confidence intervals methods and confident directions methods. Therefore, I recommend confidence intervals and confident directions methods only, which this book emphasizes. I further agree with the conclusion in the section titled 'Why confidence intervals?' in Tukey (1991) that confidence intervals are needed and irreplaceable. Therefore, I recommend confidence intervals associated with confident directions methods be given whenever such intervals have been derived. It is interesting to note that, in addition to the *British Medical Journal* editorial guideline, the *Good Clinical Practice Guideline* of the European Communities (EC-GCP 1993b, p. 30) does not consider tests of hypotheses adequate, requiring the presentation of confidence intervals instead.

Perhaps, due to the eagerness of many experimenters to find 'significance' in order to publish, a common but unfortunate practice is to infer beyond the strength of which a particular method is capable. As will be documented in Chapter 6, this practice often takes the form of executing a test of homogeneity or individual comparison method as a confident directions method instead. As will also be documented in Chapter 6, even though the method employed is typically an MCA method, often only MCB inference or MCC inference is reported. This practice leads to decision making at

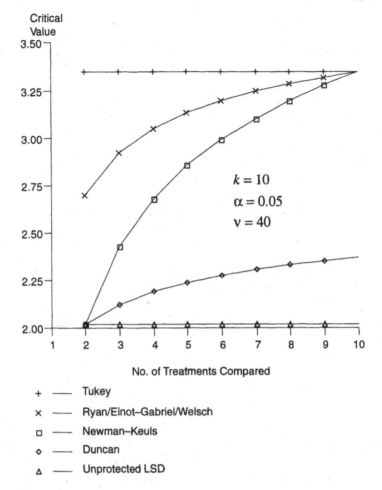

Figure 2.2 *Critical values employed by MCA methods of different strength*

an unknown error rate. A better way to achieve more 'significance,' so to speak, in situations where MCB inference or MCC inference is desired, is to use a confidence intervals MCB or MCC method, so that the error rate of consequent decision making is guaranteed to be no more than $\alpha$.

## 2.5 Exercises

1. *Example: microsurgery sutures.* It is sometimes possible surgically to reattach severed body parts. For a reattached part to survive, the surgeon must reestablish blood flow by rejoining tiny blood vessels that are fractions of a millimeter in diameter, working under a microscope with

sutures thinner than human hair. Two kinds of suture are available: the standard suture and the dissolving suture. Since swelling of the blood vessel wall after surgery reduces blood flow, it was of interest to compare the amounts of swelling associated with the two kinds of suture. An experiment took six pairs of measurements on the amount of swelling at the end of each week, where within each pair one measurement was the amount of swelling when the standard suture was used, and the other measurement was the amount of swelling when the dissolving suture was used. This experiment was carried out over a five-week period, with the measurements taken from different subjects over time. Thus, the data has exactly the same structure as the data in Table 1.1. For $i = 1, \ldots, 5$, let $\mu_i$ represent the expected difference in the amount of swelling after the $i$th week between blood vessels reattached with the standard suture and those reattached with the dissolving suture, so that the dissolving suture is better than the standard suture for the $i$th week if $\mu_i > 0$.

(a) Suppose the model (1.9) is appropriate and inference is in terms of the simultaneous confidence intervals (1.10) for $\mu_1, \ldots, \mu_5$. Show that, in this case, it is feasible to report the joint distribution of the incorrect assertion indicators $B_1, \ldots, B_5$ discussed in Section 2.2.5.

(b) Suppose, instead of simultaneous confidence intervals for $\mu_1, \ldots, \mu_5$, inference is in terms or simultaneous tests for

$$H_i : \mu_i = 0, \tag{2.2}$$

$i = 1, \ldots, 5$, with directional assertions following rejections. Note that, unlike the treatment $\times$ age scenario in Section 1.1.1, a single decision will result from the sutures experiment: either the standard suture or the dissolving suture will be used, as one can hardly perform the surgery with the standard suture and then switch mid-stream to the dissolving suture during the healing process. Thus, it is possible that a correct decision is reached even if some of the directional assertions made following the rejection of the corresponding $H_i$'s are wrong. So, for this example, controlling the error rate as defined in Definition 2.2.1 is perhaps too conservative. For such so-called 'multiple end points' situations, Benjamini and Hochberg (1995) proposed that the *false discovery rate* (FDR), which is the expected proportion of falsely rejected null hypotheses, be controlled instead. Let $V$ be the number of true null hypotheses rejected, and let $S$ be the number of false null hypotheses rejected. If one defines

$$Q = \begin{cases} 0 & \text{if } V + S = 0, \\ V/(V + S) & \text{otherwise,} \end{cases}$$

then $FDR = E(Q)$. Show that

$$FDR \leq P\{\text{reject at least one true } H_i\}$$

with equality if all $H_i$'s are true.

2. *Multiple comparisons with weighted mean.* Suppose it is of interest to compare each treatment with the (weighted) average of all the treatments, i.e., the parameters of interest are

$$\mu_i - \bar{\bar{\mu}} \text{ for } i = 1, \ldots, k$$

with

$$\bar{\bar{\mu}} = \sum_{i=1}^{k} n_i \mu_i / N$$

where

$$N = \sum_{i=1}^{k} n_i.$$

Under the model (1.1), it is natural to estimate $\mu_i - \bar{\bar{\mu}}$ by $\bar{Y}_i - \bar{\bar{Y}}$, where

$$\bar{\bar{Y}} = \sum_{i=1}^{k} n_i \bar{Y}_i / N.$$

Derive the joint distribution of

$$\bar{Y}_i - \bar{\bar{Y}}, \text{ for } i = 1, \ldots, k.$$

In particular, verify that the distribution is singular,

$$Var(\bar{Y}_i - \bar{\bar{Y}}) = \sigma^2 \left( \frac{1}{n_i} - \frac{1}{N} \right),$$

and

$$Cov(\bar{Y}_i - \bar{\bar{Y}}, \bar{Y}_j - \bar{\bar{Y}}) = -\frac{\sigma^2}{N} \text{ for all } i \neq j.$$

For this problem, simultaneous confidence intervals of the form

$$\mu_i - \bar{\bar{\mu}} \in \bar{Y}_i - \bar{\bar{Y}} \pm a\hat{\sigma} \sqrt{\frac{1}{n_i} - \frac{1}{N}}, \text{ for } i = 1, \ldots, k,$$

seem natural. While the singularity of the distribution and the negativity of the correlations make the computation of the distribution of

$$\max_{1 \leq i \leq k} \frac{|\bar{Y}_i - \bar{\bar{Y}}|}{\hat{\sigma} \sqrt{\frac{1}{n_i} - \frac{1}{N}}}$$

(which is needed to solve for the quantile $a$ in the confidence intervals) difficult relative to analogous computations for MCB and MCC, the product structure of the correlations allows the techniques developed by Nelson (1982; 1993) to apply; he has tabled $a$ for the equal sample size case

$$n_1 = \cdots = n_k.$$

However, the joint distribution of the estimators

$$\bar{Y}_i - \bar{Y} \text{ for } i = 1, \ldots, k$$

with

$$\bar{Y} = \sum_{i=1}^{k} \bar{Y}_i / k$$

for the more meaningful multiple comparisons with the (unweighted) mean (MCM) parameters

$$\mu_i - \bar{\mu} \text{ for } i = 1, \ldots, k$$

where

$$\bar{\mu} = \sum_{i=1}^{k} \mu_i / k$$

does *not* have a product correlation structure unless the sample sizes are equal. (Verify this.)

# CHAPTER 3

# Multiple comparisons with a control

Dunnett (1955) pioneered the concept that, when a control is present, the comparisons of primary interest may be the comparison of each new treatment with the control. For example, the control may be a placebo, or it may be a 'standard' treatment, or any other 'specified' treatment (such as a new drug). We call such comparisons multiple comparisons with a control (MCC). Suppose $\mu_1, \ldots, \mu_{k-1}$ are the means of the new treatments and $\mu_k$ is the mean of the control; then in MCC the parameters of primary interest are $\mu_i - \mu_k$ for $i = 1, \ldots, k-1$, the difference between each new treatment mean $\mu_i$ and the control mean $\mu_k$.

## 3.1 One-sided multiple comparisons with a control

To motivate one-sided MCC, consider the effect of smoking on pulmonary health example in Section 2.3.1. Recall that White and Froeb (1980) measured the pulmonary health of non-smokers and five groups of smokers. In that example, it is natural to take the non-smokers (NS) as the control group, and ask how much smoking might affect one's pulmonary health in terms of forced vital capacity (FVC), relative to not smoking.

### 3.1.1 Balanced one-way model

Suppose under the $i$th treatment a random sample $Y_{i1}, Y_{i2}, \ldots, Y_{in}$ of size $n$ is taken, where between the treatments the random samples are independent. Then under the usual normality and equality of variances assumptions, we have the one-way model

$$Y_{ia} = \mu_i + \epsilon_{ia}, \quad i = 1, \ldots, k, \quad a = 1, \ldots, n, \tag{3.1}$$

where $\mu_i$ is the effect of the $i$th treatment $i = 1, \ldots, k$, and $\epsilon_{11}, \ldots, \epsilon_{kn}$ are i.i.d. normal with mean 0 and variance $\sigma^2$ unknown. We use the notation

$$\hat{\mu}_i = \bar{Y}_i = \sum_{a=1}^{n} Y_{ia}/n,$$

$$\hat{\sigma}^2 = MSE = \sum_{i=1}^{k} \sum_{a=1}^{n} (Y_{ia} - \bar{Y}_i)^2 / [k(n-1)]$$

for the sample means and the pooled sample variance.

### 3.1.1.1 Dunnett's method

If a larger treatment effect is better and it is desired to infer as many new treatments as possible to be better than the control, or if a smaller treatment effect is better and it is desired to infer as many new treatments as possible to be inferior to the control, then Dunnett's method gives the following simultaneous confidence lower bounds for the difference between each new treatment mean $\mu_i$ and the control mean $\mu_k$:

$$\mu_i - \mu_k > \hat{\mu}_i - \hat{\mu}_k - d\hat{\sigma}\sqrt{2/n} \text{ for } i = 1, \dots, k-1,$$

where $d$ is the solution to the equation

$$\int_0^\infty \int_{-\infty}^{+\infty} [\Phi(z + \sqrt{2}ds)]^{k-1} d\Phi(z) \, \gamma(s) ds = 1 - \alpha. \qquad (3.2)$$

Here $\Phi$ is the standard normal distribution function and $\gamma$ is the density of $\hat{\sigma}/\sigma$. We note, however, from (3.2) that $d$ depends on $\alpha, k$ and $\nu$ only. The qmcb computer program described in Appendix D, with inputs the $(k-1) \times 1$ vector $\boldsymbol{\lambda} = (1/\sqrt{2}, \dots, 1/\sqrt{2})$, $\nu$ and $\alpha$, computes $d$. Tables of $d$ can also be found in Appendix E.

If a smaller treatment effect is better, and it is desired to infer as many new treatments as possible to be better than the control, or if a larger treatment effect is better and it is desired to infer as many new treatments as possible to be inferior to the control, then Dunnett's method gives the following simultaneous confidence upper bounds for the difference between each new treatment mean $\mu_i$ and the control mean $\mu_k$:

$$\mu_i - \mu_k < \hat{\mu}_i - \hat{\mu}_k + d\hat{\sigma}\sqrt{2/n} \text{ for } i = 1, \dots, k-1.$$

**Theorem 3.1.1** *With $d$ defined by (3.2),*

$$P\{\mu_i - \mu_k > \hat{\mu}_i - \hat{\mu}_k - d\hat{\sigma}\sqrt{2/n} \text{ for } i = 1, \dots, k-1\}$$
$$= P\{\mu_i - \mu_k < \hat{\mu}_i - \hat{\mu}_k + d\hat{\sigma}\sqrt{2/n} \text{ for } i = 1, \dots, k-1\}$$
$$= 1 - \alpha.$$

*Proof.*

$$P\{\mu_i - \mu_k > \hat{\mu}_i - \hat{\mu}_k - d\hat{\sigma}\sqrt{2/n} \text{ for } i = 1, \dots, k-1\}$$
$$= E\left[E\left[P\left\{d\sqrt{2} > \frac{\sqrt{n}(\hat{\mu}_i - \hat{\mu}_k - (\mu_i - \mu_k))/\sigma}{\hat{\sigma}/\sigma}\right.\right.\right.$$

$$\text{for } i = 1, \ldots, k-1 \Big\} \, |\hat{\mu}_k\Big] \, |\hat{\sigma}\Big]$$

$$= \int_0^\infty \int_{-\infty}^{+\infty} [\Phi(z + \sqrt{2}ds)]^{k-1} \, d\Phi(z) \, \gamma(s) ds$$

$$= 1 - \alpha.$$

Similarly for $P\{\mu_i - \mu_k < \hat{\mu}_i - \hat{\mu}_k + d\hat{\sigma}\sqrt{2/n}$ for $i = 1, \ldots, k-1\}$. $\square$

### 3.1.1.2 The stepdown method of Naik, Marcus, Peritz and Gabriel

Naik (1975) and Marcus, Peritz and Gabriel (1976) proposed a stepdown one-sided MCC method which, for the same data and at the same error rate $\alpha$ as Dunnett's method, will infer the same treatments as better than the control as Dunnett's method does, and sometimes more.

The idea of a *stepdown* MCC method is to decide, at step 1, whether there is sufficient evidence to infer that the new treatment that appears to be most significantly better than the control is indeed better. If the answer is 'no,' then proceed no further. If the answer is 'yes,' then at step 2 one decides whether the treatment that appears to be the next most significantly better than the control is indeed better, and so on.

The inference given by Naik (1975) and Marcus, Peritz and Gabriel (1976) was of the form of confident directions (i.e., $\mu_i - \mu_k > 0$ or $\mu_i - \mu_k < 0$), not confidence bounds. For a long time, it was thought that no stepwise procedure has a corresponding confidence set (e.g. Lehmann 1986, p. 388). However, recently Bofinger (1987) and Stefansson, Kim and Hsu (1988) gave two different derivations of the confidence bounds corresponding to this stepdown method. Thus, the inference given by this stepdown method can be upgraded to confidence bounds, making it a confidence interval MCC method. It is this confidence bounds version that is presented below.

Although the stepdown method has been generalized to the case of unbalanced designs by Bofinger (1987) and Dunnett and Tamhane (1991), for simplicity our description is restricted to the balanced one-way setting.

Let

$$T_i = \frac{\hat{\mu}_i - \hat{\mu}_k}{\hat{\sigma}\sqrt{2/n}}, \; i = 1, \ldots, k-1,$$

and let $[1], [2], \ldots, [k-1]$ denote the random indices such that

$$T_{[1]} \leq \cdots \leq T_{[k-1]}.$$

(Since the $T_i$'s are continuous random variables, ties occur among them with probability zero.) In other words, $[i]$ is the *anti-rank* of $T_i$ among $T_1, \ldots, T_{k-1}$. For example, suppose $k = 4$ and

$$T_2 < T_3 < T_1,$$

then
$$[1] = 2, \ [2] = 3, \ [3] = 1.$$

In the following, $d_h$ denotes the solution to (3.2) with $k-1 = h$. Note that $d_{k-1} = d$, the quantile for Dunnett's one-step method, while $d_1 = t_{\alpha,\nu}$, the upper $\alpha$ quantile of the univariate $t$ distribution with $\nu$ degrees of freedom. Further,
$$d_1 < d_2 < \cdots < d_{k-1}$$
since the integrand in (3.2) is decreasing in $k$.

If a larger treatment effect is better and it is desired to infer as many new treatments as possible to be better than the control, or if a smaller treatment effect is better and it is desired to infer as many new treatments as possible to be inferior to the control, then the confidence bounds version of the stepdown method of Naik, Marcus, Peritz and Gabriel proceeds as follows.

### Step 1

If    $T_{[k-1]} < d_{k-1}$

        then                       assert $\mu_{[i]} - \mu_k > \hat{\mu}_{[i]} - \hat{\mu}_k - d_{k-1}\hat{\sigma}\sqrt{2/n}$
                                       for $i \leq k - 1$;
                                       stop

        else                       assert $\mu_{[k-1]} > \mu_k$;
                                       go to step 2.

### Step 2

If    $T_{[k-2]} < d_{k-2}$

        then                       assert $\mu_{[i]} - \mu_k > \hat{\mu}_{[i]} - \hat{\mu}_k - d_{k-2}\hat{\sigma}\sqrt{2/n}$
                                         for $i \leq k - 2$;
                                       stop

        else                       assert $\mu_{[k-2]} > \mu_k$
                                       go to step 3.

$$\vdots$$

### Step $k - 1$

If    $T_{[1]} < t_{\alpha,\nu}$

        then             assert $\mu_{[i]} - \mu_k > \hat{\mu}_{[i]} - \hat{\mu}_k - t_{\alpha,\nu}\hat{\sigma}\sqrt{2/n}$
                                       for $i = 1$;
                                       stop

        else             assert $\mu_{[i]} - \mu_k > \hat{\mu}_{[1]} - \hat{\mu}_k - t_{\alpha,\nu}\hat{\sigma}\sqrt{2/n}$
                                     for $i \leq k - 1$;
                                       stop.

Since $d_1 < d_2 < \cdots < d_{k-1} = d$, given the same data set and error rate $\alpha$, every treatment inferred to be better than the control by Dunnett's method is guaranteed to be so inferred by the stepdown method, but the stepdown method may infer additional treatments to be better than the control. Thus, in terms of 'significant directional difference' inference, the stepdown method has an advantage over the one-step method. This has to be interpreted with care, however. The lower confidence bounds on $\mu_i - \mu_k$ of treatments asserted to be better than the control by Dunnett's one-step method are positive, while the lower confidence bounds on $\mu_i - \mu_k$ of treatments asserted to be better than the control by the stepdown MCC method are zero (except in the lucky situation when all new treatments are asserted to be better than the control). Thus, it is inappropriate to say that the stepdown method is uniformly better than the one-step method.

To gain an intuitive understanding of how the stepdown method works, consider the case of $k = 3$.

In the subset of the parameter space $\Theta_{12} = \{\mu_1 \leq \mu_3 \text{ and } \mu_2 \leq \mu_3\}$, an assertion of either $\mu_1 > \mu_3$ or $\mu_2 > \mu_3$ would be incorrect. To guard against such an incorrect assertion, for every $\boldsymbol{\mu} = (\mu_1^*, \mu_2^*, \mu_3^*)$ in $\Theta_{12}$, one performs the size-$\alpha$ test for the null hypothesis $H_0 : \mu_3 - \mu_1 \geq \mu_3^* - \mu_1^*$ and $\mu_3 - \mu_2 \geq \mu_3^* - \mu_2^*$ with acceptance region

$$\left\{ \hat{\mu}_3 - \hat{\mu}_1 > \mu_3^* - \mu_1^* - d_2\hat{\sigma}\sqrt{2/n} \text{ and } \hat{\mu}_3 - \hat{\mu}_2 > \mu_3^* - \mu_2^* - d_2\hat{\sigma}\sqrt{2/n} \right\}.$$

Stopping at step 1 corresponds to the acceptance of at least one $\boldsymbol{\mu}$ in $\Theta_{12}$.

In the subset of the parameter space $\Theta_1 = \{\mu_1 \leq \mu_3 \text{ and } \mu_2 > \mu_3\}$, an assertion of $\mu_1 > \mu_3$ would be incorrect (while an assertion of $\mu_2 > \mu_3$ would indeed be correct). To guard against such an incorrect assertion, for every $\boldsymbol{\mu} = (\mu_1^*, \mu_2^*, \mu_3^*)$ in $\Theta_1$, one performs the size-$\alpha$ test for the null hypothesis $H_0 : \mu_3 - \mu_1 \geq \mu_3^* - \mu_1^*$ with acceptance region

$$\left\{ \hat{\mu}_3 - \hat{\mu}_1 > \mu_3^* - \mu_1^* - d_1\hat{\sigma}\sqrt{2/n} \right\},$$

which incidentally is the usual one-sided size-$\alpha$ $t$-test.

In the subset of the parameter space $\Theta_2 = \{\mu_1 > \mu_3 \text{ and } \mu_2 \leq \mu_3\}$, an assertion of $\mu_2 > \mu_3$ would be incorrect (while an assertion of $\mu_1 > \mu_3$ would indeed be correct). To guard against such an incorrect assertion, for every $\boldsymbol{\mu} = (\mu_1^*, \mu_2^*, \mu_3^*)$ in $\Theta_2$, one performs the size-$\alpha$ test for the null hypothesis $H_0 : \mu_3 - \mu_2 \geq \mu_3^* - \mu_2^*$ with acceptance region

$$\left\{ \hat{\mu}_3 - \hat{\mu}_2 > \mu_3^* - \mu_2^* - d_1\hat{\sigma}\sqrt{2/n} \right\},$$

which incidentally is the usual one-sided size-$\alpha$ $t$-test.

If all $\boldsymbol{\mu} \in \Theta_{12}$ are rejected, that is, if the stepdown method proceeds to step 2, then since $d_1 < d_2$, either all $\boldsymbol{\mu} \in \Theta_1$ are rejected as well (i.e., $T_1 = T_{[2]} > d_1$), in which case $\mu_1 > \mu_3$ can be asserted, or all $\boldsymbol{\mu} \in \Theta_2$ are rejected as well (i.e., $T_2 = T_{[2]} > d_1$), in which case $\mu_2 > \mu_3$ can be asserted, or all $\boldsymbol{\mu} \in \Theta_1 \cup \Theta_2$ are rejected (i.e., $T_{[1]} > d_1$), in which case $\mu_1 > \mu_3$ and $\mu_2 > \mu_3$ can be asserted.

Note that we have defined a size-$\alpha$ test for each parameter point except for $\mu \in \Theta_0 = \{\mu_1 > \mu_3 \text{ and } \mu_2 > \mu_3\}$. Once a level-$\alpha$ test for each $\mu \in \Theta_0$ is defined, a confidence set corresponding to the stepdown method can be obtained via Lemma B.3.1. For every $\mu = (\mu_1^*, \mu_2^*, \mu_3^*)$ in $\Theta_0$, one can, for example, perform the level-$\alpha$ test for the null hypothesis

$$H_0 : \mu_3 - \min\{\mu_1, \mu_2\} \geq \mu_3^* - \min\{\mu_1^*, \mu_2^*\}$$

with acceptance region

$$\left\{ \hat{\mu}_3 - \min\{\hat{\mu}_1, \hat{\mu}_2\} > \mu_3^* - \min\{\mu_1^*, \mu_2^*\} - d_1 \hat{\sigma} \sqrt{2/n} \right\},$$

which is an example of what Roger Berger (1982) calls an *intersection-union* test. See Exercise 1. (Another possibility is given in Hayter and Hsu 1994.) The boundaries of the resulting confidence set on $\mu_1, \mu_2, \mu_3$ obtained via Lemma B.3.1 depend on $\hat{\mu}_1, \hat{\mu}_2, \hat{\mu}_3$, and are tedious to describe. But the confidence set can be plotted using Lemma B.3.2, and one can verify that the confidence intervals on $\mu_i - \mu_3, i = 1, 2$, that result are as described in the stepdown method.

To gain an understanding of the stepdown method for general $k$, let the random variable $M^\downarrow$ be the largest integer such that $T_{[i]} \leq d_i$, with the understanding that if no such integer exist, then $M^\downarrow = 0$. When $M^\downarrow > 0$, $\mu_{[1]}, \ldots, \mu_{[M^\downarrow]}$ are the treatment means that the stepdown method fails to declare to be larger than the control mean $\mu_k$. Note that, unless $M^\downarrow = 0$, the confidence bounds version of the stepdown method gives a lower bound of 0 for

$$\mu_{[i]} - \mu_k, \text{ for } i = M^\downarrow + 1, \ldots, k - 1$$

(if $M^\downarrow < k - 1$ of course) and negative lower bounds for

$$\mu_{[i]} - \mu_k, \text{ for } i = 1, \ldots, M^\downarrow.$$

Thus, the stepdown method has the disadvantage that no strictly positive lower bound on $\mu_i - \mu_k$ can be given, except in the lucky situation when all the new treatments can be inferred to be better than the control. So the stepdown method cannot be said to be uniformly better than Dunnett's method.

The proof of the validity of the stepdown method, which is rather involved, is given in Section 3.5 at the end of this chapter.

Finally, if a larger treatment effect is better and it is desired to infer as many new treatments as possible to be inferior to the control, or if a smaller treatment effect is better and it is desired to infer as many new treatments as possible to be better than the control, then the confidence bounds version of the stepdown method is as follows.

$$\boxed{\text{Step 1}}$$

If $\quad T_{[1]} > -d_{k-1}$
then
$\qquad$ assert $\mu_{[i]} - \mu_k < \hat{\mu}_{[i]} - \hat{\mu}_k + d_{k-1}\hat{\sigma}\sqrt{2/n}$ for $i \geq 1$;
$\qquad$ stop
else
$\qquad$ assert $\mu_{[1]} < \mu_k$;
$\qquad$ go to step 2.

$$\boxed{\text{Step 2}}$$

If $\quad T_{[2]} > -d_{k-2}$
then
$\qquad$ assert $\mu_{[i]} - \mu_k < \hat{\mu}_{[i]} - \hat{\mu}_k + d_{k-2}\hat{\sigma}\sqrt{2/n}$ for $i \geq 2$;
$\qquad$ stop
else
$\qquad$ assert $\mu_{[2]} < \mu_k$
$\qquad$ go to step 3.

$$\vdots$$

$$\boxed{\text{Step } k-1}$$

If $\quad T_{[k-1]} > -t_{\alpha,\nu}$
then
$\qquad$ assert $\mu_{[i]} - \mu_k < \hat{\mu}_{[i]} - \hat{\mu}_k + t_{\alpha,\nu}\hat{\sigma}\sqrt{2/n}$
$\qquad$ for $i = k-1$;
$\qquad$ stop
else
$\qquad$ assert $\mu_{[i]} - \mu_k < \hat{\mu}_{[k-1]} - \hat{\mu}_k + t_{\alpha,\nu}\hat{\sigma}\sqrt{2/n}$
$\qquad$ for $i \leq k-1$;
$\qquad$ stop.

In some situations, one may suspect

$$\mu_1 \leq \mu_2 \leq \cdots \leq \mu_k \tag{3.3}$$

as, for example, when they are mean responses under increasing dose of a substance, with $\mu_1$ corresponding to a zero dose and serving as a control. When monotonicity (3.3) is suspected, it is reasonable to use a method which is monotone in the sense that if it fails to assert $\mu_m > \mu_1$, then it does not assert $\mu_{m^*} > \mu_1$ for any $m^* < m$ either. Otherwise, what is one to make of assertions that the mean under a low dose is significantly higher than the mean under zero dose, but the mean under a higher dose is not significantly different?

Williams (1971) proposed a monotone method in the form of a stepdown test for the hypotheses

$$H_0^m : \mu_1 = \cdots = \mu_m$$

versus

$$H_a^m : \mu_1 \leq \cdots \leq \mu_m \text{ (with at least one strict inequality)}$$

with treatment 1 being the control and $m = 2, \ldots, k$. The intended interpretation is that if $H_0^m$ is rejected, then $\mu_m > \mu_1$. Williams' method is only valid if the monotonicity assumption (3.3) is true. Further, for its inference to be meaningful, one must believe the point hypotheses $H_0^m : \mu_1 = \mu_m$ can be true exactly, which not everyone does (see Sections 4.1 and Section 6.2). Therefore, when one suspects monotonicity (3.3), perhaps a better strategy is to employ a monotone method whose validity does not depend on (3.3) but which takes advantage of the suspected monotonicity if it manifests itself in the data. A simple yet powerful stepdown (and therefore monotone) MCC method for asserting $\mu_m > \mu_1$ whose validity does not depend on (3.3) is constructed in Berger and Hsu (1995).

Williams (1971) seemed to suggest, in toxicity studies, the acceptance of $H_0^m : \mu_1 = \cdots = \mu_m$ implies the substance is safe up to dose $m$. As Schoenfeld (1986) pointed out, this is not the case, as the acceptance may be due to too small a sample size relative to noise. Rather, safety up to dose $m$ can be concluded if one can assert

$$\delta_L < \mu_i - \mu_1 < \delta_U, i = 1, \ldots, m, \tag{3.4}$$

where $\delta_L$ and $\delta_U$ are prespecified constants defining practical equivalence. Assuming monotonicity in the parameters (3.3) and $\mu_1$ is known, Schoenfeld (1986) constrcuted a stepwise (and therefore monotone) method for asserting (3.4) based on isotonic regression estimates. For situations where monotonicity (3.3) is suspected but not assumed, a simple yet powerful stepwise (and therefore monotone) MCC method for asserting (3.4) whose validity does not depend on (3.3) is constructed in Berger and Hsu (1995).

### 3.1.1.3 Dunnett and Tamhane's stepup method

The idea of a *stepup* MCC method is to decide, at step 1, whether there is sufficient evidence to infer that the new treatment that appears to be least significantly better than the control is indeed better. If the answer is 'yes,' then infer all new treatments to be better than the control and stop. If the answer is 'no,' then at step 2, one uses a larger critical value to decide whether the treatment that appears to be the next least significantly better than the control is indeed better, and so on.

As before, for the balanced one-way model, let

$$T_i = \frac{\hat{\mu}_i - \hat{\mu}_k}{\hat{\sigma}\sqrt{2/n}}, \ i = 1, \ldots, k - 1,$$

and let $[1], [2], \ldots, [k - 1]$ denote the random indices such that

$$T_{[1]} \leq \cdots \leq T_{[k-1]}.$$

Define the constants $c_1, \ldots, c_{k-1}$ recursively such that

$$P\{(T_1, \ldots, T_m) < (c_1, \ldots, c_m)\} = 1 - \alpha \tag{3.5}$$

for $m = 1, \ldots, k - 1$ when $\mu_1 = \cdots = \mu_k$. Here

$$(x_1, \ldots, x_m) < (y_1, \ldots, y_m)$$

means

$$x_{1,m} < y_{1,m}, \ldots, x_{m,m} < y_{m,m}$$

where the $x_{i,m}$ and $y_{i,m}$ are the ordered $\{x_1, \ldots, x_m\}$ and $\{y_1, \ldots, y_m\}$. Thus, first $c_1 = t_{\alpha,\nu}$ is defined, and then $c_2$ is defined given $c_1$, and so on. Dunnett and Tamhane (1992) conjectured that $c_i$'s satisfying (3.5) for $m = 1, \ldots, k-1$ and the monotonicity condition

$$c_1 \leq \cdots \leq c_{k-1}$$

exist for arbitrary $k$, but stated the conjecture had only been proven for $k = 3$. For the case of independent $T_1, \ldots, T_{k-1}$, Finner, Hayter and Roters (1993) proved the conjecture to be true for $\alpha$ and $k$ in the usual ranges that occur in practice (e.g., $\alpha \leq 0.1$ and $k \leq 96$) but, as it turned out, Dalal and Mallows (1992) proved the conjecture for arbitrary $k$ in the context of software reliability testing. (I thank Helmut Finner for pointing this out.) While a proof of the conjecture for the correlated $T_1, \ldots, T_{k-1}$ case here has yet to be given, we nevertheless proceed assuming it is true.

If a larger treatment effect is better and it is desired to infer as many new treatments as possible to be better than the control, or if a smaller treatment effect is better and it is desired to infer as many new treatments as possible to be inferior to the control, then Dunnett and Tamhane's stepup method proceeds as follows.

$$\boxed{\text{Step 1}}$$

If $\quad T_{[1]} > t_{\alpha,\nu}$
then $\qquad$ assert $\mu_{[i]} > \mu_k$ for $i = 1, \ldots, k-1$;
$\qquad\qquad$ stop
else $\qquad$ go to step 2.

$$\boxed{\text{Step 2}}$$

If $\quad T_{[2]} > c_2$
then $\qquad$ assert $\mu_{[i]} > \mu_k$ for $i = 2, \ldots, k-1$;
$\qquad\qquad$ stop
else $\qquad$ go to step 3.

$$\vdots$$

$$\boxed{\text{Step } k-1}$$

If $\quad T_{[k-1]} > c_{k-1}$
then $\qquad$ assert $\mu_{[k-1]} > \mu_k$;
$\qquad\qquad$ stop
else $\qquad$ stop.

Let the random variable $M^{\dagger*}$ be the smallest integer $i$ such that $T_{[i]} > c_i$, with the understanding that if no such integer exist, then $M^{\dagger*} = k$. If we let $M^\dagger = \dot{M}^{\dagger *} - 1$, then when $M^{\dagger *} > 1$, $\mu_{[1]}, \ldots, \mu_{[M^\dagger]}$ are the treatment means that Dunnett and Tamhane's method fail to declare to be bigger than the control mean $\mu_k$.

The proof of the validity of Dunnett and Tamhane's stepup method, which is rather involved, is given in Section 3.5 at the end of this chapter. If a larger treatment effect is better and it is desired to infer as many new treatments as possible to be inferior to the control, or if a smaller treatment effect is better and it is desired to infer as many new treatments as possible to be better than the control, then Dunnett and Tamhane's stepup method proceeds as follows.

$$\boxed{\text{Step 1}}$$

If   $T_{[k-1]} < -t_{\alpha,\nu}$
    then             assert $\mu_{[i]} < \mu_k$ for $i = 1, \ldots, k-1$;
                           stop
    else             go to step 2.

$$\boxed{\text{Step 2}}$$

If   $T_{[k-2]} < -c_2$
    then             assert $\mu_{[i]} < \mu_k$ for $i = 1, \ldots, k-2$;
                           stop
    else             go to step 3.

$$\vdots$$

$$\boxed{\text{Step } k-1}$$

If   $T_{[1]} < -c_{k-1}$
    then             assert $\mu_{[1]} < \mu_k$;
                           stop
    else             stop.

Hayter and Hsu (1994) derived a confidence set corresponding to this stepup method for the case of $k = 3$. Again, in contrast to Dunnett's one-step method, the lower confidence bounds on $\mu_i - \mu_k$ of treatments asserted to be better than the control by the stepup MCC method are zero (except in the lucky situation when all new treatments are asserted to be better than the control). Thus, it cannot be said that the stepup method is uniformly better than the one-step method.

### 3.1.2 Example: smoking and pulmonary health (continued)

The smoking and pulmonary health data in White and Froeb (1980) were given in the form of summary statistics. Without access to the original data, it is impossible to perform the diagnostics illustrated in Chapter 1 to assess the appropriateness of the model (3.1). Thus, in using this data for illustration, we are assuming that such diagnostics were performed, and no obvious departure from the model assumptions was found.

To illustrate one-sided MCC methods for the balanced one-way model, let us ignore the non-inhaling smokers (NI) for now (see Table 2.4). Then

$$\hat{\sigma} = ((0.63^2 + 0.46^2 + 0.39^2 + 0.38^2 + 0.38^2)/5)^{\frac{1}{2}} = 0.46$$

with $5(200-1) = 995$ degrees of freedom. If we take the degrees of freedom to be infinite, then for $\alpha = 0.01$,

$$\begin{aligned}
d_1 &= 2.326 = c_1 \\
d_2 &= 2.558 \\
d_3 &= 2.685 \\
d_4 &= 2.772
\end{aligned}$$

For ease of identification, instead of subscripting $\mu$ by the group index $1, \ldots, 6$, we subscript $\mu$ with the group labels $NS, \ldots, HS$ (see Table 2.4). Dunnett's method infers

$$\begin{aligned}
\mu_{PS} - \mu_{NS} &< 3.23 - 3.35 + 2.772(0.46)\sqrt{\tfrac{2}{200}} = 0.007 \\
\mu_{LS} - \mu_{NS} &< 3.15 - 3.35 + 2.772(0.46)\sqrt{\tfrac{2}{200}} = -0.073 \\
\mu_{MS} - \mu_{NS} &< 2.80 - 3.35 + 2.772(0.46)\sqrt{\tfrac{2}{200}} = -0.423 \\
\mu_{HS} - \mu_{NS} &< 2.55 - 3.35 + 2.772(0.46)\sqrt{\tfrac{2}{200}} = -0.673
\end{aligned}$$

The confidence bounds version of the stepdown method proceeds as follows.

$$\boxed{\text{Step 1}}$$

Since $T_{[1]} = \frac{2.55 - 3.35}{0.46\sqrt{2/200}} = -17.46 < -2.772 = -d_4$,
assert $\mu_{HS} < \mu_{NS}$;
go to step 2.

$$\boxed{\text{Step 2}}$$

Since $T_{[2]} = \frac{2.80 - 3.35}{0.46\sqrt{2/200}} = -12.01 < -2.685 = -d_3$,
assert $\mu_{MS} < \mu_{NS}$;
go to step 3.

$$\boxed{\text{Step 3}}$$

Since $T_{[3]} = \frac{3.15 - 3.35}{0.46\sqrt{2/200}} = -4.37 < -2.558 = -d_2$,
assert $\mu_{LS} < \mu_{NS}$;
go to step 4.

$$\boxed{\text{Step 4}}$$

Since $\quad T_{[4]} = \frac{3.23-3.35}{0.46\sqrt{2/200}} = -2.62 < -2.326 = -d_1,$

assert

$$\mu_{HS} - \mu_{NS} < 3.23 - 3.35 + 2.326(0.46)\sqrt{\tfrac{2}{200}} = -0.013$$

$$\mu_{MS} - \mu_{NS} < 3.23 - 3.35 + 2.326(0.46)\sqrt{\tfrac{2}{200}} = -0.013$$

$$\mu_{LS} - \mu_{NS} < 3.23 - 3.35 + 2.326(0.46)\sqrt{\tfrac{2}{200}} = -0.013$$

$$\mu_{PS} - \mu_{NS} < 3.23 - 3.35 + 2.326(0.46)\sqrt{\tfrac{2}{200}} = -0.013$$

stop .

Dunnett and Tamhane's stepup method proceeds as follows.

$$\boxed{\text{Step 1}}$$

Since $\quad T_{[4]} = \frac{3.23-3.35}{0.46\sqrt{2/200}} = -2.62 < -2.326 = -c_1,$

assert

$\mu_{HS} < \mu_{NS},$

$\mu_{MS} < \mu_{NS},$

$\mu_{LS} < \mu_{NS},$

$\mu_{PS} < \mu_{NS},$

stop .

The female FVC data without the NI group is such that the stepdown method and Dunnett and Tamhane's stepup method are able to declare the passive smokers (PS) to be worse off than the non-smokers, while Dunnett's one-step method is not. However, Dunnett's one-step method gives better bounds on how much worse off are the light smokers (LS), moderate smokers (MS), and heavy smokers (HS), relative to non-smokers.

### 3.1.3 Unbalanced one-way model

Suppose now the sample sizes are unequal, and under the $i$th treatment a random sample $Y_{i1}, Y_{i2}, \ldots, Y_{in_i}$ is taken, where between the treatments the random samples are independent. Then under the usual normality and equality of variance assumptions, we have the one-way model (1.1)

$$Y_{ia} = \mu_i + \epsilon_{ia}, \quad i = 1, \ldots, k, \quad a = 1, \ldots, n_i, \tag{3.6}$$

where $\epsilon_{11}, \ldots, \epsilon_{kn_k}$ are i.i.d. normal with mean 0 and variance $\sigma^2$ unknown. We use the notation

$$\hat{\mu}_i \quad = \bar{Y}_i \quad = \sum_{a=1}^{n_i} Y_{ia}/n_i,$$

$$\hat{\sigma}^2 \; = MSE \; = \sum_{i=1}^{k}\sum_{a=1}^{n_i}(Y_{ia} - \bar{Y}_i)^2 / \sum_{i=1}^{k}(n_i - 1)$$

for the sample means and the pooled sample variance.

### 3.1.3.1 Dunnett's method

If a larger treatment effect is better and it is desired to infer as many new treatments as possible to be better than the control, or if a smaller treatment effect is better and it is desired to infer as many new treatments as possible to be inferior to the control, then Dunnett's method gives the following simultaneous confidence intervals for the difference between each new treatment mean $\mu_i$ and the control mean $\mu_k$:

$$\mu_i - \mu_k > \hat{\mu}_i - \hat{\mu}_k - d\hat{\sigma}\sqrt{n_i^{-1} + n_k^{-1}} \text{ for } i = 1,\dots,k-1,$$

where $d = d_{\boldsymbol{\lambda},\alpha,\nu}$ is the solution to the equation

$$\int_0^\infty \int_{-\infty}^{+\infty} \prod_{i=1}^{k-1}[\Phi((\lambda_i z + ds)/(1 - \lambda_i^2)^{1/2})]d\Phi(z)\ \gamma(s)ds = 1 - \alpha \qquad (3.7)$$

with

$$\lambda_i = \left(1 + \frac{n_k}{n_i}\right)^{-1/2}, \quad i = 1,\dots,k-1.$$

We note from (3.7) that in addition to $\alpha, k$ and $\nu$, the critical value $d$ depends on the sample size ratios $n_k/n_1,\dots,n_k/n_{k-1}$. Thus, it is not possible to tabulate $d$ in general. But it is possible to program the computer to solve for $d$ in (3.7) at execution time depending on the sample size pattern $\boldsymbol{n} = (n_1,\dots,n_k)$ of the data to be analyzed. The qmcb computer program described in Appendix D with inputs $\boldsymbol{\lambda} = (\lambda_1,\dots,\lambda_{k-1})$, $\nu$ and $\alpha$ computes $d$.

If a smaller treatment effect is better and it is desired to infer as many new treatments as possible to be better than the control, or if a larger treatment effect is better and it is desired to infer as many new treatments as possible to be inferior to the control, then Dunnett's method gives the following simultaneous confidence intervals for the difference between each new treatment mean $\mu_i$ and the control mean $\mu_k$:

$$\mu_i - \mu_k < \hat{\mu}_i - \hat{\mu}_k + d\hat{\sigma}\sqrt{n_i^{-1} + n_k^{-1}} \text{ for } i = 1,\dots,k-1.$$

**Theorem 3.1.2** *With $d$ defined by (3.7),*

$$P\left\{\mu_i - \mu_k > \hat{\mu}_i - \hat{\mu}_k - d\hat{\sigma}\sqrt{n_i^{-1} + n_k^{-1}} \text{ for } i = 1,\dots,k-1\right\}$$

$$= \; P\left\{\mu_i - \mu_k < \hat{\mu}_i - \hat{\mu}_k + d\hat{\sigma}\sqrt{n_i^{-1} + n_k^{-1}} \text{ for } i = 1,\dots,k-1\right\}$$

$$= 1 - \alpha.$$

*Proof.* If we let $Z_1, \ldots, Z_k$ be i.i.d. standard normal random variables, then

$$P\{\mu_i - \mu_k < \hat{\mu}_i - \hat{\mu}_k + d\hat{\sigma}\sqrt{n_i^{-1} + n_k^{-1}} \text{ for } i = 1, \ldots, k-1\}$$

$$= P\{\mu_i - \mu_k > \hat{\mu}_i - \hat{\mu}_k - d\hat{\sigma}\sqrt{n_i^{-1} + n_k^{-1}} \text{ for } i = 1, \ldots, k-1\}$$

$$= E\left[E\left[P\left\{\frac{\sqrt{n_i}(\hat{\mu}_i - \mu_i)}{\sigma} < \sqrt{\frac{n_i}{n_k}}\left[\frac{\sqrt{n_k}(\hat{\mu}_k - \mu_k)}{\sigma}\right] + d\left(\frac{\hat{\sigma}}{\sigma}\right)\sqrt{1 + \frac{n_i}{n_k}}\right.\right.\right.$$

$$\left.\left.\left. \text{for } i = 1, \ldots, k-1\right\} \,|\hat{\mu}_k\right] |\hat{\sigma}\right]$$

$$= E\left[E\left[P\left\{Z_i < \sqrt{\frac{n_i}{n_k}}Z_k + d\left(\frac{\hat{\sigma}}{\sigma}\right)\sqrt{1 + \frac{n_i}{n_k}}\right.\right.\right.$$

$$\left.\left.\left. \text{for } i = 1, \ldots, k-1\right\} \,|\hat{\mu}_k\right] |\hat{\sigma}\right]$$

$$= \int_0^\infty \int_{-\infty}^{+\infty} \prod_{i=1}^{k-1} [\Phi(\lambda_i z + ds)/(1 - \lambda_i^2)^{1/2}] d\Phi(z) \gamma(s) ds$$

$$= 1 - \alpha,$$

since

$$\lambda_i/(1 - \lambda_i^2)^{1/2} = \sqrt{\frac{n_i}{n_k}},$$

$$1/(1 - \lambda_i^2)^{1/2} = \sqrt{1 + \frac{n_i}{n_k}}.$$

□

### 3.1.3.2 *The Miller–Winer method*

For unbalanced designs, instead of solving for $d$ in (3.7), the so-called Miller–Winer method (implemented in program P7D of BMDP, for example) takes the harmonic mean of the new treatment sample sizes

$$\tilde{n} = \left[\left(\frac{1}{n_1} + \cdots + \frac{1}{n_{k-1}}\right)/(k-1)\right]^{-1}$$

(excluding the control) to be the 'common' sample size of the new treatments or, equivalently, takes the average of the variances of $\hat{\mu}_1, \ldots, \hat{\mu}_{k-1}$ to be the common variance, and Dunnett's method for treatment-balanced ($n_1 = \cdots = n_{k-1}$) designs is employed. This leads to invalid statistical inference in general. We will show that the probability of an incorrect assertion associated with treatments with small sample sizes can be much higher than $\alpha$.

The probability that the assertions associated with $\mu_i - \mu_k, i \in I$, are correct is

$$P\left\{\mu_i - \mu_k > \hat{\mu}_i - \hat{\mu}_k - d\hat{\sigma}\sqrt{1/\tilde{n} + 1/n_k} \text{ for all } i \in I\right\} \quad (3.8)$$

$$= P\left\{(\hat{\mu}_i - \hat{\mu}_k - (\mu_i - \mu_k))/\hat{\sigma} < d\sqrt{1/\tilde{n} + 1/n_k} \text{ for all } i \in I\right\}$$

$$= P\left\{\frac{\hat{\mu}_i - \hat{\mu}_k - (\mu_i - \mu_k)}{\hat{\sigma}\sqrt{1/n_i + 1/n_k}} < d\sqrt{\frac{1/\tilde{n} + 1/n_k}{1/n_i + 1/n_k}} \text{ for all } i \in I\right\}.$$

When the sample sizes $n_i, i \in I$, are relatively small, the factor

$$\sqrt{\frac{1/\tilde{n} + 1/n_k}{1/n_i + 1/n_k}} \quad (3.9)$$

may be so substantially less than one that the probability (3.8) becomes less than $1 - \alpha$. For example, suppose $k = 21$ and the sample sizes are $n_1 = \cdots = n_4 = 2, n_5 = \cdots = n_{21} = 20$. Then $\tilde{n} = 1/0.14$ and (3.9) equals $\sqrt{3.8/11}$ for $i = 1, \ldots, 4$. With $\alpha = 0.05$, using the qmcb program with input $\lambda = (1/\sqrt{1 + 20(0.14)}, \ldots, 1/\sqrt{1 + 20(0.14)}), \nu = 327$ and $\alpha = 0.05$, we find $d = 2.752$ and the probability (3.8) with $I = \{1, 2, 3, 4\}$ turns out to be 0.811, which is much less than 0.95. The probability that all the assertions associated with $\mu_i - \mu_k, i = 1, \ldots, k - 1$, are simultaneously correct is of course lower still.

In fact, if

$$n_1 \ll n_2 < \cdots < n_{k-1} \ll n_k, \quad (3.10)$$

where $\ll$ means 'much smaller than,' then

$$\frac{1/\tilde{n} + 1/n_k}{1/n_1 + 1/n_k}$$

is close to $1/(k - 1)$. So the probability that the inference associated with $\mu_1 - \mu_k$ will be correct is close to 50% when $k$ is large. Thus, the Miller–Winer method is not recommended.

### 3.1.3.3 Methods based on probabilistic inequalities

If you do not have access to software which computes $d$ (e.g., qmcb), but have access to software which computes univariate $t$ quantiles, then the following two methods based on probabilistic inequalities can be considered conservative alternatives of last resort.

Let

$$E_i = \{(\hat{\mu}_i - \hat{\mu}_k - (\mu_i - \mu_k))/\hat{\sigma}\sqrt{n_i^{-1} + n_k^{-1}} > -d\} \quad (3.11)$$

or

$$E_i = \{(\hat{\mu}_i - \hat{\mu}_k - (\mu_i - \mu_k))/\hat{\sigma}\sqrt{n_i^{-1} + n_k^{-1}} < d\}, \quad (3.12)$$

depending on whether upper confidence bounds or lower confidence bounds for $\mu_i - \mu_k$ are desired, respectively. The Bonferroni inequality (A.3) states

$$P(\bigcup_{i=1}^{k-1} E_i^c) \leq \sum_{i=1}^{k-1} P(E_i^c).$$

If each $E_i^c$ is such that

$$P(E_i^c) = \alpha/(k-1),$$

then

$$P(\bigcup_{i=1}^{k-1} E_i^c) \leq \alpha.$$

Thus a conservative approximation to $d$ is

$$d_{Bonferroni} = t_{\frac{\alpha}{k-1},\nu}. \tag{3.13}$$

A slightly better approximation is given by Slepian's inequality (Theorem A.3.1), which states

$$P(\bigcap_{i=1}^{k-1} E_i \mid \hat{\sigma}) \geq \prod_{i=1}^{k-1} P(E_i \mid \hat{\sigma}).$$

Now $P(E_i|\hat{\sigma}), i = 1,\ldots,k-1$, are monotone in $\hat{\sigma}$ in the same direction, thus one can further apply Corollary A.1.1 to get

$$P(\bigcap_{i=1}^{k-1} E_i) \geq E_{\hat{\sigma}}[\prod_{i=1}^{k-1} P(E_i \mid \hat{\sigma})] \geq \prod_{i=1}^{k-1} E_{\hat{\sigma}}[P(E_i|\hat{\sigma})] = \prod_{i=1}^{k-1} P(E_i).$$

If each $E_i$ is such that

$$P(E_i) = (1-\alpha)^{1/(k-1)},$$

then

$$P(\bigcap_{i=1}^{k-1} E_i) \geq 1 - \alpha.$$

So a conservative approximation $d_{Slepian}$ to $d$ results if one pretends $E_1,\ldots,E_{k-1}$ were independent:

$$d_{Slepian} = t_{1-(1-\alpha)^{1/(k-1)},\nu}.$$

Note that this approximation for one-sided MCC is analogous to the product inequality approximation for inference on the means in Section 1.3.3. However, relative to exact critical values, this approximation for one-sided MCC is worse, due to the additional inequality (Slepian's) employed in arriving at the approximation. Also, as explained in Section 1.3.5, this approximation is slightly less conservative than the Bonferroni approximation. Therefore, unless the software you have access to is only capable of univariate $t$ quantiles, neither the approximation based on the Bonferroni inequality nor the approximation based on Slepian's inequality is recommended.

One could also obtain a better (less conservative) approximation than $d_{\text{Bonferroni}}$ by using the Hunter–Worsley inequality (A.6) instead of the Bonferroni inequality. However, implementing the Hunter–Worsley inequality method involves no less coding and computation than the exact method. One might as well use the exact method.

### 3.1.4 Example: smoking and pulmonary health (continued)

To illustrate one-sided MCC for unbalanced designs, let us include the non-inhaling smokers (NI) in the analysis. Then

$$
\hat{\sigma} = \left( \frac{199(0.63^2 + 0.46^2 + 0.39^2 + 0.38^2 + 0.38^2) + 49(0.52^2)}{1044} \right)^{\frac{1}{2}}
$$
$$
= 0.46
$$

with $5(200 - 1) + (50 - 1) = 1044$ degrees of freedom. If we take the degrees of freedom to be infinity, then

$$
\begin{aligned}
d &= 2.848 \\
d_{\text{Slepian}} &= 2.883 \\
d_{\text{Bonferroni}} &= 2.885
\end{aligned}
$$

for $\alpha = 0.01$.

Therefore, Dunnett's method infers

$$
\begin{aligned}
\mu_{PS} - \mu_{NS} &< 3.23 - 3.35 + 2.848(0.46)\sqrt{\tfrac{2}{200}} &&= 0.011 \\
\mu_{NI} - \mu_{NS} &< 3.19 - 3.35 + 2.848(0.46)\sqrt{\tfrac{1}{50} + \tfrac{1}{200}} &&= 0.048 \\
\mu_{LS} - \mu_{NS} &< 3.15 - 3.35 + 2.848(0.46)\sqrt{\tfrac{2}{200}} &&= -0.069 \\
\mu_{MS} - \mu_{NS} &< 2.80 - 3.35 + 2.848(0.46)\sqrt{\tfrac{2}{200}} &&= -0.419 \\
\mu_{HS} - \mu_{NS} &< 2.55 - 3.35 + 2.848(0.46)\sqrt{\tfrac{2}{200}} &&= -0.669
\end{aligned}
$$

Inference based on the Bonferroni inequality or Slepian's inequality will be less precise (more conservative). Note that, with the degrees of freedom $\nu = 1044$ associated with $\hat{\sigma}$ being essentially infinity, the method based on Slepian's inequality would give the exact critical value 2.848 if the only source of dependence among $E_1, \ldots, E_{k-1}$ were the common $\hat{\sigma}$. But the common $\hat{\mu}_k$ also contributes to dependence, making the approximation based on Slepian's inequality a conservative 2.883 instead.

## 3.2 Two-sided multiple comparisons with a control

To motivate two-sided MCC, consider the bovine growth hormone safety study example described in Section 2.3.2. Recall that a subset of the data from one experiment reported in Juskevich and Guyer (1990) gave weight changes in rbGH (recombinant bovine growth hormone) treated rats. The treatments included a negative control diet or placebo treatment (labeled dose group 1), a positive control treatment which was rbGH by injection (labeled dose group 2), and four different doses of bovine growth hormone given orally (labeled dose groups 3–6). Changes in weights of male rats after 85 days were given in Table 2.5, which is reproduced in Table 3.1 for convenience.

Table 3.1 *Body weight changes (in grams) of male rats*

| Level | Method | Dosage (mg/kg per day) | Sample size | Mean weight change | Std. dev. weight change |
|-------|--------|------------------------|-------------|--------------------|-------------------------|
| 1 | oral | 0 | 30 | 324 | 39.2 |
| 2 | injection | 1.0 | 30 | 432 | 60.3 |
| 3 | oral | 0.1 | 30 | 327 | 39.1 |
| 4 | oral | 0.5 | 30 | 318 | 53.0 |
| 5 | oral | 5 | 30 | 325 | 46.3 |
| 6 | oral | 50 | 30 | 328 | 43.0 |

For this experiment, it is of interest to compare the positive control with the negative control, to verify that the measurement process was capable of detecting expected difference. It is also of interest to compare the four oral bovine growth hormone doses with the negative control, to see whether weight gains given any dose are close to weight gains given the placebo.

### 3.2.1 Balanced one-way model

### 3.2.1.1 Dunnett's method

Consider the balanced one-way model (3.1). Dunnett's (1955) two-sided method provides the following simultaneous confidence intervals for the difference between each new treatment mean $\mu_i$ and the control mean $\mu_k$:

$$\mu_i - \mu_k \in \hat{\mu}_i - \hat{\mu}_k \pm |d|\hat{\sigma}\sqrt{2/n} \text{ for } i = 1, \ldots, k-1,$$

where $|d|$ is the solution to the equation

$$\int_0^\infty \int_{-\infty}^{+\infty} [\Phi(z + \sqrt{2}|d|s) - \Phi(z - \sqrt{2}|d|s)]^{k-1} d\Phi(z) \, \gamma(s)ds = 1 - \alpha. \text{ (3.14)}$$

**Theorem 3.2.1** *With $|d|$ defined by (3.14)*

$$P\{\mu_i - \mu_k \in \hat{\mu}_i - \hat{\mu}_k \pm |d| \, \hat{\sigma}\sqrt{2/n} \text{ for } i = 1, \ldots, k-1\} = 1 - \alpha.$$

*Proof.*

$$
\begin{aligned}
&P\left\{\mu_i - \mu_k \in \hat{\mu}_i - \hat{\mu}_k \pm |d| \, \hat{\sigma}\sqrt{2/n} \text{ for } i = 1, \ldots, k-1\right\} \\
&= E\left[E\left[P\left\{-|d|\sqrt{2} < \frac{\sqrt{n}(\hat{\mu}_i - \hat{\mu}_k - (\mu_i - \mu_k))/\sigma}{\hat{\sigma}/\sigma} < |d|\sqrt{2}\right.\right.\right. \\
&\qquad\qquad \left.\left.\left. \text{for } i = 1, \ldots, k-1\right\} \Big| \hat{\mu}_k\right] \Big| \hat{\sigma}\right] \\
&= \int_0^\infty \int_{-\infty}^{+\infty} [\Phi(z + \sqrt{2}|d|s) - \Phi(z - \sqrt{2}|d|s)]^{k-1} d\Phi(z)\gamma(s)ds \\
&= 1 - \alpha. \quad \square
\end{aligned}
$$

The qmcc computer program described in Appendix D, with inputs the $(k-1) \times 1$ vector $\boldsymbol{\lambda} = (1/\sqrt{2}, \ldots, 1/\sqrt{2})$, $\nu$ and $\alpha$, computes $|d|$. Tables of $|d|$ can also be found in Appendix E.

The two-sided stepdown MCC method of Dunnett and Tamhane (1991) and the two-sided stepup MCC method of Dunnett and Tamhane (1992) are confident inequalities methods. Even though the technique of Finner (1990a) can perhaps be adapted to show that, for comparing two treatments with a control (i.e., $k = 3$), the two-sided stepdown MCC method is a confident directions method, there is yet no proof that either is a confident directions method in general. Therefore, instead of discussing them here, we refer the reader to the original papers.

### 3.2.2 Example: bovine growth hormone safety (continued)

The bovine growth hormone safety study data in Juskevich and Guyer (1990) was given in the form of summary statistics. Without access to the original data, it is impossible to perform the diagnostics illustrated in Chapter 1 to assess the appropriateness of the model (3.1). Thus, in using this data to illustrate Dunnett's two-sided MCC methods for balanced designs, we are assuming that such diagnostics were performed, and no obvious departure from the model assumptions was found.

For this data,

$$\hat{\sigma} = ((39.2^2 + 60.3^2 + 39.1^2 + 53.0^2 + 46.3^2 + 43.0^2)/6)^{\frac{1}{2}} = 47.44$$

with $6(30 - 1) = 174$ degrees of freedom. For $\alpha = 0.05$, $|d| = 2.536$. Therefore, Dunnett's two-sided MCC method infers

$$76.89 \ < \ \mu_2 - \mu_1 \ < \ 139.11$$
$$-28.11 \ < \ \mu_3 - \mu_1 \ < \ 34.11$$
$$-37.11 \ < \ \mu_4 - \mu_1 \ < \ 25.11$$
$$-30.11 \ < \ \mu_5 - \mu_1 \ < \ 32.11$$
$$-27.11 \ < \ \mu_6 - \mu_1 \ < \ 35.11$$

Production statistical packages generally expect raw input data. To illustrate computer analysis, an artificial data set with summary statistics matching those in Juskevich and Guyer (1990) was generated. The MCC subcommand of the ONEWAY command of MINITAB was then invoked to produce the output displayed in Figure 3.1.

```
MTB > Oneway 'change' 'level';
SUBC>    Dunnett 5 1.

ANALYSIS OF VARIANCE ON change
SOURCE      DF       SS        MS        F        p
level        5    291280     58256     25.89    0.000
ERROR      174    391594      2251
TOTAL      179    682874
                                  INDIVIDUAL 95 PCT CI'S FOR MEAN
                                  BASED ON POOLED STDEV
   LEVEL      N      MEAN      STDEV   ----------+---------+---------+------
     1       30    324.00     39.20   (---*--)
     2       30    432.00     60.30                            (--*---)
     3       30    327.00     39.10   (--*---)
     4       30    318.00     53.00  (---*--)
     5       30    325.00     46.30   (--*--)
     6       30    328.00     43.00   (---*--)
                                      ----------+---------+---------+------
POOLED STDEV =     47.44              350       400       450

Dunnett's intervals for treatment mean minus control mean

     Family error rate = 0.0500
Individual error rate = 0.0121

Critical value = 2.54

Control = level 1 of level

Level    Lower    Center    Upper  --------+---------+---------+---------+---
  2      76.89    108.00    139.11                       (------*-----)
  3     -28.11      3.00     34.11   (------*-----)
  4     -37.11     -6.00     25.11   (-----*-----)
  5     -30.11      1.00     32.11   (-----*-----)
  6     -27.11      4.00     35.11    (-----*-----)
                                    --------+---------+---------+---------+---
                                            0        50       100       150
```

Figure 3.1 *Dunnett's two-sided MCC of bovine growth hormone* ($\alpha = 0.05$)

Since the confidence interval for $\mu_2 - \mu_1$ is to the right of 0, we have established that the experiment is sufficiently sensitive to distinguish between the positive control and the negative control. Since the confidence intervals for $\mu_i - \mu_1, i = 3, 4, 5, 6$, all cover zero, there is no evidence that orally fed growth hormone changes the rate of weight gain at any dose. (Such information can be deduced by noting the $p$-values reported in Juskevich and Guyer 1990 are larger than $\alpha$.) Whether the experiment has established that the effects of doses 3, 4, 5 and 6 are practically equivalent to the negative control or not depends on what can be considered practically equivalent. If two doses with a difference within 50 can be considered equivalent (say), then practical equivalence has been established. If two doses must have a difference within 25 to be considered equivalent (say), then the present data are insufficient to support a claim of practical equivalence, at least by Dunnett's method. (The reader is reminded that this illustrative analysis is based on a subset of the data from one of a number of experiments described in Juskevich and Guyer 1990.)

### 3.2.3 Unbalanced one-way model

Consider the unbalanced one-way model (3.6).

### 3.2.3.1 Dunnett's method

Dunnett's (1955) two-sided method provides the following simultaneous confidence intervals for the difference between each new treatment mean $\mu_i$ and the control mean $\mu_k$:

$$\mu_i - \mu_k \in \hat{\mu}_i - \hat{\mu}_k \pm |d|\hat{\sigma}\sqrt{n_i^{-1} + n_k^{-1}} \text{ for } i = 1, \ldots, k-1,$$

where $|d|$ is the solution to the equation

$$\int_0^\infty \int_{-\infty}^{+\infty} \prod_{i=1}^{k-1} [\Phi((\lambda_i z + |d|s)/(1 - \lambda_i^2)^{1/2})$$
$$- \Phi((\lambda_i z - |d|s)/(1 - \lambda_i^2)^{1/2})] d\Phi(z) \, \gamma(s) ds = 1 - \alpha \qquad (3.15)$$

with

$$\lambda_i = \left(1 + \frac{n_k}{n_i}\right)^{-1/2}, \quad i = 1, \ldots, k-1.$$

Note that, in addition to $\alpha$, $k$ and $\nu$, the quantile $|d|$ depends on the sample size ratios $n_k/n_1, \ldots, n_k/n_{k-1}$. Thus, it is not possible to tabulate $|d|$ in general. But it is possible to program the computer to solve for $|d|$ in (3.15) at execution time depending on the sample size pattern $n = (n_1, \ldots, n_k)$ of the data to be analyzed, which is the implementation on the Fit Y by X platform of JMP Version 2 and in the ONEWAY command of MINITAB Release 8. The qmcc computer program described in Appendix D with inputs $\lambda = (\lambda_1, \ldots, \lambda_{k-1})$, $\nu$ and $\alpha$ computes $|d|$.

**Theorem 3.2.2** *With $|d|$ defined by (3.15),*

$$P\{\mu_i - \mu_k \in \hat{\mu}_i - \hat{\mu}_k \pm |d| \hat{\sigma}\sqrt{n_i^{-1} + n_k^{-1}} \text{ for } i = 1, \ldots, k-1\} = 1 - \alpha.$$

*Proof.* If we let $Z_1, \ldots, Z_k$ be i.i.d. standard normal random variables, then

$$P\left\{\mu_i - \mu_k \in \hat{\mu}_i - \hat{\mu}_k \pm |d| \hat{\sigma}\sqrt{n_i^{-1} + n_k^{-1}} \text{ for } i = 1, \ldots, k-1\right\}$$

$$= E\left[E\left[P\left\{\left|\frac{\sqrt{n_i}(\hat{\mu}_i - \mu_i)}{\sigma} - \sqrt{\frac{n_i}{n_k}}\left[\frac{\sqrt{n_k}(\hat{\mu}_k - \mu_k)}{\sigma}\right]\right| < \right.\right.\right.$$

$$\left.\left.\left. |d|\left(\frac{\hat{\sigma}}{\sigma}\right)\sqrt{1 + \frac{n_i}{n_k}} \text{ for } i = 1, \ldots, k-1\right\}\Big|\hat{\mu}_k\right]\Big|\hat{\sigma}\right]$$

$$= E\left[E\left[P\left\{\left|Z_i - \sqrt{\frac{n_i}{n_k}}Z_k\right| < |d|\left(\frac{\hat{\sigma}}{\sigma}\right)\sqrt{1 + \frac{n_i}{n_k}}\right.\right.\right.$$

$$\left.\left.\left. \text{ for } i = 1, \ldots, k-1\right\}\Big|\hat{\mu}_k\right]\Big|\hat{\sigma}\right]$$

$$= \int_0^\infty \int_{-\infty}^{+\infty} \prod_{i=1}^{k-1}[\Phi((\lambda_i z + |d|s)/(1 - \lambda_i^2)^{1/2}) -$$

$$\Phi((\lambda_i z - |d|s)/(1 - \lambda_i^2)^{1/2})]d\Phi(z)\gamma(s)ds$$

$$= 1 - \alpha,$$

since

$$\lambda_i/(1 - \lambda_i^2)^{1/2} = \sqrt{\frac{n_i}{n_k}}$$

$$1/(1 - \lambda_i^2)^{1/2} = \sqrt{1 + \frac{n_i}{n_k}}.$$

□

### 3.2.3.2 The Miller–Winer method

For unbalanced designs, instead of solving for $|d|$ in (3.15), the so-called Miller–Winer method (implemented in program P7D of BMDP, for example), takes the harmonic mean of the new treatment sample sizes

$$\tilde{n} = \left(\left(\frac{1}{n_1} + \cdots + \frac{1}{n_{k-1}}\right)/(k-1)\right)^{-1}$$

(excluding the control) to be the 'common' sample size of the new treatments, and Dunnett's method for treatment-balanced ($n_1 = \cdots = n_{k-1}$) designs is employed. As for one-sided MCC inference, this leads to invalid statistical inference in general, and the Miller–Winer method is not recommended.

### 3.2.3.3 Methods based on probabilistic inequalities

If you do not have access to a computer package which correctly implements two-sided MCC (e.g., JMP or MINITAB), or software which computes $|d|$ (e.g., qmcc), but have access to software which computes univariate $t$ quantiles, then the following two methods based on probabilistic inequalities can be considered conservative alternatives of last resort.

Let

$$E_i = \{|\hat{\mu}_i - \hat{\mu}_k - (\mu_i - \mu_k)|/\hat{\sigma}\sqrt{n_i^{-1} + n_k^{-1}} < |d|\}. \tag{3.16}$$

The Bonferroni inequality (A.3) states

$$P(\bigcup_{i=1}^{k-1} E_i^c) \leq \sum_{i=1}^{k-1} P(E_i^c).$$

If each $E_i^c$ is such that

$$P(E_i^c) = \alpha/(k-1),$$

then

$$P(\bigcup_{i=1}^{k-1} E_i^c) \leq \alpha.$$

Thus, a conservative approximation to $|d|$ is

$$|d|_{\text{Bonferroni}} = t_{\frac{\alpha}{2(k-1)},\nu}. \tag{3.17}$$

Šidák's inequality (Theorem A.4.1) states

$$P(\bigcap_{i=1}^{k-1} E_i \mid \hat{\sigma}) \geq \prod_{i=1}^{k-1} P(E_i \mid \hat{\sigma}).$$

Now $P(E_i|\hat{\sigma}), i = 1, \ldots, k-1$, are monotone in $\hat{\sigma}$ in the same direction, so one can further apply Corollary A.1.1 to give

$$P(\bigcap_{i=1}^{k-1} E_i) \geq E_{\hat{\sigma}}[\prod_{i=1}^{k-1} P(E_i \mid \hat{\sigma})] \geq \prod_{i=1}^{k-1} E_{\hat{\sigma}}[P(E_i|\hat{\sigma})] = \prod_{i=1}^{k-1} P(E_i).$$

If each $E_i$ is such that

$$P(E_i) = (1 - \alpha)^{1/(k-1)},$$

then

$$P(\bigcap_{i=1}^{k-1} E_i) \geq 1 - \alpha.$$

Thus, a conservative approximation $|d|_{\text{Šidák}}$ to $|d|$ results if one pretends $E_1, \ldots, E_{k-1}$ were independent:

$$|d|_{\text{Šidák}} = t_{\frac{1-(1-\alpha)^{1/(k-1)}}{2},\nu}.$$

In other words, Šidák's approximation for two-sided MCC is analogous to the product inequality approximation for inference on the means in Section 1.3.3. However, relative to exact critical values, this approximation for two-sided MCC is worse, due to the additional inequality (Šidák's) employed in arriving at the approximation. Also, as explained in Section 1.3.5, this approximation is slightly less conservative than the Bonferroni approximation. Therefore, unless the software you have access to is only capable of univariate $t$ quantiles, neither the approximation based on the Bonferroni inequality nor the approximation based on Šidák's inequality is recommended.

One could also obtain a better (less conservative) approximation than $|d|_{\text{Bonferroni}}$ by using the Hunter–Worsley inequality (A.6) instead of the Bonferroni inequality. However, implementing the Hunter–Worsley inequality method involves no less coding and computation than the exact method. One might as well use the exact method.

### 3.2.4 Example: human food safety of recombinant insulin-like growth factor-I (rIGF-I)

Juskevich and Guyer (1990) also included data from an experiment in which absolute weights of various organs were measured from control hypophysectomized rats and hypophysectomized rats treated orally with the peptide hormone rIGF-I. In addition to groups given rIGF-I orally, one group was given a saline control; another was given bovine serum albumin (BSA) as a negative 'oral protein' control; yet another group was given rIGF-I via a subcutaneously implanted osmotic minipump as a positive control. Spleen weights of rats treated for either 17 days by gavage or 15 days by continuous subcutaneous (SC) infusion are given in Table 3.2. While comparisons with both the saline control and the oral protein control are of interest, for illustration, we concentrate on comparisons with the saline control.

Table 3.2 *Spleen weight (in grams) of male rats*

| Hormone | Dosage (mg/kg per day) | Sample size | Mean weight | SEM weight |
|---|---|---|---|---|
| 1 = None | 0 | 20 | 147.6 | 8.8 |
| 2 = Oral BSA | 1.0 | 20 | 151.3 | 8.6 |
| 3 = Oral rIGF-I | 0.01 | 20 | 147.2 | 5.7 |
| 4 = Oral rIGF-I | 0.1 | 20 | 149.6 | 5.8 |
| 5 = Oral rIGF-I | 1.0 | 20 | 147.1 | 6.6 |
| 6 = SC infusion rIGF-I | 1.0 | 10 | 239.6 | 17.9 |

After converting the reported standard errors of the mean (SEMs) to standard deviations, we find

$$\hat{\sigma} = \left( \frac{19(39.35^2 + 38.46^2 + 25.49^2 + 25.94^2 + 29.52^2) + 9(56.60^2)}{104} \right)^{\frac{1}{2}}$$

$$= 35.09$$

with $\nu = 5(20 - 1) + (10 - 1) = 104$ degrees of freedom. For $\alpha = 0.05$,

$$|d| = 2.561$$
$$|d|_{\text{Šidák}} = 2.617$$
$$|d|_{\text{Bonferroni}} = 2.624.$$

Therefore, Dunnett's two-sided MCC method infers

$$-24.71 < \mu_2 - \mu_1 < 32.11$$
$$-28.81 < \mu_3 - \mu_1 < 28.01$$
$$-26.41 < \mu_4 - \mu_1 < 30.41$$
$$-28.91 < \mu_5 - \mu_1 < 27.91$$
$$57.21 < \mu_6 - \mu_1 < 126.79$$

Inferences based on the Bonferroni inequality or Šidák's inequality will be less precise (more conservative).

We again use an artificially generated data set with summary statistics matching those in Juskevich and Guyer (1990) to illustrate computer analysis. Figure 3.2 displays Dunnett's two-sided method performed by the MCC subcommand of the ONEWAY command of MINITAB.

Since the confidence interval for $\mu_6 - \mu_1$ is to the right of 0, we have established that the experiment is sufficiently sensitive to distinguish between the positive control and the saline control. Since the confidence intervals for $\mu_i - \mu_1, i = 3, 4, 5$, all cover zero, there is no evidence that orally fed rIGF-I changes spleen weight. (Such information can be deduced by noting that the $p$-values reported in Juskevich and Guyer 1990 are larger than $\alpha$.) Whether the experiment has established that rIGF-I doses 3, 4, and 5 are practically equivalent to the saline control or not depends on what can be considered practically equivalent. If mean spleen weights with a difference within 40 can be considered equivalent (say), then practical equivalence has been established. If mean spleen weights must have a difference within 30 to be considered equivalent (say), then the present data are insufficient to support a claim of practical equivalence (at least by Dunnett's method).

## 3.3 Nonparametric methods

Suppose that the one-way model (3.6) holds but that the error distribution may not be normal:

$$Y_{ia} = \mu_i + \epsilon_{ia}, \quad i = 1, \ldots, k, \quad a = 1, \ldots, n_i, \tag{3.18}$$

```
MTB > Oneway 'Weight' 'Hormone';
SUBC>    Dunnett 5 1.

ANALYSIS OF VARIANCE ON Weight
SOURCE      DF        SS        MS       F        p
Hormone      5      75618     15124    12.28    0.000
ERROR      104     128051      1231
TOTAL      109     203669
                                  INDIVIDUAL 95 PCT CI'S FOR MEAN
                                  BASED ON POOLED STDEV
LEVEL        N      MEAN     STDEV   --------+---------+---------+--------
  1         20    147.60     39.35   (---*---)
  2         20    151.30     38.46    (---*---)      /
  3         20    147.20     25.49   (---*---)
  4         20    149.60     25.94    (--*---)
  5         20    147.10     29.52   (---*---)
  6         10    239.60     56.60                         (-----*----)
                                  --------+---------+---------+--------
POOLED STDEV =      35.09               160       200       240

Dunnett's intervals for treatment mean minus control mean

     Family error rate = 0.0500
Individual error rate = 0.0119

Critical value = 2.56

Control = level 1 of Hormone

Level     Lower    Center    Upper  --------+---------+---------+---------+---
  2      -24.71      3.70    32.11   (------*------)
  3      -28.81     -0.40    28.01  (------*------)
  4      -26.41      2.00    30.41   (-------*------)
  5      -28.91     -0.50    27.91  (------*------)
  6       57.21     92.00   126.79                   (--------*--------)
                                  --------+---------+---------+---------+---
                                          0        40        80       120
```

Figure 3.2 *Dunnett's two-sided MCC of recombinant insulin-like growth factor-I*
*($\alpha = 0.05$)*

where $\epsilon_{11}, \ldots, \epsilon_{kn_k}$ are i.i.d. with distribution $F$, which is absolutely continuous but otherwise unknown. By the central limit theorem, under suitable
conditions the methods based on sample means discussed in this chapter
will hold their error rates at approximately the nominal error rate when
the sample sizes are large. However, these methods are inefficient for many
possible $F$ in the sense that they give relatively wide confidence intervals,
or are relatively ineffective in discovering which treatments are better (or
worse) than the control. One way to construct statistical methods that hold
their error rates at the nominal level for all possible $F$ and any sample size,
as well as provide more efficient inference for a large class of possible $F$s,
is to use the fact that, under

$$H_0 : \mu_1 = \cdots = \mu_k,$$

all rankings of

$$Y_{11}, \ldots, Y_{1n_1}, Y_{21}, \ldots, Y_{2n_2}, \ldots, Y_{k1}, \ldots, Y_{kn_k}$$

are equally likely.

### 3.3.1 Pairwise ranking methods

Let $R_{ia}^k(\delta_i)$ denote the rank of $Y_{ia} - \delta_i$ in the combined sample of size $n_i + n_k$

$$Y_{i1} - \delta_i, \ldots, Y_{in_i} - \delta_i, Y_{k1}, \ldots, Y_{kn_k},$$

and let

$$R_i^k(\delta_i) = \sum_{a=1}^{n_i} R_{ia}^k(\delta_i).$$

Let

$$D_{i[1]} \leq \cdots \leq D_{i[n_i n_k]}$$

denote the ordered $n_i n_k$ differences

$$Y_{ia} - Y_{kb}, 1 \leq a \leq n_i, \ 1 \leq b \leq n_k$$

with the additional understanding that

$$D_{i[0]} = -\infty$$

and

$$D_{i[n_i n_k + 1]} = +\infty.$$

There is a well-known relationship between $R_i^k(\delta_i)$ and the $D_{i[a]}$'s, the proof of which can be found in Lehmann (1975, p. 87), for example.

**Lemma 3.3.1** *For any $l$ $(0 \leq l \leq n_i n_k + 1)$ and any $\delta_i$,*

$$D_{i[l]}^k \leq \delta_i \text{ if and only if } R_i^k(\delta_i) \leq n_i n_k + \frac{n_i(n_i + 1)}{2} - l$$

*and*

$$D_{i[l]}^k > \delta_i \text{ if and only if } R_i^k(\delta_i) \geq n_i n_k + \frac{n_i(n_i + 1)}{2} - l + 1.$$

Using this relationship, level $1 - \alpha$ simultaneous MCC confidence intervals can be constructed as follows.

For one-sided MCC inference under the one-way location model (3.18) with balanced sample sizes $n_1 = \cdots = n_k = n$, let $r$ be the smallest integer such that, under $H_0$,

$$P_{H_0}\{R_i^k(0) \leq r \text{ for } i = 1, \ldots, k - 1\} \geq 1 - \alpha.$$

Then since under $H_0$ all rankings of

$$Y_{11}, \ldots, Y_{1n}, Y_{21}, \ldots, Y_{2n}, \ldots, Y_{k1}, \ldots, Y_{kn}$$

are equally likely, $r$ can be computed without knowledge of $F$. Note that, by the symmetry of the distribution of $R_1^k(0), \ldots, R_{k-1}^k(0)$ under $H_0$, the critical value $r$ is also the smallest integer such that, under $H_0$,

$$P\{R_i^k(0) \geq n(2n+1) - r \text{ for } i = 1, \ldots, k-1\} \geq 1 - \alpha.$$

**Theorem 3.3.1**

$$P\{\mu_i - \mu_k < D_{i[r-n(n+1)/2+1]} \text{ for } i = 1, \ldots, k-1\} \geq 1 - \alpha$$

*and*

$$P\{\mu_i - \mu_k \geq D_{i[n(3n+1)/2-r]} \text{ for } i = 1, \ldots, k-1\} \geq 1 - \alpha.$$

*Proof.* We first note that changing the value of $\delta_i$ in $R_i^k(\delta_i)$ may change the value of $R_i^k(\delta_i)$, but not the value of any other $R_j^k(\delta_j), j \neq i$. Therefore

$$
\begin{aligned}
1 - \alpha \quad &\leq \quad P_{H_0}\{R_i^k(0) \leq r \text{ for } i = 1, \ldots, k-1\} \\
&= \quad P\{R_i^k(\mu_i - \mu_k) \leq r \text{ for } i = 1, \ldots, k-1\} \\
&= \quad P\{\mu_i - \mu_k \geq D_{i[n(3n+1)/2-r]} \\
&\qquad \text{for } i = 1, \ldots, k-1\},
\end{aligned}
$$

$$
\begin{aligned}
1 - \alpha \quad &\leq \quad P_{H_0}\{R_i^k(0) \geq n(2n+1) - r \text{ for } i = 1, \ldots, k-1\} \\
&= \quad P\{R_i^k(\mu_i - \mu_k) \geq n(2n+1) - r \text{ for } i = 1, \ldots, k-1\} \\
&= \quad P\{\mu_i - \mu_k < D_{i[r-n(n+1)/2+1]} \text{ for } i = 1, \ldots, k-1\}. \quad \square
\end{aligned}
$$

For two-sided MCC inference under the one-way model (3.18) with balanced sample sizes $n_1 = \cdots = n_k = n$, let $|r|$ be the smallest integer such that, under $H_0$,

$$P_{H_0}\{n(2n+1) - |r| \leq R_i^k(0) \leq |r| \text{ for } i = 1, \ldots, k-1\} \geq 1 - \alpha,$$

then again $|r|$ can be computed without knowledge of $F$.

**Theorem 3.3.2**

$$
\begin{aligned}
P\{D_{i[n(3n+1)/2-|r|]} \leq \mu_i - \mu_k &< D_{i[|r|-n(n+1)/2+1]} \\
\text{for } i = 1, \ldots, k-1\} \quad &\geq \quad 1 - \alpha.
\end{aligned}
$$

*Proof.* Noting that changing the value of $\delta_i$ in $R_i^k(\delta_i)$ may change the value of $R_i^k(\delta_i)$, but not the value of any other $R_j^k(\delta_j), j \neq i$, we have

$$
\begin{aligned}
1 - \alpha \quad &\leq \quad P_{H_0}\{n(2n+1) - |r| \leq R_i^k(0) \leq |r| \text{ for } i = 1, \ldots, k-1\} \\
&= \quad P\{n(2n+1) - |r| \leq R_i^k(\mu_i - \mu_k) \leq |r| \text{ for } i = 1, \ldots, k-1\} \\
&= \quad P\{D_{i[n(3n+1)/2-|r|]} \leq \mu_i - \mu_k < D_{i[|r|-n(n+1)/2+1]} \\
&\qquad \text{for } i = 1, \ldots, k-1\}. \quad \square
\end{aligned}
$$

Steel (1959a), who first proposed these methods, gave exact tables of $r$ and $|r|$ only for $k = 3, 4$, and $n = 2, 3, 4$. But for large $n$, using the asymptotic multivariate normality of $R_1^k(0), \ldots, R_{k-1}^k(0)$ under $H_0$, the critical value $r$ can be approximated by

$$r \approx \frac{n(2n+1)}{2} + d_\infty \sqrt{\frac{n^2(2n+1)}{12}} + 0.5,$$

where $d_\infty$ is the critical value of Dunnett's one-sided method for the same one-way model, but with degrees of freedom $\nu = \infty$, and the critical value $|r|$ can be approximated by

$$|r| \approx \frac{n(2n+1)}{2} + |d|_\infty \sqrt{\frac{n^2(2n+1)}{12}} + 0.5,$$

where $|d|_\infty$ is the critical value of Dunnett's two-sided method for the same one-way model, except with degrees of freedom $\nu = \infty$.

Steel's method (one-sided or two-sided) can be thought of as the pairwise Wilcoxon–Mann–Whitney statistic analog of Dunnett's corresponding sample-means-based method. Let the asymptotic efficiency of confidence set A relative to confidence set B be defined as the reciprocal of the limiting ratio of sample sizes needed so that the two confidence sets have the same asymptotic probability of excluding the same sequence of false parameters $\mu_i^{(n)} - \mu_k^{(n)}, i = 1, \ldots, k-1$, approaching $(0, \ldots, 0)$ at the rate of $n^{1/2}$, i.e.

$$\mu_i^{(n)} - \mu_k^{(n)} = \frac{\Delta_i}{\sqrt{n}}, \ i = 1, \ldots, k-1,$$

where $\Delta_1, \ldots, \Delta_{k-1}$ are fixed positive constants. The asymptotic relative efficiency (ARE) of Steel's method relative to Dunnett's method, under certain regularity conditions, is the same as the Pitman ARE of the Wilcoxon–Mann–Whitney test relative to the $t$-test:

$$e_{Steel, Dunnett}(F) = 12\sigma^2 \left[ \int_{-\infty}^{\infty} f^2(y) dy \right]^2, \tag{3.19}$$

where $f$ is the density of $F$ and $\sigma^2$ is the variance of a typical $Y_{ia}$. When $F$ is a normal distribution,

$$e_{Steel, Dunnett}(F) = 3/\pi \approx 0.955.$$

When $F$ is not a normal distribution, $e_{Steel, Dunnett}(F)$ is never less than 0.864 for all $F$ with finite variance, and can be considerably greater than 1 (equaling 3 if $F$ is an exponential distribution, for example). Thus, if the normality assumption in the location model (3.18) is in doubt, then Steel's method is preferable to Dunnett's method.

For one-sided MCC inference under the unbalanced one-way location model (3.18), in analogy to Theorem 3.3.1, it is easy to deduce from Lemma

3.3.1 that, if $r_1, \ldots, r_{k-1}$ are integers satisfying

$$P_{H_0}\{R_i^k(0) \le r_i \text{ for } i = 1, \ldots, k-1\} \ge 1 - \alpha, \qquad (3.20)$$

then

$$P\{\mu_i - \mu_k < D_{i[r_i - n_i(n_i+1)/2+1]} \text{ for } i = 1, \ldots, k-1\} \ge 1 - \alpha$$

and

$$P\{\mu_i - \mu_k \ge D_{i[n_i n_k + n_i(n_i+1)/2 - r_i]} \text{ for } i = 1, \ldots, k-1\} \ge 1 - \alpha.$$

Software to compute the probability

$$P_{H_0}\{R_i^k(0) \le r_i \text{ for } i = 1, \ldots, k-1\} \qquad (3.21)$$

exactly given a set of candidate $r_1, \ldots, r_{k-1}$ does not seem readily available. But, for large $n_1, \ldots, n_k$, one can use the asymptotic multivariate normality of $R_1^k(0), \ldots, R_{k-1}^k(0)$ under $H_0$ to approximate the probability (3.21). In general, there are many combinations of $r_1, \ldots, r_{k-1}$ satisfying (3.20). If, in analogy with Dunnett's method, one standardizes the individual comparisons to approximately the same level, then for large $n_1, \ldots, n_k$, one can approximate the $r_i$'s by

$$r_i \approx \frac{n_i(n_i + n_k + 1)}{2} + d_\infty \sqrt{\frac{n_i n_k (n_i + n_k + 1)}{12}} + 0.5,$$

where $d_\infty$ is the critical value of Dunnett's one-sided method for the same one-way model, but with degrees of freedom $\nu = \infty$.

For two-sided MCC inference under the unbalanced one-way location model (3.18), in analogy to Theorem 3.3.2, it is easy to deduce from Lemma 3.3.1 that if $|r_1|, \ldots, |r_{k-1}|$ are integers such that

$$P_{H_0}\{n_i(n_i+n_k+1) - |r_i| \le R_i^k(0) \le |r_i| \text{ for } i = 1, \ldots, k-1\} \ge 1-\alpha, \quad (3.22)$$

then

$$P\{D_{i[n_i n_k + n_i(n_i+1)/2 - |r_i|]} \le \mu_i - \mu_k$$
$$< D_{i[|r_i| - n_i(n_i+1)/2+1]} \text{ for } i = 1, \ldots, k-1\} \ge 1 - \alpha.$$

Software to compute the probability

$$P_{H_0}\{n_i(n_i + n_k + 1) - |r_i| \le R_i^k(0) \le |r_i| \text{ for } i = 1, \ldots, k-1\} \qquad (3.23)$$

exactly given a set of candidate $|r_1|, \ldots, |r_{k-1}|$ does not seem readily available. But, for large $n_1, \ldots, n_k$, one can use the asymptotic multivariate normality of $R_1^k(0), \ldots, R_{k-1}^k(0)$ under $H_0$ to approximate the probability (3.23). In general, there are many combinations of $|r_1|, \ldots, |r_{k-1}|$ satisfying (3.22). If, in analogy with Dunnett's method, one standardizes the individual comparisons to approximately the same level, then for large $n_1, \ldots, n_k$, one can approximate the $|r_i|$'s by

$$|r_i| \approx \frac{n_i(n_i + n_k + 1)}{2} + |d|_\infty \sqrt{\frac{n_i n_k (n_i + n_k + 1)}{12}} + 0.5,$$

where $|d|_\infty$ is the critical value of Dunnett's two-sided method for the same one-way model, but with degrees of freedom $\nu = \infty$.

### 3.3.2 Joint ranking methods

Under the one-way location model (3.18) with equal sample sizes $n_1 = \cdots = n_k = n$, let $\boldsymbol{\delta} = (\delta_1, \ldots, \delta_{k-1})$, let $R_{ia}(\boldsymbol{\delta})$ denote the rank of $Y_{ia} - \delta_i$ in the sample combining all $kn$ observations

$$Y_{11} - \delta_1, \ldots, Y_{1n} - \delta_1, \ldots, Y_{(k-1)1} - \delta_{k-1}, \ldots, Y_{(k-1)n} - \delta_{k-1}, Y_{k1}, \ldots, Y_{kn}$$

and let $R_i(\boldsymbol{\delta})$ denote the rank sum of the $i$th treatment (in this joint ranking of all $k$ treatments):

$$R_i(\boldsymbol{\delta}) = \sum_{a=1}^{n} R_{ia}(\boldsymbol{\delta}).$$

Suppose $r_J$ is the smallest integer such that

$$P_{H_0}\{R_i - R_k \le r_J \text{ for } i = 1, \ldots, k-1\} \ge 1 - \alpha \qquad (3.24)$$

under

$$H_0 : \mu_1 = \cdots = \mu_k,$$

where $R_i = R_i(0, \ldots, 0), i = 1, \ldots, k$; then since

$$(3.24) = P_{\boldsymbol{\mu}}\{R_i(\boldsymbol{\delta}^*) - R_k(\boldsymbol{\delta}^*) \le r_J \text{ for } i = 1, \ldots, k-1\},$$

where $\boldsymbol{\delta}^* = (\mu_1 - \mu_k, \ldots, \mu_{k-1} - \mu_k)$, a $100(1 - \alpha)\%$ confidence set for $\mu_1 - \mu_k, \ldots, \mu_{k-1} - \mu_k$ is

$$C_1 = \{(\delta_1, \ldots, \delta_{k-1}) : R_i(\boldsymbol{\delta}) - R_k(\boldsymbol{\delta}) \le r_J \text{ for } i = 1, \ldots, k-1\}.$$

Suppose $|r|_J$ is the smallest integer such that

$$P_{H_0}\{|R_i - R_k| \le |r|_J \text{ for } i = 1, \ldots, k-1\} \ge 1 - \alpha \qquad (3.25)$$

under

$$H_0 : \mu_1 = \cdots = \mu_k,$$

then since

$$(3.25) = P_{\boldsymbol{\mu}}\{|R_i(\boldsymbol{\delta}^*) - R_k(\boldsymbol{\delta}^*)| \le |r|_J \text{ for } i = 1, \ldots, k-1\},$$

a $100(1 - \alpha)\%$ confidence set for $\mu_1 - \mu_k, \ldots, \mu_{k-1} - \mu_k$ is

$$C_2 = \{(\delta_1, \ldots, \delta_{k-1}) : |R_i(\boldsymbol{\delta}) - R_k(\boldsymbol{\delta})| \le |r|_J \text{ for } i = 1, \ldots, k-1\}.$$

Unfortunately, no general technique for computing $C_1$ or $C_2$ is known. The difficulty is that, in contrast to the pairwise ranking method, changes in the value of $\delta_i$ not only induce changes in $R_i(\delta) - R_k(\delta)$, but changes in $R_j(\delta) - R_k(\delta), j \neq i$, as well.

A popular one-sided (0-1 decision) MCC method (Hollander and Wolfe, 1973, p. 130) asserts

$$\mu_i > \mu_k \text{ for all } i \ (1 \leq i \leq k-1) \text{ such that } R_i - R_k > r_J.$$

When $H_0$ is true, clearly

$$P_{H_0}\{\text{at least one incorrect assertion}\} \leq \alpha. \tag{3.26}$$

However, as we indicate below, at least for some $k$,

$$\lim_{n \to \infty} \sup_{F,\mu} P_{F,\mu}\{\text{at least one incorrect assertion}\} > \alpha$$

at the usual $\alpha$ levels.

Suppose $\mu_1 = \delta, \mu_2 = \cdots = \mu_k = 0$, so that an assertion of $\mu_i \neq \mu_k$ for any $i, i = 2, \ldots, k-1$, is incorrect. The distributional results in Oude Voshaar (1980) showed that, for this parameter configuration, asymptotically $(n^3 b)^{-1/2}(R_2 - R_k, \ldots, R_{k-1} - R_k)$ and $Z_2 + Z_k, \ldots, Z_{k-1} + Z_k$ have the same distribution, where $Z_1, \ldots, Z_k$ are i.i.d. standard normal random variables, and $b = b(F,\delta) = k^2/12 + (2Cov(F(X), F(X-\delta)) - 1/6)(k-1) + Var(F(X-\delta)) - 1/12$. Oude Voshaar (1980) further showed that

$$\sup_{F,\delta} b(F,\delta) \geq (k^2 + k/2 + 5/16)/12.$$

Thus, noting (Hollander and Wolfe 1973, p.130) that

$$\lim_{n \to \infty} r_J = \sqrt{2} d_{k-1} n(k(kn+1)/12)^{1/2}$$

where $d_{k-1}$ is the critical value for Dunnett's one-sided method for comparing $k-1$ new treatments with a control with infinite $MSE$ degrees of freedom ($\nu = \infty$), we have

$$\lim_{n \to \infty} \sup_{F,\mu} P_{F,\mu}\{\text{at least one incorrect assertion}\}$$
$$\geq \quad P\{Z_i - Z_k > \sqrt{2} d_{k-1} k/(k^2 + k/2 + 5/16)^{1/2} \text{ for some } i,$$
$$i = 2, \ldots, k-1\}.$$

Table 3.3 tabulates this probability which, as can be seen, can be greater than $\alpha$. So this 'nonparametric' MCC method is not even a confident inequalities method, and is not recommended.

A popular two-sided MCC method (Hollander and Wolfe, 1973, p. 130) asserts that

$$\mu_i \neq \mu_k \text{ for all } i \ (1 \leq i \leq k-1) \text{ such that } |R_i - R_k| > |r|_J.$$

When $H_0$ is true, clearly (3.26) holds. However, in analogy to the one-sided

Table 3.3 *Lower bound on asymptotic error rate of one-sided joint ranking method*

| | | | $k$ | | | |
|---|---|---|---|---|---|---|
| $\alpha$ | 3 | 5 | 7 | 10 | 15 | 20 |
| 0.10 | 0.0751 | 0.0982 | 0.1022 | 0.1035 | 0.1034 | 0.1030 |
| 0.05 | 0.0402 | 0.0516 | 0.0531 | 0.0532 | 0.0528 | 0.0523 |
| 0.01 | 0.0098 | 0.0118 | 0.0117 | 0.0114 | 0.0111 | 0.0109 |

case, Fligner (1984) showed that

$$\lim_{n \to \infty} \sup_{F,\mu} P_{F,\mu}\{\text{at least one incorrect assertion}\}$$

$$\geq \quad P\{|Z_i - Z_k| > \sqrt{2}|d|_{k-1}k/(k^2 + k/2 + 5/16)^{1/2} \text{ for some } i,$$
$$i = 2, \ldots, k-1\}$$

where $|d|_{k-1}$ is the critical value of Dunnett's two-sided method for comparing $k-1$ new treatments with a control with infinite $MSE$ degrees of freedom ($\nu = \infty$). Table 3.4 tabulates this probability which, as can be seen, can be greater than $\alpha$. So this 'nonparametric' MCC method is also not a confident inequalities method, and is not recommended.

Table 3.4 *Lower bound on asymptotic error rate of two-sided joint ranking method*

| | | | $k$ | | | |
|---|---|---|---|---|---|---|
| $\alpha$ | 3 | 5 | 7 | 10 | 15 | 20 |
| 0.10 | 0.0804 | 0.1032 | 0.1062 | 0.1065 | 0.1055 | 0.1047 |
| 0.05 | 0.0436 | 0.0545 | 0.0554 | 0.0549 | 0.0539 | 0.0532 |
| 0.01 | 0.0108 | 0.0125 | 0.0123 | 0.0118 | 0.0113 | 0.0111 |

## 3.4 Other approaches to stepwise testing

The derivation of stepwise methods in this book follows the idea of Stefansson, Kim and Hsu (1988), which is to partition the parameter space and choose a different (typically equivariant) family of tests for each member of the partition in accordance with the decision appropriate to that member. In particular, for one-sided multiple comparisons with a control, the approach presented here partitions the parameter space $\Theta = \Re^k$ into $2^{k-1}$ parts as

$$\Theta = \bigcup_B \{\mu \mid \{i : \mu_i \leq \mu_k\} = B\}.$$

A different approach is to test multiple hypotheses (each at level $\alpha$) and make decisions using the so-called *closure* principle described by Marcus, Peritz and Gabriel (1976) which, in the setting of one-sided multiple comparisons with a control, rejects

$$H_i : \mu_i \leq \mu_k \tag{3.27}$$

and infers

$$\mu_i > \mu_k$$

if and only if all hypotheses $H_B : \mu_j \leq \mu_k \ \forall j \in B$ implying (3.27) are rejected. In Section 5.1.7, a general description of the closure technique is given, and its application to all-pairwise comparisons is discussed.

Since the closure technique is usually described in terms of tests of hypotheses, it is sometimes stated in terms of $p$-values associated with the tested hypotheses. Placed in the context of multiple comparisons with a control, the 'sequentially rejective' method of Holm (1979a), for example, is a stepdown method analogous to the Naik/Marcus–Peritz–Gabriel stepdown method, except that the critical value of Holm's method at step $j$ is $t_{\alpha/(k-j),\nu}$, which is larger than the critical value $d_{k-j}$ of the stepdown method at all steps except the $(k-1)$th. In essence, Holm's method uses the Bonferroni inequality to guarantee

$$E_B^{\downarrow} = \left\{ \mu_i - \mu_k > \hat{\mu}_i - \hat{\mu}_k - t_{\alpha/|B|,\nu}\hat{\sigma}\sqrt{2/n} \text{ for all } i, i \in B \right\}$$

occurs with a probability of at least $1 - \alpha$. As another example, in the context of one-sided multiple comparisons with a control, the method of Hochberg (1988) is a stepup method analogous to Dunnett and Tamhane's, except that the critical value of Hochberg's method at step $j$ is $t_{\alpha/j,\nu}$, which Dunnett and Tamhane conjecture to be larger than the critical value $c_j$ of their method. Currently, the adjustment of $p$-values in testing multiple hypotheses is an area of very active research. See, for example, Simes (1986), Hommel (1988; 1989) and Rom (1990). Two excellent articles surveying this area are Shaffer (1995) and Tamhane (1995).

## 3.5 Appendix: detailed proofs

### 3.5.1 Proof of the validity of the stepdown method

To prove the validity of the stepdown method, let

$$B = \{i : \mu_i \leq \mu_k\},$$

the set of indices of treatments worse than the control, and let $|B|$ be the number of elements in $B$.

Suppose $B \neq \emptyset$ for the true $\mu$; then the event

$$E_B^{\downarrow} = \left\{ \mu_i - \mu_k > \hat{\mu}_i - \hat{\mu}_k - d_{|B|}\hat{\sigma}\sqrt{2/n} \text{ for all } i, i \in B \right\}$$

occurs with a probability of $1 - \alpha$ by the definition of $d_{|B|}$.

**Lemma 3.5.1** *When $E_B^\downarrow$ occurs,*

$$\{[1],\ldots,[M^\downarrow]\} \supseteq B.$$

*Proof.* When $E_B^\downarrow$ occurs,

$$T_i < d_{|B|} \text{ for all } i \in B$$

since $\mu_i - \mu_k \leq 0$ for all $i \in B$. To illustrate why this implies

$$\{[1],\ldots,[M^\downarrow]\} \supseteq B,$$

suppose, for example, $k = 5$, $B = \{1,2\}$, and

$$T_4 > T_2 > T_3 > T_1.$$

Then the stepdown method proceeds as follows:

$$\begin{array}{ccccc} Step & 1 & 2 & 3 & 4 \\ & T_4 \overset{?}{>} d_4 & T_2 \overset{?}{>} d_3 & T_3 \overset{?}{>} d_2 & T_1 \overset{?}{>} d_1 \end{array}$$

Since $T_2 < d_2$ and $T_1 < d_2$ when $E_B^\downarrow$ occurs, the stepdown method stops at or before step 2, and infers at most $\mu_4 > \mu_5$. One can likewise see that any arrangement of $T_1,\ldots,T_{k-1}$ will still lead to $\{[1],\ldots,[M^\downarrow]\} \supseteq B$ when $E_B^\downarrow$ occurs. The result follows. $\square$

**Corollary 3.5.1 (Naik 1975; Marcus, Peritz and Gabriel 1976)**
*The probability that the stepdown method will incorrectly assert at least one new treatment worse than the control to be better than the control is at most $\alpha$.*

*Proof.* Suppose $B = \emptyset$, that is, every new treatment is in fact better than the control; then obviously the probability that any method will incorrectly assert at least one new treatment worse than the control to be better than the control is 0. Thus assume $B \neq \emptyset$, that is, there exists at least one new treatment worse than the control. Noting

$$P\{E_B^\downarrow\} = 1 - \alpha,$$

the result follows from Lemma 3.5.1. $\square$

**Theorem 3.5.1 (Stefansson, Kim and Hsu 1988)** *The simultaneous coverage probability of the confidence bounds version of the stepdown method is at least $100(1 - \alpha)\%$.*

*Proof.* First suppose that $B \neq \emptyset$ for the true $\mu$; then by Lemma 3.5.1,

$$\{[1],\ldots,[M^\downarrow]\} \supseteq B$$

when $E_B^\downarrow$ occurs. Therefore, when $E_B^\downarrow$ occurs,

$$\begin{aligned} M^\downarrow &> 0, \\ d_{M^\downarrow} &\geq d_{|B|}. \end{aligned}$$

Noting that

$$\hat{\mu}_{[i]} - \hat{\mu}_k - d_{M^\downarrow}\hat{\sigma}\sqrt{2/n} \;<\; 0 \text{ for all } i \le M^\downarrow,$$

we see

$$
\begin{aligned}
E_B^\downarrow &= \{\mu_i - \mu_k > \hat{\mu}_i - \hat{\mu}_k - d_{|B|}\hat{\sigma}\sqrt{2/n} \text{ for all } i, i \in B\} \\
&= \{\mu_i - \mu_k > \hat{\mu}_i - \hat{\mu}_k - d_{|B|}\hat{\sigma}\sqrt{2/n} \text{ for all } i, i \in B \\
&\quad \text{ and } \mu_i - \mu_k > 0 \text{ for all } i, i \notin B\} \\
&\subseteq \{\mu_{[i]} - \mu_k > \hat{\mu}_{[i]} - \hat{\mu}_k - d_{M^\downarrow}\hat{\sigma}\sqrt{2/n} \text{ for all } i \le M^\downarrow \\
&\quad \text{ and } \mu_{[i]} - \mu_k > 0 \text{ for all } i > M^\downarrow\}.
\end{aligned}
$$

Therefore, the confidence level is at least $1 - \alpha$ when $B \ne \emptyset$.

Now suppose that $B = \emptyset$ for the true $\mu$, and $\mu_{i^*} = \min_{1 \le i \le k-1} \mu_i$. Then

$$
\begin{aligned}
E_\emptyset &= \{\mu_i - \mu_k > \min_{1 \le i \le k-1} \hat{\mu}_i - \hat{\mu}_k - t_{\alpha,\nu}\hat{\sigma}\sqrt{2/n} \text{ for all } i = 1,\dots,k-1\} \\
&= \{\min_{1 \le i \le k-1} \hat{\mu}_i - \hat{\mu}_k - \min_{1 \le i \le k-1}(\mu_i - \mu_k) < t_{\alpha,\nu}\hat{\sigma}\sqrt{2/n}\} \\
&\supseteq \{\hat{\mu}_{i^*} - \hat{\mu}_k - (\mu_{i^*} - \mu_k) < t_{\alpha,\nu}\hat{\sigma}\sqrt{2/n}\}.
\end{aligned}
$$

Thus $P(E_\emptyset) \ge 1 - \alpha$. Since $\mu_i - \mu_k > 0$ for all $i$ when $B = \emptyset$, noting that the lower bounds on $\mu_i - \mu_k, i = 1,\dots,k-1$, given by the confidence bounds version of the stepdown method are either all non-positive (when $M^\downarrow > 0$) or all equal to the positive lower bounds in $E_\emptyset$ (when $M^\downarrow = 0$), we see that the confidence level is at least $1 - \alpha$ when $B = \emptyset$ for the true $\mu$ also.   □

### 3.5.2 Proof of the validity of the stepup method

To prove the validity of Dunnett and Tamhane's stepup method, recall $B$ is the the set of indices of treatments worse than the control, and $|B|$ is the number of elements in $B$.

Note that if $B \ne \emptyset$ for the true $\mu$, then the event

$$E_B^\uparrow = \{(\hat{\mu}_i - \hat{\mu}_k - (\mu_i - \mu_k), i \in B) < (c_1,\dots,c_{|B|})\hat{\sigma}\sqrt{2/n}\}$$

occurs with a probability of $1 - \alpha$ by the definition of $c_1,\dots,c_{|B|}$.

**Lemma 3.5.2** *When $E_B^\uparrow$ occurs,*

$$\{[1],\dots,[M^\uparrow]\} \supseteq B.$$

*Proof.* When $E_B^\uparrow$ occurs,

$$(T_i, i \in B) < (c_1,\dots,c_{|B|})$$

since $\mu_i - \mu_k < 0$ for all $i \in B$. To illustrate why this implies

$$\{[1],\dots,[M^\uparrow]\} \supseteq B,$$

suppose, for example, $k = 5$, $B = \{1, 2\}$, and

$$T_3 < T_1 < T_2 < T_4.$$

Then Dunnett and Tamhane's stepup method proceeds as follows:

| *Step* | 1 | 2 | 3 | 4 |
|---|---|---|---|---|
| | $\overset{?}{T_3 > c_1}$ | $\overset{?}{T_1 > c_2}$ | $\overset{?}{T_2 > c_3}$ | $\overset{?}{T_4 > c_4}$ |

Since $(T_1, T_2) < (c_1, c_2)$ when $E_B^\uparrow$ occurs, the stepup method does not stop until it reaches step 4 and infers at most $\mu_4 > \mu_5$. One can likewise see that any arrangement of $T_1, \ldots, T_{k-1}$ will still lead to $\{[1], \ldots, [M^\uparrow]\} \supseteq B$ when $E_B^\uparrow$ occurs. The result follows. $\square$

**Corollary 3.5.2 (Dunnett and Tamhane 1992)** *The probability that Dunnett and Tamhane's method will incorrectly infer at least one new treatment worse than the control to be better than the control is at most $\alpha$.*

*Proof.* Suppose $B = \emptyset$, that is, every new treatment is in fact better than the control; then obviously the probability that any method will incorrectly infer at least one new treatment worse than the control to be better than the control is 0. Thus assume $B \neq \emptyset$, that is, there exists at least one new treatment worse than the control. Noting

$$P\{E_B^\uparrow\} = 1 - \alpha,$$

the result follows from Lemma 3.5.2. $\square$

## 3.6 Exercises

1. Prove that a level-$\alpha$ test for

$$H_0 : \mu_3 \geq \mu_1 \text{ or } \mu_3 \geq \mu_2 \text{ vs. } H_a : \mu_3 < \mu_1 \text{ and } \mu_3 < \mu_2$$

is to reject if a size-$\alpha$ test for

$$H_0 : \mu_3 \geq \mu_1 \text{ vs. } H_a : \mu_3 < \mu_1$$

(with the usual rejection region $\hat{\mu}_3 - \hat{\mu}_1 \leq -d_1 \hat{\sigma} \sqrt{2/n}$, say) and a size-$\alpha$ test for

$$H_0 : \mu_3 \geq \mu_2 \text{ vs. } H_a : \mu_3 < \mu_2$$

(with the usual rejection region $\hat{\mu}_3 - \hat{\mu}_2 \leq -d_1 \hat{\sigma} \sqrt{2/n}$, say) both reject. Conclude that a level-$\alpha$ test for the null hypothesis

$$H_0 : \mu_3 - \min\{\mu_1, \mu_2\} \geq \mu_3^* - \min\{\mu_1^*, \mu_2^*\}$$

(for prespecified constants $\mu_1^*, \mu_2^*, \mu_3^*$) is to accept if

$$\hat{\mu}_3 - \min\{\hat{\mu}_1, \hat{\mu}_2\} > \mu_3^* - \min\{\mu_1^*, \mu_2^*\} - d_1 \hat{\sigma} \sqrt{2/n}.$$

2. In their retrospective study, White and Froeb (1980) also recorded forced mid-expiratory flow (FEF) of 200 randomly selected males from each group, except for the non-inhaling smokers group which, due to the small number of such people available, was limited to 50 males, as shown in Table 3.5.

Table 3.5  *FEF Data for smoking and non-smoking males*

| Group | Sample size | Mean FEF | Std. dev. FEF |
|-------|-------------|----------|---------------|
| NS | 200 | 3.78 | 0.79 |
| PS | 200 | 3.30 | 0.77 |
| NI | 50 | 3.32 | 0.86 |
| LS | 200 | 3.23 | 0.78 |
| MS | 200 | 2.73 | 0.81 |
| HS | 200 | 2.59 | 0.82 |

(a) Apply Dunnett's one-sided MCC method at $\alpha = 0.01$ to compare the FEF of each of the smoking groups with the non-smoking group (NS) ($d = 2.848$).

(b) Ignoring the non-inhaling smokers (NI) group, apply the confidence bounds version of the stepdown method at $\alpha = 0.01$ to compare each of the smoking groups with the non-smoking group (NS) ($d_1 = 2.326$, $d_2 = 2.558$, $d_3 = 2.685$, $d_4 = 2.772$).

(c) Ignoring the non-inhaling smokers (NI) groups, apply Dunnett and Tamhane's stepup method at $\alpha = 0.05$ to compare each of the smoking groups with the non-smoking group (NS) ($c_1 = 1.645$, $c_2 = 1.933$, $c_3 = 2.071$, $c_4 = 2.165$).

# CHAPTER 4

# Multiple comparisons with the best

To motivate multiple comparisons with the best (MCB), consider the following (somewhat morbid) example. Suppose, among five treatments being compared, two treatments are so bad that given either one most patients die within a short period of time. Then it is possibly not of primary interest to know which of those two treatments is worse; the inference that neither is best suffices. Suppose the second best treatment (among the three remaining treatments) is almost as good as the true best treatment. Then statistical inference identifying both as practically the best may be of interest, for there may be other considerations (e.g. cost, toxicity) impacting on the choice of the treatment. Thus, in these situations not all pairwise comparisons are of interest. The question is then 'What comparisons are of primary interest?'

One could characterize the comparisons of primary interest in these situations as 'multiple comparisons with the best.' Thus, if a larger treatment effect is better, then even though which treatment is best is unknown, one could define the parameters of primary interest to be

$$\max_{j=1,\ldots,k} \mu_j - \mu_i, i = 1, \ldots, k, \qquad (4.1)$$

the difference between the (unknown) true best treatment effect and each of the $k$ treatment effects. Indeed, this was the parametrization in three early papers (Hsu 1981; 1982; Edwards and Hsu 1983) on multiple comparisons with the best.

However, it turns out in most cases to be advantageous to compare each treatment with the best of the *other* treatments instead. Suppose a larger treatment effect (e.g. survival time) implies a better treatment. Then the parameters

$$\mu_i - \max_{j \neq i} \mu_j, \ i = 1, \ldots, k \qquad (4.2)$$

contain all the information that the parameters (4.1) contain, for if

$$\mu_i - \max_{j \neq i} \mu_j > 0,$$

then treatment $i$ is the best treatment. On the other hand, if

$$\mu_i - \max_{j \neq i} \mu_j < 0,$$

then treatment $i$ is not the best treatment. Further, even if the $i$th treatment is not the best, but nevertheless

$$\mu_i - \max_{j \neq i} \mu_j > -\delta$$

where $\delta$ is a small positive number, then the $i$th treatment is at least close to the best. Note that whereas the range of the parameters (4.1) is $[0, \infty)$, the range of the parameters (4.2) is $(-\infty, \infty)$. Thus, the reference point to which confidence intervals for the parameters (4.2) should be compared is the usual 0. This is one advantage of the parametrization (4.2) over the parametrization (4.1). Another is that mathematical derivations are easier with the (4.2) parametrization than with the (4.1) parametrization in most cases.

Naturally, if a smaller treatment effect (e.g. CPU time) implies a better treatment, then by symmetry the parameters of primary interest are

$$\mu_i - \min_{j \neq i} \mu_j, \ i = 1, \dots, k.$$

Note that, in contrast to the all-pairwise comparisons (MCA) parameters

$$\mu_i - \mu_j, \ i \neq j$$

and the multiple comparisons with a control (MCC) parameters

$$\mu_i - \mu_k, \ i \neq k,$$

the multiple comparisons with the best (MCB) parameters (4.2) are non-linear functions of $\mu_1, \dots, \mu_k$. Consequently, the derivation and operating characteristics of MCB confidence intervals are different from those of MCA and MCC confidence intervals.

## 4.1 Constrained multiple comparison with the best

By constrained MCB inference we mean simultaneous confidence intervals on $\mu_i - \max_{j \neq i} \mu_j$ which are constrained to contain 0. For those agreeing with John W. Tukey (1992) as I do:

> Our experience with the real world teaches us – if we are willing learners – that, provided we measure to enough decimal places, no two 'treatments' ever have identically the same long-run value.[*]

(see also Tukey 1991; Berger 1985, pp.20–21), a confidence interval for $\mu_i - \max_{j \neq i} \mu_j$ whose lower limit is 0 indicates the $i$th treatment is the best, and a confidence interval for $\mu_i - \max_{j \neq i} \mu_j$ whose upper limit is 0 indicates the $i$th treatment is not the best. By taking this view and forgoing the ability to put lower bounds on *how much* treatments identified not to be the best are worse than the true best, constrained MCB achieves sharper inference than unconstrained MCB.

---

[*] Reprinted by courtesy of Marcel Dekker Inc.

Not everyone takes the position that no two treatments ever have identical long-run average treatment effects. However, even if you believe $\mu_i = \mu_j$ is a possibility, constrained MCB still provides useful inference because a confidence interval for $\mu_i - \max_{j \neq i} \mu_j$ with a lower bound of 0 means the $i$th treatment is *one of* the best, and a lower confidence bound for $\mu_i - \max_{j \neq i} \mu_j$ close to 0 indicates the $i$th treatment is close to the best (assuming a larger treatment effect is better).

For situations where one desires lower bounds on how much treatments identified not to be the best are worse than the true best, unconstrained MCB inference, discussed in Section 4.2, is available.

### 4.1.1 Balanced one-way model

Suppose that under the $i$th treatment a random sample $Y_{i1}, Y_{i2}, \ldots, Y_{in}$ of size $n$ is taken, where the observations between the treatments are independent. Then under the usual normality and equality of variances assumptions, we have the balanced one-way model (3.1)

$$Y_{ia} = \mu_i + \epsilon_{ia}, \quad i = 1, \ldots, k, \quad a = 1, \ldots, n, \tag{4.3}$$

where $\mu_i$ is the effect of the $i$th treatment, $i = 1, \ldots, k$, and $\epsilon_{11}, \ldots, \epsilon_{kn}$ are i.i.d. normal errors with mean 0 and variance $\sigma^2$ unknown. We again use the notation

$$\hat{\mu}_i = \bar{Y}_i = \sum_{a=1}^{n} Y_{ia}/n,$$

$$\hat{\sigma}^2 = MSE = \sum_{i=1}^{k} \sum_{a=1}^{n} (Y_{ia} - \bar{Y}_i)^2 / [k(n-1)]$$

for the sample means and the pooled sample variance.

Let $d$ be the critical value for Dunnett's one-sided MCC method, that is, $d$ is the solution to the equation

$$\int_0^\infty \int_{-\infty}^{+\infty} [\Phi(z + \sqrt{2}ds)]^{k-1} d\Phi(z) \, \gamma(s)ds = 1 - \alpha, \tag{4.4}$$

so that, for any $i$,

$$P\{\hat{\mu}_i - \mu_i \geq \hat{\mu}_j - \mu_j - d\hat{\sigma}\sqrt{2/n} \text{ for all } j, j \neq i\} = 1 - \alpha.$$

Recall that the quantile $d$ depends on $\alpha, k$ and $\nu$, which can be computed using the qmcb computer program, with inputs the $(k-1) \times 1$ vector $\lambda = (1/\sqrt{2}, \ldots, 1/\sqrt{2})$, $\nu$ and $\alpha$, or found in Appendix E. Suppose a larger treatment effect implies a better treatment. We will show that the closed intervals

$$[-(\hat{\mu}_i - \max_{j \neq i} \hat{\mu}_j - d\hat{\sigma}\sqrt{2/n})^-, (\hat{\mu}_i - \max_{j \neq i} \hat{\mu}_j + d\hat{\sigma}\sqrt{2/n})^+], i = 1, \ldots, k \tag{4.5}$$

form a set of $100(1-\alpha)\%$ simultaneous confidence intervals for $\mu_i - \max_{j \neq i} \mu_j$. Here

$$-x^- = \min\{0, x\} = \begin{cases} x & \text{if } x < 0 \\ 0 & \text{otherwise} \end{cases}$$

and

$$x^+ = \max\{0, x\} = \begin{cases} x & \text{if } x > 0 \\ 0 & \text{otherwise.} \end{cases}$$

**Theorem 4.1.1 (Hsu 1984b)** *Let*

$$\begin{aligned} D_i^- &= -(\hat{\mu}_i - \max_{j \neq i} \hat{\mu}_j - d\hat{\sigma}\sqrt{2/n})^-, \\ D_i^+ &= (\hat{\mu}_i - \max_{j \neq i} \hat{\mu}_j + d\hat{\sigma}\sqrt{2/n})^+. \end{aligned}$$

*Then for all* $\boldsymbol{\mu} = (\mu_1, \ldots, \mu_k)$ *and* $\sigma^2$,

$$P_{\boldsymbol{\mu},\sigma^2}\{\mu_i - \max_{j \neq i} \mu_j \in [D_i^-, D_i^+] \text{ for } i = 1, \ldots, k\} \geq 1 - \alpha. \qquad (4.6)$$

*Proof.* It will be convenient to let $(1), \ldots, (k)$ denote the unknown indices such that

$$\mu_{(1)} \leq \mu_{(2)} \leq \cdots \leq \mu_{(k)}.$$

In other words, $(i)$ is the *anti-rank* of $\mu_i$ among $\mu_1, \ldots, \mu_k$. For example, suppose $k = 3$ and $\mu_2 < \mu_3 < \mu_1$; then $(1) = 2$, $(2) = 3$, $(3) = 1$. Ties among $\mu_i$ can be broken in any fashion without affecting the validity of the derivation of the MCB simultaneous confidence intervals below. For example, suppose $k = 3$ and $\mu_2 < \mu_3 = \mu_1$; then it is immaterial whether one chooses $(2) = 3$ and $(3) = 1$, or $(2) = 1$ and $(3) = 3$.

Define the event $E$ as

$$E = \{\hat{\mu}_{(k)} - \mu_{(k)} > \hat{\mu}_i - \mu_i - d\hat{\sigma}\sqrt{2/n} \text{ for all } i, i \neq (k)\},$$

which is a pivotal event in the sense that its probability content does not depend on any unknown parameter. In fact, by the definition of the critical value $d$,

$$P\{E\} = P\{\hat{\mu}_{(k)} - \mu_{(k)} > \hat{\mu}_i - \mu_i - d\hat{\sigma}\sqrt{2/n} \text{ for all } i, i \neq (k)\} = 1 - \alpha.$$

If the index $(k)$ were known, then $E$ could be pivoted to provide exact simultaneous confidence intervals for $\mu_i - \max_{j \neq i} \mu_j$, $i = 1, \cdots, k$. But $(k)$ is not known. Therefore, in order to obtain observable confidence bounds on the parameters of interest $\mu_i - \max_{j \neq i} \mu_j, i = 1, \cdots, k$, the derivation below includes several enlargements of the event $E$.

*Derivation of upper confidence bounds*

$$\begin{aligned} E &= \{\hat{\mu}_{(k)} - \mu_{(k)} > \hat{\mu}_i - \mu_i - d\hat{\sigma}\sqrt{2/n} \text{ for all } i, i \neq (k)\} \\ &= \{\mu_{(k)} - \mu_i < \hat{\mu}_{(k)} - \hat{\mu}_i + d\hat{\sigma}\sqrt{2/n} \text{ for all } i, i \neq (k)\} \\ &\subseteq \{\mu_{(k)} - \max_{j \neq (k)} \mu_j < \hat{\mu}_{(k)} - \hat{\mu}_i + d\hat{\sigma}\sqrt{2/n} \text{ for all } i, i \neq (k)\} \end{aligned}$$

$$= \quad \{\mu_{(k)} - \max_{j \neq (k)} \mu_j < \hat{\mu}_{(k)} - \max_{j \neq (k)} \hat{\mu}_j + d\hat{\sigma}\sqrt{2/n}\}$$

$$= \quad \{\mu_{(k)} - \max_{j \neq (k)} \mu_j < \hat{\mu}_{(k)} - \max_{j \neq (k)} \hat{\mu}_j + d\hat{\sigma}\sqrt{2/n}$$

$$\text{and } \mu_i - \max_{j \neq i} \mu_j \leq 0 \text{ for all } i, i \neq (k)\}$$

$$\subseteq \quad \{\mu_i - \max_{j \neq i} \mu_j \leq (\hat{\mu}_i - \max_{j \neq i} \hat{\mu}_j + d\hat{\sigma}\sqrt{2/n})^+ \text{ for all } i\}$$

$$= \quad E_1$$

*Derivation of lower confidence bounds*

$$E \quad = \quad \{\hat{\mu}_{(k)} - \mu_{(k)} > \hat{\mu}_i - \mu_i - d\hat{\sigma}\sqrt{2/n} \text{ for all } i, i \neq (k)\}$$

$$= \quad \{\mu_i - \mu_{(k)} > \hat{\mu}_i - \hat{\mu}_{(k)} - d\hat{\sigma}\sqrt{2/n} \text{ for all } i, i \neq (k)\}$$

$$= \quad \{\mu_i - \max_{j \neq i} \mu_j > \hat{\mu}_i - \hat{\mu}_{(k)} - d\hat{\sigma}\sqrt{2/n} \text{ for all } i, i \neq (k)\}$$

$$\subseteq \quad \{\mu_i - \max_{j \neq i} \mu_j > \hat{\mu}_i - \max_{j \neq i} \hat{\mu}_j - d\hat{\sigma}\sqrt{2/n} \text{ for all } i, i \neq (k)\}$$

$$= \quad \{\mu_i - \max_{j \neq i} \mu_j > \hat{\mu}_i - \max_{j \neq i} \hat{\mu}_j - d\hat{\sigma}\sqrt{2/n} \text{ for all } i, i \neq (k)$$

$$\text{and } \mu_i - \max_{j \neq i} \mu_j \geq 0 \text{ for } i = (k)\}$$

$$\subseteq \quad \{\mu_i - \max_{j \neq i} \mu_j \geq -(\hat{\mu}_i - \max_{j \neq i} \hat{\mu}_j - d\hat{\sigma}\sqrt{2/n})^- \text{ for all } i\}$$

$$= \quad E_2$$

We have shown $E \subseteq E_1 \cap E_2$. Therefore,

$$1 - \alpha \quad = \quad P\{E\}$$

$$\leq \quad P_{\mu,\sigma^2}\{E_1 \cap E_2\}$$

$$= \quad P_{\mu,\sigma^2}\{D_i^- \leq \mu_i - \max_{j \neq i} \mu_j \leq D_i^+ \text{ for } i = 1, \cdots, k\}. \quad \square$$

You may wonder whether the inequality (4.6) is ever an equality, that is, whether the coverage probability is exactly $1 - \alpha$ for some $(\mu, \sigma^2)$. Indeed it is. For example, suppose

$$\mu_{(k)} > \mu_{(k-1)} = \cdots = \mu_{(1)};$$

then

$$P\{\mu_i - \max_{j \neq i} \mu_j \in [D_i^-, D_i^+] \text{ for } i = 1, \ldots, k\}$$

$$\leq \quad P\{\mu_{(k)} - \mu_{(k-1)} \leq D_{(k)}^+\}$$

$$= \quad P\{\mu_{(k)} - \mu_{(k-1)} \leq \hat{\mu}_{(k)} - \max_{j \neq (k)} \hat{\mu}_j + d\hat{\sigma}\sqrt{2/n}\}$$

$$= \quad 1 - \alpha.$$

This was shown by Yu and Lam (1991). Hsu (1984a) also noted that equality is attained when $\mu_{(k)} - \mu_{(k-1)} > d\hat{\sigma}/\sqrt{n}$ in the case $\nu = \infty$ (i.e., $\hat{\sigma} = \sigma$).

You may also wonder whether it is possible to construct a confidence set with coverage probability exactly equal to $1 - \alpha$ for all $\mu = (\mu_1, \cdots, \mu_k)$ and $\sigma^2$ which is contained within the constrained MCB confidence set. The answer is 'yes.' This confidence set was discovered by Gunnar Stefansson, and shown in Stefansson, Kim and Hsu (1988). For $k > 3$ the boundaries of this exact confidence set are not easy to describe and visualize, but when this confidence set is 'projected' on the contours $\mu_i - \max_{j \neq i} \mu_j$, the constrained MCB confidence intervals (4.5) result. Section 4.1.6 does show the construction of the confidence set with coverage probability exactly $1 - \alpha$ for the simple case of $k = 2$, in the context of bioequivalence testing.

Now suppose either a smaller treatment effect implies a better treatment, or a larger treatment effect is better but multiple comparisons with the *worst* treatment are of interest; then the parameters of interest become $\mu_i - \min_{j \neq i} \mu_j$, $i = 1, \ldots, k$. Defining

$$
\begin{aligned}
D_i^- &= -(\hat{\mu}_i - \min_{j \neq i} \hat{\mu}_j - d\hat{\sigma}\sqrt{2/n})^-, \\
D_i^+ &= (\hat{\mu}_i - \min_{j \neq i} \hat{\mu}_j + d\hat{\sigma}\sqrt{2/n})^+,
\end{aligned}
$$

one can analogously prove that, for all $\mu$ and $\sigma^2$,

$$
P_{\mu, \sigma^2}\{\mu_i - \min_{j \neq i} \mu_j \in [D_i^-, D_i^+] \text{ for } i = 1, \ldots, k\} \geq 1 - \alpha.
$$

### 4.1.2 Example: insect traps

The presence of harmful insects in farm fields can be detected by examining insects trapped on boards covered with a sticky material and erected in the fields. Wilson and Shade (1967) reported on the number of cereal leaf beetles trapped when six boards of each of four colors were placed in a field of oats in July. A hypothetical data set, patterned after their experiment, is shown in Table 4.1. To illustrate MCB inference for balanced designs, we compare the attractiveness of the four colors at $\alpha = 0.01$.

Table 4.1 *Number of insects trapped on boards of various colors*

| Color | Label | Insects trapped | | | | | |
|-------|-------|----|----|----|----|----|----|
| Yellow | 1 | 45 | 59 | 48 | 46 | 38 | 47 |
| White | 2 | 21 | 12 | 14 | 17 | 13 | 17 |
| Red | 3 | 37 | 32 | 15 | 25 | 39 | 41 |
| Blue | 4 | 16 | 11 | 20 | 21 | 14 | 7 |

The sample means are

$$\hat{\mu}_1 = 47.17,$$
$$\hat{\mu}_2 = 15.67,$$
$$\hat{\mu}_3 = 31.50,$$
$$\hat{\mu}_4 = 14.83,$$

with sample standard deviation $\hat{\sigma} = 6.784$. For four treatments $(k = 4)$ and 20 error degrees of freedom $(\nu = 20)$, we have $d = 2.973$. Thus, with a common sample size of six $(n = 6)$,

$$d\hat{\sigma}\sqrt{2/n} = 11.64.$$

Therefore 99% constrained MCB confidence intervals are

$$
\begin{aligned}
0 &= -(47.17 - 31.50 - 11.64)^- &\leq& \quad \mu_1 - \max_{j\neq 1} \mu_j \\
&\leq \quad (47.17 - 31.50 + 11.64)^+ &=& \quad 27.30
\end{aligned}
$$

$$
\begin{aligned}
-43.13 &= -(15.67 - 47.17 - 11.64)^- &\leq& \quad \mu_2 - \max_{j\neq 2} \mu_j \\
&\leq \quad (15.67 - 47.17 + 11.64)^+ &=& \quad 0
\end{aligned}
$$

$$
\begin{aligned}
-27.30 &= -(31.50 - 47.17 - 11.64)^- &\leq& \quad \mu_3 - \max_{j\neq 3} \mu_j \\
&\leq \quad (31.50 - 47.17 + 11.64)^+ &=& \quad 0
\end{aligned}
$$

$$
\begin{aligned}
-43.97 &= -(14.83 - 47.17 - 11.64)^- &\leq& \quad \mu_4 - \max_{j\neq 4} \mu_j \\
&\leq \quad (14.83 - 47.17 + 11.64)^+ &=& \quad 0
\end{aligned}
$$

Figure 4.1 illustrates how the MCB subcommand of the ONEWAY command of MINITAB can be used to analyze the traps data.

If one takes the position that the attractiveness of two different colors cannot be exactly the same, then one can infer yellow to be the most attractive at a confidence level of 99%. Even if different colors can be exactly equally attractive to beetles, one can infer yellow is at least as attractive as any other color at a confidence level of 99%. Note, however, constrained MCB does not provide a positive lower confidence bound on *how much* yellow is better than the other colors. Nor does MCB infer that red is more attractive than either *white* or blue.

### 4.1.3 Comparison with the sample best

If the only inference of primary interest is whether the *sample* best treatment is indeed the best, that is, if the only parameter of primary interest is $\mu_{[k]} - \max_{j\neq[k]} \mu_j$, where $[k]$ is the random index such that

$$\hat{\mu}_{[k]} = \max_{1,\dots,k} \hat{\mu}_i,$$

```
MTB > Oneway 'Insects' 'Color';
SUBC>    MCB 1.0001 +1.

ANALYSIS OF VARIANCE ON Insects
SOURCE       DF        SS        MS         F        p
Color         3     4218.5    1406.2     30.55    0.000
ERROR        20      920.5      46.0
TOTAL        23     5139.0
                                         INDIVIDUAL 95 PCT CI'S FOR MEAN
                                         BASED ON POOLED STDEV
LEVEL         N      MEAN     STDEV    ---+---------+---------+---------+---
  1           6     47.167    6.795                                (----*----)
  2           6     15.667    3.327    (----*----)
  3           6     31.500    9.915                  (----*----)
  4           6     14.833    5.345    (---*----)
                                       ---+---------+---------+---------+---
POOLED STDEV =       6.784             12        24        36        48
```

Hsu's MCB (Multiple Comparisons with the Best)

    Family error rate = 0.0100

Critical value = 2.97

Intervals for level mean minus largest of other level means

```
Level    Lower    Center    Upper  --+---------+---------+---------+---------
  1      0.000    15.667    27.300                              (-------*-----)
  2    -43.133   -31.500     0.000  (-----*---------------)
  3    -27.300   -15.667     0.000            (-----*-------)
  4    -43.966   -32.333     0.000  (-----*---------------)
                                    --+---------+---------+---------+---------
                                    -40       -20        0        20
```

Figure 4.1 *Multiple comparison with the best trap color* ($\alpha = 0.01$)

then a sharper inference than deduction from constrained MCB is possible. For this *comparison with the sample best* formulation, Kim (1988) showed that if one asserts

$$0 \leq \mu_{[k]} - \max_{j \neq [k]} \mu_j \tag{4.7}$$

if and only if

$$0 < \hat{\mu}_{[k]} - \max_{j \neq [k]} \hat{\mu}_j - d_1^* \hat{\sigma} \sqrt{2/n} \tag{4.8}$$

where $d_1^*$ is the critical value such that

$$\int_0^\infty \max_{1 \leq r \leq k-1} \left[ r \int_{-\infty}^\infty \Phi^r(z - \sqrt{2}d_1^* s) d\Phi(z) \right] \gamma(s) ds = 1 - \alpha, \tag{4.9}$$

then

$$P\{\text{incorrect assertion}\} \leq \alpha.$$

While a detailed proof is beyond the scope of this book, his technique was to show that

$$P\{\text{incorrect assertion} \mid \hat{\sigma}\}$$

$$= P_{\mu,\sigma^2}\{\mu_{[k]} < \max_{j\neq[k]} \mu_j \text{ and } \hat{\mu}_{[k]} > \max_{j\neq[k]} \hat{\mu}_j + d_1^\star \hat{\sigma}\sqrt{2/n} \mid \hat{\sigma}\}$$

is maximized when

$$\mu_{(k-1)} < \mu_{(k)},$$
$$\mu_{(k-1)} \rightarrow \mu_{(k)},$$

and for some $r$ $(1 \leq r \leq k-3)$,

$$\mu_{(1)}, \ldots, \mu_{(r)} \rightarrow -\infty,$$
$$\mu_{(r+1)}, \ldots, \mu_{(k-2)} \rightarrow \mu_{(k)}.$$

Interestingly, at the usual error rate of $\alpha = 0.10, 0.05, 0.01$, for the vast majority (but not all) of $k, \nu$ combinations, the critical value $d_1^\star$ is the one-sided univariate $t$ critical value $t_{\alpha,\nu}$. See Bofinger (1988) for related work. It should be noted, however, that their techniques do not seem readily generalizable to unbalanced designs.

The advantage this comparison with the sample best formulation has over constrained MCB is it infers the sample best treatment to be the true best more often than constrained MCB, since clearly $d_1^\star \leq d$ always and $d_1^\star < d$ for $k > 2$ generally (compare (4.7) with the MCB lower confidence bound for $\mu_{[k]} - \max_{j\neq[k]} \mu_j$). The disadvantage of comparison with the sample best formulation is it gives no inference when (4.8) does not hold, while constrained MCB may still be able to infer some treatment or treatments to be close to the best, and others not to be the best.

### 4.1.4 Unbalanced one-way model

Suppose under the $i$th treatment a random sample $Y_{i1}, Y_{i2}, \ldots, Y_{in_i}$ of size $n_i$ is taken, where between the treatments the random samples are independent. Then under the usual normality and equality of variances assumptions, we have the one-way model (1.1)

$$Y_{ia} = \mu_i + \epsilon_{ia}, \quad i = 1, \ldots, k, \quad a = 1, \ldots, n_i, \tag{4.10}$$

where $\mu_i$ is the effect of $i$th treatment $i = 1, \ldots, k$, and $\epsilon_{11}, \ldots, \epsilon_{kn_k}$ are i.i.d. normal random variables with mean 0 and variance $\sigma^2$ unknown. We continue to use the notation

$$\hat{\mu}_i = \bar{Y}_i = \sum_{a=1}^{n_i} Y_{ia}/n_i,$$

$$\hat{\sigma}^2 = MSE = \sum_{i=1}^{k}\sum_{a=1}^{n_i}(Y_{ia} - \bar{Y}_i)^2 / \sum_{i=1}^{k}(n_i - 1)$$

for the sample means and the pooled sample variance.

For each $i$, let $d^i$ be the critical value for Dunnett's method of one-sided

MCC if treatment $i$ is the control. In other words, $d^i$ is the solution to

$$\int_0^\infty \int_{-\infty}^{+\infty} \prod_{j \neq i} [\Phi((\lambda_{ij}z + d^i s)/(1 - \lambda_{ij}^2)^{1/2})]d\Phi(z)\,\gamma(s)ds = 1 - \alpha \quad (4.11)$$

with $\lambda_{ij} = (1 + \frac{n_i}{n_j})^{-1/2}$, so that

$$P\{\hat{\mu}_i - \mu_i > \hat{\mu}_j - \mu_j - d^i \hat{\sigma}\sqrt{n_i^{-1} + n_j^{-1}} \text{ for all } j, j \neq i\} = 1 - \alpha.$$

Recall that, in addition to $\alpha$, $k$ and $\nu$, the quantile $d^i$ depends on the sample size ratios $n_i/n_j$, $j \neq i$. Thus, it is not possible to tabulate $d^i$ in general. But it is possible to program a computer to solve for $d^i$ in (4.11) at execution time depending on the sample size pattern $n = (n_1, \ldots, n_k)$ of the data to be analyzed. The qmcb computer program with inputs $\lambda = (\lambda_{ij}, j \neq i)$, $\nu$, and $\alpha$ computes $d^i$. Suppose a larger treatment effect implies a better treatment. We will show that the closed intervals

$$[D_i^-, D_i^+], \quad i = 1, \ldots, k, \tag{4.12}$$

where

$$D_i^+ = +\left(\min_{j \neq i}\{\hat{\mu}_i - \hat{\mu}_j + d^i \hat{\sigma}\sqrt{n_i^{-1} + n_j^{-1}}\}\right)^+, \tag{4.13}$$

$$G = \{i : D_i^+ > 0\}, \tag{4.14}$$

$$D_i^- = \begin{cases} 0 \text{ if } G = \{i\} \\ \min_{j \in G, j \neq i}\{\hat{\mu}_i - \hat{\mu}_j - d^j \hat{\sigma}\sqrt{n_i^{-1} + n_j^{-1}}\} \text{ otherwise,} \end{cases} \tag{4.15}$$

form a set of $100(1 - \alpha)\%$ simultaneous confidence intervals for

$$\mu_i - \max_{j \neq i} \mu_j, \quad i = 1, \ldots, k.$$

Note that it is impossible for $D_i^-$ to be strictly positive, for if $D_i^-$ were positive, then there exists a $j$, $j \in G, j \neq i$, such that

$$\hat{\mu}_j - \hat{\mu}_i + d^j \hat{\sigma}\sqrt{n_i^{-1} + n_j^{-1}} < 0,$$

which contradicts the requirement for $j \in G$.

The computation of MCB confidence intervals can be described as follows. Let an arbitrary treatment, the $i$th treatment say, be the 'control' treatment for the first step. We first compute the critical value $d^i$ for comparing the other treatments with the $i$th treatment using Dunnett's one-sided MCC method. The next step is to calculate the upper confidence bound for $\mu_i - \max_{j \neq i} \mu_j$. First computer the MCC upper confidence bounds

$$\hat{\mu}_i - \hat{\mu}_j + d^i \hat{\sigma}\sqrt{n_i^{-1} + n_j^{-1}}$$

for all $j \neq i$. Then let (the positive part of) the minimum of such bounds

$$D_i^+ = + \left( \min_{j \neq i} \{ \hat{\mu}_i - \hat{\mu}_j + d^i \hat{\sigma} \sqrt{n_i^{-1} + n_j^{-1}} \} \right)^+$$

be the upper confidence bound for $\mu_i - \max_{j \neq i} \mu_j$. This process is repeated using each treatment as the 'control,' thereby producing upper confidence bounds $D_i^+, i = 1, \ldots, k$.

If any $\hat{\mu}_i - \hat{\mu}_j + d^i \hat{\sigma} \sqrt{n_i^{-1} + n_j^{-1}}$ is negative, which indicates the $i$th treatment is not the best, then $D_i^+ = 0$. Thus $G = \{ i : D_i^+ > 0 \}$ is the set of possible treatments with the largest mean. If $G = \{ M \}$, then the $M$th treatment alone is declared the best treatment. Therefore, the lower confidence bound is 0 for the $M$th treatment and

$$D_j^- = \hat{\mu}_j - \hat{\mu}_M - d^M \hat{\sigma} \sqrt{n_j^{-1} + n_M^{-1}}$$

for the $j$th treatment, $j \neq M$. If $G$ contains more than one element, then compute MCC lower confidence bounds using only treatments $j$ that are candidates for the best treatment,

$$\hat{\mu}_i - \hat{\mu}_j - d^j \hat{\sigma} \sqrt{n_i^{-1} + n_j^{-1}},$$

and let (the negative part of) the minimum of such bounds

$$D_i^- = \min_{j \in G, j \neq i} \{ \hat{\mu}_i - \hat{\mu}_j - d^j \hat{\sigma} \sqrt{n_i^{-1} + n_j^{-1}} \}$$

be the lower confidence bound for $\mu_i - \max_{j \neq i} \mu_j$.

**Theorem 4.1.2 (Hsu 1985)** *For all $\mu = (\mu_1, \ldots, \mu_k)$ and $\sigma^2$,*

$$P_{\mu, \sigma^2} \{ \mu_i - \max_{j \neq i} \mu_j \in [D_i^-, D_i^+] \text{ for } i = 1, \ldots, k \} \geq 1 - \alpha. \tag{4.16}$$

*Proof.* It will again be convenient to let $(k)$ denote the unknown index such that

$$\mu_{(k)} = \max_{1 \leq i \leq k} \mu_i.$$

Ties among $\mu_i$ can be broken in any fashion without affecting the validity of the derivation below.

Define the event $E$ as

$$E = \{ \hat{\mu}_{(k)} - \mu_{(k)} > \hat{\mu}_i - \mu_i - d^{(k)} \hat{\sigma} \sqrt{n_i^{-1} + n_{(k)}^{-1}} \text{ for all } i, i \neq (k) \}.$$

By the definition of the critical value $d^{(k)}$, $P\{E\} = 1 - \alpha$.

*Derivation of upper confidence bounds*

$$\begin{aligned} E &= \{ \hat{\mu}_{(k)} - \mu_{(k)} > \hat{\mu}_i - \mu_i - d^{(k)} \hat{\sigma} \sqrt{n_i^{-1} + n_{(k)}^{-1}} \text{ for all } i, i \neq (k) \} \\ &= \{ \mu_{(k)} - \mu_i < \hat{\mu}_{(k)} - \hat{\mu}_i + d^{(k)} \hat{\sigma} \sqrt{n_i^{-1} + n_{(k)}^{-1}} \text{ for all } i, i \neq (k) \} \end{aligned}$$

$$\subseteq \quad \{\mu_{(k)} - \max_{j \neq (k)} \mu_j < \hat{\mu}_{(k)} - \hat{\mu}_i + d^{(k)} \hat{\sigma} \sqrt{n_i^{-1} + n_{(k)}^{-1}} \text{ for all } i, i \neq (k)\}$$

$$= \quad \{\mu_{(k)} - \max_{j \neq (k)} \mu_j < \min_{j \neq (k)} (\hat{\mu}_{(k)} - \hat{\mu}_j + d^{(k)} \hat{\sigma} \sqrt{n_j^{-1} + n_{(k)}^{-1}})\}$$

$$= \quad \{\mu_{(k)} - \max_{j \neq (k)} \mu_j < \min_{j \neq (k)} (\hat{\mu}_{(k)} - \hat{\mu}_j + d^{(k)} \hat{\sigma} \sqrt{n_j^{-1} + n_{(k)}^{-1}}) \text{ and}$$

$$\mu_i - \max_{j \neq i} \mu_j \leq 0 \text{ for all } i, i \neq (k)\}$$

$$\subseteq \quad \{\mu_i - \max_{j \neq i} \mu_j \leq (\min_{j \neq i} \{\hat{\mu}_i - \hat{\mu}_j + d^i \hat{\sigma} \sqrt{n_i^{-1} + n_j^{-1}}\})^+ \text{ for all } i\}$$

$$= \quad E_1.$$

Note also that $E \subseteq \{(k) \in G\}$ because $\mu_{(k)} - \max_{j \neq (k)} \mu_j \geq 0$.

*Derivation of lower confidence bounds*

$$E \quad = \quad \{(k) \in G \text{ and } \hat{\mu}_{(k)} - \mu_{(k)} > \hat{\mu}_i - \mu_i - d^{(k)} \hat{\sigma} \sqrt{n_i^{-1} + n_{(k)}^{-1}} \text{ for all } i,$$

$$i \neq (k)\}$$

$$= \quad \{(k) \in G \text{ and } \mu_i - \mu_{(k)} > \hat{\mu}_i - \hat{\mu}_{(k)} - d^{(k)} \hat{\sigma} \sqrt{n_i^{-1} + n_{(k)}^{-1}} \text{ for all } i,$$

$$i \neq (k)\}$$

$$= \quad \{(k) \in G \text{ and } \mu_i - \max_{j \neq i} \mu_j > \hat{\mu}_i - \hat{\mu}_{(k)} - d^{(k)} \hat{\sigma} \sqrt{n_i^{-1} + n_{(k)}^{-1}}$$

$$\text{for all } i, i \neq (k)\}$$

$$\subseteq \quad \{\mu_i - \max_{j \neq i} \mu_j > \min_{j \in G, j \neq i} (\hat{\mu}_i - \hat{\mu}_j - d^j \hat{\sigma} \sqrt{n_i^{-1} + n_j^{-1}}) \text{ for all } i,$$

$$i \neq (k) \text{ and } (k) \in G\}$$

$$= \quad \{\mu_i - \max_{j \neq i} \mu_j > \min_{j \in G, j \neq i} (\hat{\mu}_i - \hat{\mu}_j - d^j \hat{\sigma} \sqrt{n_i^{-1} + n_j^{-1}}) \text{ for all } i,$$

$$i \neq (k) \text{ and } (k) \in G \text{ and } \mu_i - \max_{j \neq i} \mu_j \geq 0 \text{ for } i = (k)\}$$

$$\subseteq \quad \{\mu_i - \max_{j \neq i} \mu_j \geq D_i^- \text{ for all } i\}$$

$$= \quad E_2.$$

We have shown $E \subseteq E_1 \cap E_2$. Therefore,

$$1 - \alpha \quad = \quad P\{E\}$$

$$\leq \quad P_{\mu, \sigma^2}\{E_1 \cap E_2\}$$

$$= \quad P_{\mu, \sigma^2}\{D_i^- \leq \mu_i - \max_{j \neq i} \mu_j \leq D_i^+ \text{ for all } i\}. \quad \square$$

Since $d^i = d$ when $n_1 = \cdots = n_k = n$, it can be verified that the confidence intervals (4.12) reduce to (4.5) when the sample sizes are equal.

Note that, in contrast to traditional tests of hypothesis and multiple comparison methods, if there are $m$ different sample sizes then MCB utilizes $m$ critical values. When $m$ is large, depending on the speed of the computing platform, computing all the critical values may be too time-consuming. When that is the case, one can note that if $n_1 > n_2$, for example, then the vector $(\lambda_{12}, \lambda_{13}, \ldots, \lambda_{1k})$ is smaller than the vector $(\lambda_{21}, \lambda_{23}, \ldots, \lambda_{2k})$ component-wise. Therefore, by the corollary to Slepian's inequality (see Appendix A), $d^1 > d^2$. In other words, the larger the sample size $n_i$, the larger the critical value $d^i$. Thus, if

$$n_M = \max\{n_1, \ldots, n_k\},$$

then

$$d^M = \max\{d^1, \ldots, d^k\}$$

so if $d^i$ in (4.13)–(4.15) are replaced by $d^M$ for all $i$, then a set of conservative simultaneous MCB confidence intervals is obtained. This is, in fact, the implementation of MCB in the ONEWAY command of MINITAB. The Compare with Best option on the Fit y by x platform in JMP computes all $d^i$ exactly. However, instead of reporting the simultaneous confidence intervals $[D_i^-, D_i^+], i = 1, \ldots, k$, its comparison circles graphical user interface indicates only which treatments have $D_i^+ = 0$, those being the treatments which can be inferred to not be the best. Therefore, compared to MINITAB, JMP has the advantage of potentially being able to infer more treatments not to be the best when the sample sizes are unequal, but the disadvantage of not being able to indicate how close a promising treatment may be to the true best treatment.

Now suppose either a smaller treatment effect implies a better treatment, or a larger treatment effect is better but multiple comparisons with the worst treatment are of interest; then the parameters of interest become $\mu_i - \min_{j \neq i} \mu_j$, $i = 1, \ldots, k$. Defining

$$D_i^- = -(\max_{j \neq i}\{\hat{\mu}_i - \hat{\mu}_j - d^i \hat{\sigma} \sqrt{n_i^{-1} + n_j^{-1}}\})^-,$$

$$G = \{i : D_i^- < 0\},$$

$$D_i^+ = \begin{cases} 0 \text{ if } G = \{i\} \\ \max_{j \in G, j \neq i}\{\hat{\mu}_i - \hat{\mu}_j + d^j \hat{\sigma} \sqrt{n_i^{-1} + n_j^{-1}}\} \text{ otherwise,} \end{cases}$$

one can analogously prove that, for all $\mu$ and $\sigma^2$,

$$P_{\mu, \sigma^2}\{\mu_i - \min_{j \neq i} \mu_j \in [D_i^-, D_i^+] \text{ for } i = 1, \ldots, k\} \geq 1 - \alpha.$$

*Extensions.* Competing systems are sometimes compared by simulation. Constrained MCB has been extended by Yuan and Nelson (1993) to the setting of steady-state simulation, in which outcomes may be dependent within systems but not between systems.

### 4.1.5 Example: SAT scores

The computer science department of a large university finds many of its students change their major after the first year. Campbell and McCabe (1984) report on a study of 216 students conducted in an attempt to understand this phenomenon. Students were classified based on academic status at the beginning of their second year. One variable measured was Scholastic Aptitude Test (SAT) mathematics scores, which are summarized in Table 4.2.

Table 4.2 *SAT mathematics scores*

| Second-year major | Label | Sample size | Sample mean | Sample standard deviation |
|---|---|---|---|---|
| Computer science | 1 | 103 | 619 | 86 |
| Engineering | 2 | 31 | 629 | 67 |
| Other | 3 | 122 | 575 | 83 |

For the sample size pattern $n_1 = 103, n_2 = 31, n_3 = 122$, error degrees of freedom $\nu = 253$ (large enough to be taken as $\infty$ by some computer packages and the qmcb program described in Appendix D) and $\alpha = 0.10$, we have

$$
\begin{aligned}
d^1 &= 1.605, \\
d^2 &= 1.505, \\
d^3 &= 1.612.
\end{aligned}
$$

The pooled sample standard deviation $\hat{\sigma} = 82.52$. (Had the computer codes not taken $\nu$ as $\infty$, the critical values would have been slightly larger.)

It is convenient first to compute MCB upper bounds, since the set $G$ consists of those treatments whose upper MCB bounds are positive.

$$
\begin{aligned}
D_1^+ &= +(\min\{619 - 629 + 1.605(82.52)\sqrt{1/103 + 1/31}, \\
&\quad 619 - 575 + 1.605(82.52)\sqrt{1/103 + 1/122}\})^+ = 17.14
\end{aligned}
$$

$$
\begin{aligned}
D_2^+ &= +(\min\{629 - 619 + 1.505(82.52)\sqrt{1/31 + 1/103}, \\
&\quad 629 - 575 + 1.505(82.52)\sqrt{1/31 + 1/122}\})^+ = 35.44
\end{aligned}
$$

$$
\begin{aligned}
D_3^+ &= +(\min\{575 - 619 + 1.612(82.52)\sqrt{1/122 + 1/103}, \\
&\quad 575 - 629 + 1.612(82.52)\sqrt{1/122 + 1/31}\})^+ = 0.
\end{aligned}
$$

Therefore, $G = \{1, 2\}$.

We proceed to compute lower MCB confidence bounds.

$$D_1^- = 619 - 629 - 1.505(82.52)\sqrt{1/103 + 1/31} = -35.44$$

$$D_2^- = 629 - 619 - 1.605(82.52)\sqrt{1/31 + 1/103} = -17.14$$

$$D_3^- = \min\{575 - 619 - 1.605(82.52)\sqrt{1/122 + 1/103},$$
$$575 - 629 - 1.505(82.52)\sqrt{1/122 + 1/31}\} = -78.98.$$

We thus conclude

$$-35.44 \leq \mu_1 - \max_{j \neq 1} \mu_j \leq 17.14$$

$$-17.14 \leq \mu_2 - \max_{j \neq 2} \mu_j \leq 35.44$$

$$-78.98 \leq \mu_3 - \max_{j \neq 3} \mu_j \leq 0$$

Figures 4.2 and 4.3 illustrate how the MCB subcommand of the ONEWAY command of MINITAB and the MCB option of the Fit Y by X platform of JMP can be used to analyze the SAT data. Note that the confidence intervals computed by MINITAB are somewhat wider than the ones reported above, since MINITAB replaces $d^1$ and $d^2$ by $d^3 = 1.612$.

### 4.1.6 Connection with bioequivalence

There is an interesting connection between constrained MCB confidence intervals and bioequivalence testing.

Consider the case of two treatments ($k = 2$), where $\mu_1$ represents the mean effect of a reference drug and $\mu_2$ is the mean effect of a test drug. In bioequivalence testing, the objective is to see whether there is sufficient evidence that the mean effect $\mu_2$ of the test drug is 'practically equivalent' to the mean effect $\mu_1$ of the reference drug, where 'practically equivalent' means $-\delta_1 < \mu_1 - \mu_2 < \delta_2$, for some prespecified positive $\delta_1$ and $\delta_2$.

*The two one-sided tests procedure* To establish bioequivalence between $\mu_1$ and $\mu_2$, Westlake (1981) proposed the *two one-sided tests* procedure (cf. Schuirmann 1987). This procedure tests

$$H_0 : \mu_1 - \mu_2 \leq -\delta_1 \text{ or } \mu_1 - \mu_2 \geq \delta_2 \tag{4.17}$$

vs.

$$H_a : -\delta_1 < \mu_1 - \mu_2 < \delta_2 \tag{4.18}$$

using the ordinary one-sided size-$\alpha$ $t$-test for

$$H_{01} : \mu_1 - \mu_2 \leq -\delta_1 \tag{4.19}$$

vs.

$$H_{a1} : -\delta_1 < \mu_1 - \mu_2 \tag{4.20}$$

```
MTB > Oneway 'SAT' 'Major';
SUBC>    MCB 10 +1.

ANALYSIS OF VARIANCE ON SAT
SOURCE      DF         SS         MS        F         p
Major        2      139357      69679     10.23     0.000
ERROR      253     1722631       6809
TOTAL      255     1861988
                                     INDIVIDUAL 95 PCT CI'S FOR MEAN
                                     BASED ON POOLED STDEV
  LEVEL      N       MEAN      STDEV  ----+---------+---------+---------+--
    1       103     619.00     86.00                  (----*-----)
    2        31     629.00     67.00                  (---------*--------)
    3       122     575.00     83.00    (----*----)
                                     ----+---------+---------+---------+--
POOLED STDEV =      82.52            570       600       630       660
```

Hsu's MCB (Multiple Comparisons with the Best)

    Family error rate = 0.100

Critical value = 1.61

Intervals for level mean minus largest of other level means

```
  Level    Lower    Center    Upper -------+---------+---------+---------+----
    1      -37.22   -10.00    17.22                  (--------*--------)
    2      -17.22    10.00    37.22                         (--------*--------)
    3      -80.72   -54.00     0.00  (--------*-----------------)
                                    -------+---------+---------+---------+----
                                         -60       -30        0        30
```

Figure 4.2 *Multiple comparison with the best SAT score using MINITAB*

and the ordinary one-sided size-$\alpha$ $t$-test for

$$H_{02} : \mu_1 - \mu_2 \geq \delta_2 \qquad (4.21)$$

vs.

$$H_{a2} : \mu_1 - \mu_2 < \delta_2. \qquad (4.22)$$

If both $H_{01}$ and $H_{02}$ are rejected, then $H_0$ is rejected in favor of $H_a$, establishing bioequivalence between the test drug and the reference drug. The two one-sided tests procedure is an example of what Berger (1982) calls *intersection-union* tests, and is a level-$\alpha$ test for $H_0$. To see that, define

$$E_1 = \{\text{reject } H_{01}\}$$
$$E_2 = \{\text{reject } H_{02}\}.$$

Suppose the test drug and the reference drug are not bioequivalent, that is, $H_0$ is true. Without loss of generality, suppose $\mu_1 - \mu_2 \geq \delta_2$, that is,

**SAT By Major**

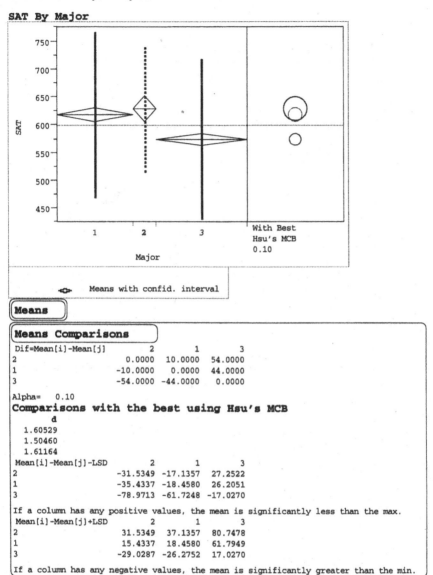

Figure 4.3 *Multiple comparison with the best SAT score using JMP*

$H_{02}$ is true. Then

$$P\{\text{reject } H_0 \mid H_{02} \text{ is true}\}$$
$$= \quad P\{\text{reject } H_{01} \text{ and reject } H_{02} \mid H_{02} \text{ is true}\}$$
$$\leq \quad P\{\text{reject } H_{02} \mid H_{02} \text{ is true}\}$$

$$\leq \; \alpha.$$

One would expect the two one-sided tests procedure to be identical to some confidence interval procedure: For some appropriate $100(1 - \alpha)\%$ confidence interval $[C^-, C^+]$ for $\mu_1 - \mu_2$, declare the test drug to be bioequivalent to the reference drug if and only if $[C^-, C^+] \subset (-\delta_1, \delta_2)$.

It has been noted (e.g., Westlake 1981, p. 593; Schuirmann 1987, p. 661) that the two one-sided tests procedure is operationally identical to the procedure of declaring equivalence only if the ordinary $100(1 - 2\alpha)\%$, not $100(1 - \alpha)\%$, two-sided confidence interval for $\mu_1 - \mu_2$

$$(\hat{\mu}_1 - \hat{\mu}_2 - t_{\alpha,\nu}\hat{\sigma}\sqrt{n_1^{-1} + n_2^{-1}}, \hat{\mu}_1 - \hat{\mu}_2 + t_{\alpha,\nu}\hat{\sigma}\sqrt{n_1^{-1} + n_2^{-1}}) \qquad (4.23)$$

is completely contained in the interval $(-\delta_1, \delta_2)$. Here $t_{\gamma,\nu}$ again denotes the upper $\gamma$ quantile of the $t$ distribution with $\nu$ degrees of freedom. In fact, both the Food and Drug Administration bioequivalence guideline (FDA 1992) and the European Community guideline (EC-GCP 1993a) specify that the two one-sided tests procedure should be executed in this fashion.

We will show that the $100(1 - \alpha)\%$ constrained MCB confidence interval

$$[-(\hat{\mu}_1 - \hat{\mu}_2 - t_{\alpha,\nu}\hat{\sigma}\sqrt{n_1^{-1} + n_2^{-1}})^-, \; (\hat{\mu}_1 - \hat{\mu}_2 + t_{\alpha,\nu}\hat{\sigma}\sqrt{n_1^{-1} + n_2^{-1}})^+] \quad (4.24)$$

corresponds to testing (4.17) versus (4.18). (Note that when $k = 2$, $d^1 = d^2 = t_{\alpha,\nu}$.) In other words, the two one-sided tests procedure is identical to the procedure which declares bioequivalence when the $100(1 - \alpha)\%$ constrained MCB confidence interval (4.24) is contained in $(-\delta_1, \delta_2)$. To see that, we use the connection between tests and confidence sets (Lemma B.3.1) which, in the setting of bioequivalence, can be stated as follows.

**Lemma 4.1.1** *Let* $\delta = \mu_1 - \mu_2$. *If* $\{\phi_\delta(\hat{\mu}_1 - \hat{\mu}_2) : \delta \in \Re\}$ *is a family of size-$\alpha$ tests for $\delta$, that is,*

$$P_\delta\{\phi_\delta(\hat{\mu}_1 - \hat{\mu}_2) = 0\} = 1 - \alpha$$

*for all* $\delta \in \Re$, *then*

$$C(\hat{\mu}_1 - \hat{\mu}_2) = \{\delta : \phi_\delta(\hat{\mu}_1 - \hat{\mu}_2) = 0\}$$

*is a* $100(1 - \alpha)\%$ *confidence set for $\delta$.*

The usual family of tests for this situation is

$$\phi_\delta(\hat{\mu}_1 - \hat{\mu}_2) = \begin{cases} 1 & \text{if } |\hat{\mu}_1 - \hat{\mu}_2 - \delta| \geq t_{\alpha/2,\nu}\hat{\sigma}\sqrt{n_1^{-1} + n_2^{-1}} \\ 0 & \text{otherwise.} \end{cases} \qquad (4.25)$$

As this family of tests is obtained by applying equivariance to the test for $H_0 : \delta = 0$ over the real line, the corresponding confidence set obtained via Lemma 4.1.1 is the usual equivariant confidence interval (4.23) but with $t_{\alpha,\nu}$ replaced by $t_{\alpha/2,\nu}$.

However, no current law or regulation states one must employ confidence sets that are equivariant over the entire real line. Consider the family of tests such that, for $\delta \geq 0$,

$$\phi_\delta(\hat{\mu}_1 - \hat{\mu}_2) = \begin{cases} 1 & \text{if } \hat{\mu}_1 - \hat{\mu}_2 - \delta \leq -t_{\alpha,\nu}\hat{\sigma}\sqrt{n_1^{-1} + n_2^{-1}} \\ 0 & \text{otherwise} \end{cases} \quad (4.26)$$

and, for $\delta < 0$,

$$\phi_\delta(\hat{\mu}_1 - \hat{\mu}_2) = \begin{cases} 1 & \text{if } \hat{\mu}_1 - \hat{\mu}_2 - \delta \geq t_{\alpha,\nu}\hat{\sigma}\sqrt{n_1^{-1} + n_2^{-1}} \\ 0 & \text{otherwise.} \end{cases} \quad (4.27)$$

Note that this family of tests is obtained by applying equivariance to two different one-sided tests for $H_0 : \delta = 0$ for the two halves of the real line. The confidence set obtained from this family of tests via Lemma 4.1.1 is

$$[-(\hat{\mu}_1 - \hat{\mu}_2 - t_{\alpha,\nu}\hat{\sigma}\sqrt{n_1^{-1} + n_2^{-1}})^-, (\hat{\mu}_1 - \hat{\mu}_2 + t_{\alpha,\nu}\hat{\sigma}\sqrt{n_1^{-1} + n_2^{-1}})^+) \quad (4.28)$$

which differs from the constrained MCB confidence interval $[D_1^-, D_1^+]$ in (4.24) only in that the upper bound is open rather than closed. (Technically, the lower confidence limit can be open *except* when it is 0.) This point is inconsequential in bioequivalence testing. The only value of the upper bound with positive probability is 0 and, in bioequivalence testing, the inference $\mu_1 \neq \mu_2$ is not of interest. (In terms of operating characteristics, the confidence interval (4.28) has coverage probability $100(1-\alpha)\%$ everywhere, which the confidence interval (4.24) has also except at $\mu_1 - \mu_2 = 0$ where it has 100% coverage probability.)

Note that the family of tests (4.26) contains the one-sided size-$\alpha$ $t$-test for (4.21), and the family of tests (4.27) contains the one-sided size-$\alpha$ $t$-test for (4.19), in contrast to the family of tests (4.25). The two one-sided tests procedure is equivalent to asserting bioequivalence

$$-\delta_1 < \mu_1 - \mu_2 < \delta_2 \quad (4.29)$$

if and only if

$$[D_1^-, D_1^+] \subset (-\delta_1, \delta_2),$$

which clearly has a probability of an incorrect assertion no more than $\alpha$. Therefore, as pointed out by Hsu *et al.* (1994), it is more appropriate to say that the $100(1-\alpha)\%$ constrained MCB confidence interval $[D_1^-, D_1^+]$, rather than the ordinary $100(1 - 2\alpha)\%$ confidence interval (4.23), corresponds to the two one-sided tests procedure. In a way, it is not surprising that constrained MCB and bioequivalence testing are closely related. Constrained MCB strives to identify treatment(s) close to the true best treatment. In the case of two treatments ($k = 2$), a treatment close to the other treatment is either the true best itself or close to the true best treatment.

Hsu *et al.* (1994) noted that this confidence interval is the Studentized version of a confidence interval derived by Pratt (1961), who showed that for the $\nu = \infty$ case (i.e. $\hat{\sigma} = \sigma$), when $\mu_1 = \mu_2$, $[D_1^-, D_1^+]$ has the smallest expected length among all $100(1 - \alpha)\%$ confidence intervals for $\mu_1 - \mu_2$. In other words, for large sample sizes, when the test drug is indeed equivalent to the reference drug, the confidence interval $[D_1^-, D_1^+]$ is optimal for the purpose of inferring $\mu_1$ and $\mu_2$ are indeed close to each other. They also showed that the equivalence confidence interval (4.24) is always contained in Westlake's (1976) symmetric confidence interval.

### 4.1.7 Connection with ranking and selection

Multiple comparison with the best has its roots in ranking and selection, which has two principal formulations: subset selection, and indifference zone selection.

Subset selection inference, due to Gupta (1956; 1965), gives a confidence set $G$ for the index $(k)$ of the true best treatment. (Recall that, in the case of ties for $\max_{i=1,\cdots,k} \mu_i$, a specific $i$ can be arbitrarily chosen to be $(k)$ or, in the language of subset selection, the $i$th population is 'tagged' to be the best.) If one discounts the possibility of ties for $\max_{i=1,\cdots,k} \mu_i$ or, equivalently, assumes $\mu_{(k)} > \mu_{(k-1)}$, then the implied inference is treatments with indices not in $G$ are not the best treatment.

For the balanced one-way model (4.3), subset selection in fact gives the set

$$G = \left\{ i : \hat{\mu}_i - \max_{j \neq i} \hat{\mu}_j + d\hat{\sigma}\sqrt{2/n} > 0 \right\}$$

in (4.14) as a $100(1 - \alpha)\%$ confidence set for $(k)$, which indeed it is, as can be seen in the proof of Theorem 4.1.2, since $E \subseteq \{(k) \in G\}$ and $P\{E\} = 1 - \alpha$. If one assumes $\mu_{(k)} > \mu_{(k-1)}$, then the subset selection confidence statement

$$\inf_{\mu,\sigma^2} P_{\mu,\sigma^2}\{(k) \in G\} \geq 1 - \alpha$$

is implied by the confidence statement associated with constrained MCB upper bounds

$$\inf_{\mu,\sigma^2} P_{\mu,\sigma^2} \left\{ \mu_i - \max_{j \neq i} \mu_j \leq \left( \hat{\mu}_i - \max_{j \neq i} \hat{\mu}_j + d\hat{\sigma}\sqrt{2/n} \right)^+ \right\} \geq 1 - \alpha,$$

since a non-positive upper bound on $\mu_i - \max_{j \neq i} \mu_j$ implies $i \neq (k)$.

Indifference zone selection, due to Bechhofer (1954), has a *design* aspect and an *inference* aspect.

Let $[1], \ldots, [k]$ denote the random indices such that

$$\hat{\mu}_{[1]} < \hat{\mu}_{[2]} < \ldots < \hat{\mu}_{[k]}.$$

(Since $\hat{\mu}_i$ are continuous random variables, ties occur in them with probability zero.) In other words, $[i]$ is the anti-rank of $\hat{\mu}_i$ among $\hat{\mu}_1, \dots, \hat{\mu}_k$. For example, suppose $k = 3$ and $\hat{\mu}_2 < \hat{\mu}_1 < \hat{\mu}_3$; then $[1] = 2$, $[2] = 1$, $[3] = 3$.

For the balanced one-way model (4.3) with $\sigma$ known, the design aspect of indifference zone selection sets the common sample size $n$ to be the smallest integer such that

$$d\sigma\sqrt{2/n} \leq \delta^*, \tag{4.30}$$

where $\delta^*(> 0)$ represents the quantity of *indifference* to the user, that is, treatment means within $\delta^*$ of the best are considered to be equivalent to the best.

Once data has been collected, the inference aspect of indifference zone selection then 'selects' the $[k]$th treatment as the best treatment. The indifference zone confidence statement is that if $\boldsymbol{\mu} = (\mu_1, \dots, \mu_k)$ is in the so-called *preference zone*

$$\{\mu_{(k)} - \mu_{(k-1)} > \delta^*\},$$

then with a probability of at least $1 - \alpha$ the true best treatment will be selected. In other words, the indifference zone confidence statement is

$$\inf_{\mu_{(k)} - \mu_{(k-1)} > \delta^*} P_{\boldsymbol{\mu}, \sigma^2}\{\mu_{[k]} = \mu_{(k)}\} = 1 - \alpha. \tag{4.31}$$

This confidence statement is implied by the confidence statement

$$\inf_{\boldsymbol{\mu}, \sigma^2} P_{\boldsymbol{\mu}, \sigma^2}\{-\delta^* \leq \mu_{[k]} - \max_{j \neq [k]} \mu_j\} \geq 1 - \alpha \tag{4.32}$$

because, for $\boldsymbol{\mu}$ in the preference zone, the only treatment mean $\mu_i$ with

$$-\delta^* \leq \mu_i - \max_{j \neq i} \mu_j$$

is $\mu_{(k)}$. While Fabian (1962) gave a direct proof of (4.32), we can see that the confidence statement (4.32) is implied by the confidence statement

$$\inf_{\boldsymbol{\mu}, \sigma^2} P_{\boldsymbol{\mu}, \sigma^2}\{-d\sigma\sqrt{2/n} \leq \mu_{[k]} - \max_{j \neq [k]} \mu_j\} \geq 1 - \alpha$$

because

$$d\sigma\sqrt{2/n} \leq \delta^*$$

by indifference zone sample size design (4.30). This last confidence statement in turn is implied by

$$\inf_{\boldsymbol{\mu}, \sigma^2} P_{\boldsymbol{\mu}, \sigma^2}\{-(\hat{\mu}_{[k]} - \max_{j \neq [k]} \hat{\mu}_j - d\sigma\sqrt{2/n})^- \leq \mu_{[k]} - \max_{j \neq [k]} \mu_j\} \geq 1 - \alpha,$$

because

$$\hat{\mu}_{[k]} - \max_{j \neq [k]} \hat{\mu}_j \geq 0$$

and

$$-d\sigma\sqrt{2/n} < 0.$$

The last confidence bound now looks familiar. It is one of the constrained lower MCB confidence bounds on $\mu_i - \max_{j \neq i} \mu_j$, $i = 1, \ldots, k$, namely, the one on $\mu_{[k]} - \max_{j \neq [k]} \mu_j$, in the special case of $\sigma$ known, which can be thought of as the case where the degree of freedom of $\hat{\sigma}$ equals infinity.

In essence, the design aspect of indifference zone selection guarantees a desired accuracy of the MCB lower bound for $\mu_{[k]} - \max_{j \neq [k]} \mu_j$, so that after experimentation, the useful level $1 - \alpha$ confidence statement

$$-\delta^* \leq \mu_{[k]} - \max_{j \neq [k]} \mu_j \tag{4.33}$$

can be made with probability one. For single-stage experiments, this can be achieved only when $\sigma$ is known. When $\sigma$ is unknown and must be estimated, there is a positive probability that the lower confidence bound

$$\hat{\mu}_{[k]} - \max_{j \neq [k]} \hat{\mu}_j - d\hat{\sigma}\sqrt{2/n}$$

on $\mu_{[k]} - \max_{j \neq [k]} \mu_j$ is less than $-\delta^*$. However, in analogy with sample size computation based on power of tests, one can, as will be discussed in Appendix C, design a single-stage experiment so that, with a probability $1 - \beta < 1 - \alpha$, the lower bound on $\mu_{[k]} - \max_{j \neq [k]} \mu_j$ will be greater than $-\delta^*$. Otherwise, to guarantee that the useful statement (4.33) will be made with certainty, one must conduct the experiment in two or more stages. Matejcik and Nelson (1995) have developed 2-stage MCB methods which guarantee this useful statement will be made. Their methods are in fact valid under quite mild conditions (including some models with serial dependence and unequal variances), thereby extending the range of applicability of MCB, especially to the setting of computer simulations. Note that, for multi-stage experiments that eliminate apparently bad treatments at the end of each stage (which is ethically desirable in clinical trials, for example), proving that the more useful confidence statement (4.31) holds can be more challenging than merely proving that the confidence statement (4.32) holds. See Hsu and Edwards (1983) and Edwards (1987), for example.

For book-length discussions on ranking and selection, see Gibbons, Olkin and Sobel (1977), Gupta and Panchapakesan (1979), and Bechhofer, Santner and Goldsman (1995).

Having shown subset selection and indifference zone selection correspond to upper and lower MCB confidence bounds, a most important observation to make at this point is that, since the MCB confidence intervals are guaranteed to cover the parameters $\mu_i - \max_{j \neq i} \mu_j$, $i = 1, \ldots, k$, simultaneously with a probability of at least $1 - \alpha$, subset selection inference and indifference zone selection inference can be given simultaneously with the guarantee that both inferences are correct with a probability of at least $1 - \alpha$.

## 4.2 Unconstrained multiple comparisons with the best

For situations where one desires lower bounds on how much treatments identified not to be the best are worse than the true best, two kinds of unconstrained MCB method are available. By unconstrained MCB inference we mean simultaneous confidence intervals for $\mu_i - \max_{j \neq i} \mu_j$ or $\mu_i - \max_{1 \leq j \leq k} \mu_j$ which are not constrained to contain 0.

### 4.2.1 Balanced One-way Model

Consider the balanced one-way model (4.3)

$$Y_{ia} = \mu_i + \epsilon_{ia}, \quad i = 1, \ldots, k, \quad a = 1, \ldots, n,$$

where $\mu_i$ is the effect of $i$th treatment, $i = 1, \ldots, k$, and $\epsilon_{11}, \ldots, \epsilon_{kn}$ are i.i.d. normal with mean 0 and variance $\sigma^2$ unknown. We again use the following notation

$$\hat{\mu}_i = \bar{Y}_i = \sum_{a=1}^{n} Y_{ia}/n,$$

$$\hat{\sigma}^2 = MSE = \sum_{i=1}^{k} \sum_{a=1}^{n} (Y_{ia} - \bar{Y}_i)^2 / [k(n-1)]$$

for the sample means and the pooled sample variance.

### 4.2.1.1 Deduction from all-pairwise comparisons

As will be discussed in more detail in Section 5.1.1, Tukey (1953) pivoted the probabilistic statement

$$P\{\hat{\mu}_i - \mu_i - (\hat{\mu}_j - \mu_j) < |q^*|\hat{\sigma}\sqrt{2/n} \text{ for all } i, j, \ i \neq j\} = 1 - \alpha, \quad (4.34)$$

where $|q^*|$ is the solution to the equation

$$P\left\{ \frac{|Z_i - Z_j|}{\sqrt{2}\hat{\sigma}} \leq |q^*| \text{ for all } i > j \right\} = 1 - \alpha$$

in which $Z_1, \ldots, Z_k$ are i.i.d. standard normal random variables, to give

$$(\hat{\mu}_i - \hat{\mu}_j - |q^*|\hat{\sigma}\sqrt{2/n}, \hat{\mu}_i - \hat{\mu}_j + |q^*|\hat{\sigma}\sqrt{2/n}) \text{ for all } i \neq j \quad (4.35)$$

as a set of $100(1 - \alpha)\%$ simultaneous confidence intervals for all pairwise differences $\mu_i - \mu_j, i \neq j$. We will show that if one keeps the form of the constrained MCB confidence intervals, that is, one centers the confidence intervals for

$$\mu_i - \max_{j \neq i} \mu_j$$

at

$$\hat{\mu}_i - \max_{j \neq i} \hat{\mu}_j$$

but removes the 'constraint' and forms the confidence intervals

$$(\hat{\mu}_i - \max_{j \neq i} \hat{\mu}_j - c\hat{\sigma}\sqrt{2/n}, \hat{\mu}_i - \max_{j \neq i} \hat{\mu}_j + c\hat{\sigma}\sqrt{2/n}), \quad i = 1, \ldots, k, \quad (4.36)$$

then to achieve confidence level $1 - \alpha$, the critical value $c$ must be equal to $|q^*|$.

**Theorem 4.2.1** *For all $\mu = (\mu_1, \ldots, \mu_k)$ and $\sigma^2$,*

$$P_{\mu,\sigma^2}\{\mu_i - \max_{j \neq i} \mu_j \in \hat{\mu}_i - \max_{j \neq i} \hat{\mu}_j \pm |q^*|\hat{\sigma}\sqrt{2/n} \text{ for } i = 1, \ldots, k\}$$
$$\geq \; 1 - \alpha,$$

*with equality when $\mu_1 = \cdots = \mu_k$.*

*Proof.* Consider the pivotal event

$$E \; = \; \{\hat{\mu}_i - \mu_i - (\hat{\mu}_j - \mu_j) < |q^*|\hat{\sigma}\sqrt{2/n} \text{ for all } j \neq i, \text{ for } i = 1, \ldots, k\}.$$

$$E \; = \; \{\hat{\mu}_i - \hat{\mu}_j - |q^*|\hat{\sigma}\sqrt{2/n} < \mu_i - \mu_j \text{ for all } j \neq i, \text{ for } i = 1, \ldots, k\}$$

$$\subseteq \; \{\hat{\mu}_i - \max_{m \neq i} \hat{\mu}_m - |q^*|\hat{\sigma}\sqrt{2/n} < \mu_i - \mu_j \text{ for all } j \neq i,$$

$$\text{for } i = 1, \ldots, k\}$$

$$= \; \{\hat{\mu}_i - \max_{j \neq i} \hat{\mu}_j - |q^*|\hat{\sigma}\sqrt{2/n} < \mu_i - \max_{j \neq i} \mu_j \text{ for } i = 1, \ldots, k\}$$

$$= \; E_1 \text{(say)}.$$

$$E \; = \; \{\mu_i - \mu_j < \hat{\mu}_i - \hat{\mu}_j + |q^*|\hat{\sigma}\sqrt{2/n} \text{ for all } j \neq i, \text{ for } i = 1, \ldots, k\}$$

$$\subseteq \; \{\mu_i - \max_{m \neq i} \mu_m < \hat{\mu}_i - \hat{\mu}_j + |q^*|\hat{\sigma}\sqrt{2/n} \text{ for all } j \neq i,$$

$$\text{for } i = 1, \ldots, k\}$$

$$= \; \{\mu_i - \max_{j \neq i} \mu_j < \hat{\mu}_i - \max_{j \neq i} \hat{\mu}_j + |q^*|\hat{\sigma}\sqrt{2/n} \text{ for } i = 1, \ldots, k\}$$

$$= \; E_2 \text{(say)}.$$

Since $E \subseteq E_1 \cap E_2$ for all $\mu$ and $P\{E\} = 1 - \alpha$,

$$P_{\mu,\sigma^2}\{E_1 \cap E_2\} \geq 1 - \alpha$$

for all $\mu$ and $\sigma^2$. When $\mu_1 = \cdots = \mu_k$, $E = E_2$ so

$$P_{\mu,\sigma^2}(E_1 \cap E_2) = P\{E\} = 1 - \alpha. \quad \square$$

The astute reader will notice that one can actually prove

**Theorem 4.2.2** *For all $\mu = (\mu_1, \ldots, \mu_k)$ and $\sigma^2$,*

$$P_{\mu,\sigma^2}\{\mu_i - \max_{j \neq i} \mu_j \in \hat{\mu}_i - \max_{j \neq i} \hat{\mu}_j \pm |q^*|\hat{\sigma}\sqrt{2/n} \text{ and}$$

$$\mu_i - \min_{j \neq i} \mu_j \in \hat{\mu}_i - \min_{j \neq i} \hat{\mu}_j \pm |q^*|\hat{\sigma}\sqrt{2/n} \text{ for } i = 1, \ldots, k\} \geq 1 - \alpha.$$

That is, one can deduce multiple comparison with the best and multiple comparison with the worst inferences from Tukey's all-pairwise comparison method simultaneously. (See Exercise 1.)

### 4.2.1.2 The Edwards–Hsu method

This method is more simply stated in terms of the parameters

$$\mu_i - \max_{1 \le j \le k} \mu_j, i = 1, \ldots, k$$

instead of $\mu_i - \max_{j \ne i} \mu_j$.

Let $|d|$ be the critical value for Dunnett's two-sided MCC method, that is, $|d|$ is the solution to the equation

$$\int_0^\infty \int_{-\infty}^{+\infty} [\Phi(z + \sqrt{2}|d|s) - \Phi(z - \sqrt{2}|d|s)]^{k-1} d\Phi(z) \, \gamma(s) ds = 1 - \alpha, \quad (4.37)$$

so that taking arbitrarily the $i$th treatment as the control, we have

$$P\{|\hat{\mu}_i - \mu_i - (\hat{\mu}_j - \mu_j)| < |d|\hat{\sigma}\sqrt{2/n} \text{ for all } j, j \ne i\} = 1 - \alpha.$$

Recall that the quantile $|d|$ depends on $\alpha, k$, and $\nu$. The qmcc computer program with inputs the $(k-1) \times 1$ vector $\boldsymbol{\lambda} = (1/\sqrt{2}, \ldots, 1/\sqrt{2})$, $\nu$ and $\alpha$ computes $|d|$, and tables of $|d|$ can also be found in Appendix E, for example. We will show that the intervals

$$[L_i, U_i], \quad i = 1, \ldots, k, \quad (4.38)$$

where

$$
\begin{aligned}
S &= \{i : \hat{\mu}_i - \hat{\mu}_j + |d|\hat{\sigma}\sqrt{2/n} > 0 \text{ for all } j, j \ne i\}, \\
L_{ij} &= \begin{cases} 0 \text{ if } i = j, \\ \hat{\mu}_i - \hat{\mu}_j - |d|\hat{\sigma}\sqrt{2/n} \text{ otherwise}, \end{cases} \\
L_i &= \min_{j \in S} L_{ij}, \\
U_{ij} &= \begin{cases} 0 \text{ if } i = j, \\ -(\hat{\mu}_i - \hat{\mu}_j + |d|\hat{\sigma}\sqrt{2/n})^- \text{ otherwise}, \end{cases} \\
U_i &= \max_{j \in S} U_{ij},
\end{aligned}
$$

form a set of $100(1 - \alpha)\%$ simultaneous confidence intervals for

$$\mu_i - \max_{1 \le j \le k} \mu_j, \quad i = 1, \ldots, k.$$

Note that if $i \in S$, then $U_i = 0$. Further, if $S = \{M\}$, then $L_M = U_M = 0$ and the confidence intervals $[L_i, U_i], i \ne M$, are Dunnett's two-sided MCC confidence intervals with the $M$th treatment as the control.

**Theorem 4.2.3** *For all* $\boldsymbol{\mu} = (\mu_i, \ldots, \mu_k)$ *and* $\sigma^2$,

$$P_{\boldsymbol{\mu},\sigma^2}\{\mu_i - \max_{1 \le j \le k} \mu_j \in [L_i, U_i] \text{ for } i = 1, \ldots, k\} \ge 1 - \alpha. \quad (4.39)$$

*Proof.* It will again be convenient to let $(k)$ denote the unknown index such that

$$\mu_{(k)} = \max_{1 \le i \le k} \mu_i.$$

Ties among $\mu_i$ can be broken in any fashion without affecting the validity of the derivation below.

Define the event $E$ as

$$E = \{\hat{\mu}_i - \mu_i - |d|\hat{\sigma}\sqrt{2/n} < \hat{\mu}_{(k)} - \mu_{(k)} < \hat{\mu}_i - \mu_i + |d|\hat{\sigma}\sqrt{2/n} \text{ for all } i, i \ne (k)\}.$$

By the definition of the critical value $|d|$, $P\{E\} = 1 - \alpha$. Since $\mu_{(k)} \ge \mu_i$ for all $i$,

$$\begin{aligned} E \quad &\subseteq \quad \{\hat{\mu}_{(k)} > \hat{\mu}_i - |d|\hat{\sigma}\sqrt{2/n} \text{ for all } i, i \ne (k)\} \\ &= \quad \{(k) \in S\}. \end{aligned}$$

*Derivation of upper confidence bounds*

$$\begin{aligned} E \quad &\subseteq \quad \{(k) \in S \text{ and } \mu_{(k)} - \mu_{(k)} = 0 \text{ and} \\ & \qquad \mu_i - \mu_{(k)} \le -(\hat{\mu}_i - \hat{\mu}_{(k)} + |d|\hat{\sigma}\sqrt{2/n})^- \text{ for all } i, i \ne (k)\} \\ &\subseteq \quad \{\mu_i - \mu_{(k)} \le U_i \text{ for } i = 1, \dots, k\} \\ &= \quad E_1 \text{ (say)}. \end{aligned}$$

*Derivation of lower confidence bounds*

$$\begin{aligned} E \quad &\subseteq \quad \{(k) \in S \text{ and } 0 = \mu_{(k)} - \mu_{(k)} \text{ and} \\ & \qquad \hat{\mu}_i - \hat{\mu}_{(k)} - |d|\hat{\sigma}\sqrt{2/n} < \mu_i - \mu_{(k)} \text{ for all } i, i \ne (k)\} \\ &\subseteq \quad \{L_i \le \mu_i - \mu_{(k)} \text{ for } i = 1, \dots, k\} \\ &= \quad E_2 \text{ (say)}. \end{aligned}$$

We have shown $E \subseteq E_1 \cap E_2$. Thus,

$$\begin{aligned} 1 - \alpha \quad &= \quad P\{E\} \\ &\le \quad P_{\mu,\sigma^2}\{E_1 \cap E_2\} \\ &= \quad P_{\mu,\sigma^2}\{L_i \le \mu_i - \max_{1 \le j \le k} \mu_j \le U_i \text{ for all } i\}. \quad \square \end{aligned}$$

Starting with the same pivotal event $E$ as in Theorem 4.2.3, it is possible to derive unconstrained simultaneous confidence intervals for $\mu_i - \max_{j \ne i} \mu_j, i = 1, \dots, k$. However, the expressions for the confidence bounds get complicated, and their derivation is left as an exercise for fanatics. (See Exercise 3(a).) Instead, we indicate in the following example how confidence intervals for $\mu_i - \max_{j \ne i} \mu_j, i = 1, \dots, k$, can be given after confidence intervals for $\mu_i - \max_{1 \le j \le k} \mu_j, i = 1, \dots, k$, have been computed.

### 4.2.2 Example: insect traps (continued)

For four treatments ($k = 4$), 20 error degrees of freedom ($\nu = 20$), and $\alpha = 0.01$, we have $|q^*| = 3.549$ and $|d| = 3.286$. Thus, with a common sample size of six ($n = 6$),

$$|q^*|\hat{\sigma}\sqrt{2/n} = 13.90,$$

$$|d|\hat{\sigma}\sqrt{2/n} = 12.87.$$

*Deduction from Tukey's method*   Deduction from Tukey's method gives the following 99% simultaneous confidence intervals:

$$
\begin{aligned}
1.77 &= 47.17 - 31.50 - 13.90 &<& \quad \mu_1 - \max_{j \neq 1} \mu_j \\
&< 47.17 - 31.50 + 13.90 &=& \quad 29.57
\end{aligned}
$$

$$
\begin{aligned}
-45.40 &= 15.67 - 47.17 - 13.90 &<& \quad \mu_2 - \max_{j \neq 2} \mu_j \\
&< 15.67 - 47.17 + 13.90 &=& \quad -17.60
\end{aligned}
$$

$$
\begin{aligned}
-29.57 &= 31.50 - 47.17 - 13.90 &<& \quad \mu_3 - \max_{j \neq 3} \mu_j \\
&< 31.50 - 47.17 + 13.90 &=& \quad -1.77
\end{aligned}
$$

$$
\begin{aligned}
-46.24 &= 14.83 - 47.17 - 13.90 &<& \quad \mu_4 - \max_{j \neq 4} \mu_j \\
&< 14.83 - 47.17 + 13.90 &=& \quad -18.44.
\end{aligned}
$$

*The Edwards–Hsu method*   The only $i$ with $\hat{\mu}_i - \hat{\mu}_j + |d|\hat{\sigma}\sqrt{2/n} > 0$ is $i = 1$, so, $S = \{1\}$. Therefore, the Edwards–Hsu method gives the following 99% confidence intervals for $\mu_i - \max_{1 \leq j \leq k} \mu_j$:

$$\mu_1 = \max_{1 \leq j \leq k} \mu_j$$

$$
\begin{aligned}
-44.39 &= 15.67 - 47.17 - 12.87 &\leq& \quad \mu_2 - \max_{1 \leq j \leq 4} \mu_j \\
&\leq -(15.67 - 47.17 + 12.87)^- &=& \quad -18.61
\end{aligned}
$$

$$
\begin{aligned}
-28.55 &= 31.50 - 47.17 - 12.87 &\leq& \quad \mu_3 - \max_{1 \leq j \leq 4} \mu_j \\
&\leq -(31.50 - 47.17 + 12.87)^- &=& \quad -2.78
\end{aligned}
$$

$$
\begin{aligned}
-45.22 &= 14.83 - 47.17 - 12.87 &\leq& \quad \mu_4 - \max_{1 \leq j \leq 4} \mu_j \\
&\leq -(14.83 - 47.17 + 12.87)^- &=& \quad -19.45.
\end{aligned}
$$

Using the fact that if $\mu_i - \max_{1 \leq j \leq k} \mu_j$ is negative then $\mu_i - \max_{1 \leq j \leq k} \mu_j =$

$\mu_i - \max_{j\neq i}\mu_j$, we can restate these confidence intervals as:

$$2.78 \;\leq\; \mu_1 - \max_{j\neq 1}\mu_j \;\leq\; 28.55$$

$$\begin{aligned}
-44.39 &= & 15.67 - 47.17 - 12.87 & \;\leq\; \mu_2 - \max_{j\neq 2}\mu_j \\
&\leq & -(15.67 - 47.17 + 12.87)^- & \;=\; -18.61
\end{aligned}$$

$$\begin{aligned}
-28.55 &= & 31.50 - 47.17 - 12.87 & \;\leq\; \mu_3 - \max_{j\neq 3}\mu_j \\
&\leq & -(31.50 - 47.17 + 12.87)^- & \;=\; -2.78
\end{aligned}$$

$$\begin{aligned}
-45.22 &= & 14.83 - 47.17 - 12.87 & \;\leq\; \mu_4 - \max_{j\neq 4}\mu_j \\
&\leq & -(14.83 - 47.17 + 12.87)^- & \;=\; -19.45.
\end{aligned}$$

So, deduction from both Tukey's method and the Edwards–Hsu method infers yellow to be the most attractive color. However, in contrast to constrained MCB, both methods provide positive lower confidence bounds on *how much* yellow is better than the other colors.

### 4.2.3 Comparison with the sample best

Let $[1], [2], \ldots, [k]$ denote the random indices such that

$$\hat{\mu}_{[1]} \leq \cdots \leq \hat{\mu}_{[k]}.$$

If the only inference of primary interest is whether the sample best treatment is indeed the best, that is, if the only parameter of primary interest is $\mu_{[k]} - \max_{j\neq[k]}\mu_j$, then a sharper inference than deduction from unconstrained MCB is possible. For this *comparison with the sample best* formulation, using very different techniques, Gutmann and Maymin (1987) and Stefansson, Kim and Hsu (1988) showed that if one asserts

$$0 < \mu_{[k]} - \max_{j\neq[k]}\mu_j \tag{4.40}$$

if and only if

$$0 < \hat{\mu}_{[k]} - \max_{j\neq[k]}\hat{\mu}_j - t_{\alpha/2,\nu}\hat{\sigma}\sqrt{2/n}, \tag{4.41}$$

where $t_{\alpha/2,\nu}$ is the upper $\alpha/2$ quantile of the $t$ distribution with $\nu$ degrees of freedom, then for all $\boldsymbol{\mu} = (\mu_1,\ldots,\mu_k)$ and $\sigma^2$,

$$P_{\boldsymbol{\mu},\sigma^2}\{\text{incorrect assertion}\} \leq \alpha.$$

Comparing (4.40) and (4.41) with (4.7) and (4.8), we have the following interpretation. One can assert

$$\mu_{[k]} \geq \max_{j\neq[k]}\mu_j$$

if (essentially) the one-sided two-sample size-$\alpha$ $t$-test for

$$H_0 : \mu_{[k]} \leq \mu_{[k-1]}$$

rejects, while one can assert

$$\mu_{[k]} > \max_{j \neq [k]} \mu_j$$

if the two-sided two-sample size-$\alpha$ $t$-test for

$$H_0 : \mu_{[k]} = \mu_{[k-1]}$$

rejects.

While a detailed proof is beyond the scope of this book, the technique used by Stefansson, Kim and Hsu (1988) was to partition the parameter space with the contours

$$\mu_{i,\delta} = \{(\mu_1, \ldots, \mu_k) : \mu_i - \max_{j \neq i} \mu_j = \delta\}$$

so that

$$\Re^k \subseteq \bigcup_{i=1}^{k} \bigcup_{\delta \geq 0} \mu_{i,\delta}.$$

They then showed that a level-$\alpha$ test for the composite hypothesis

$$H_{i,\delta} : \mu_i - \max_{j \neq i} = \delta$$

is to reject if

$$\hat{\mu}_{[k]} - \hat{\mu}_{[k-1]} - \delta > t_{\alpha/2,\nu} \hat{\sigma} \sqrt{2/n}$$

by showing that the supremum of the probability of a Type I error occurs when $\mu_i - \mu_j = \delta$ for one $j, j \neq i$, while all the other $\mu_j, j \neq i$, approach $-\infty$. Gutmann and Maymin's (1987) proof, on the other hand, followed a conditional retrospective testing approach. Neither technique seems readily generalizable to unbalanced designs.

The advantage this comparison with the sample best formulation has over unconstrained MCB is it infers the sample best treatment to be the true best more often than unconstrained MCB, since it uses the critical value for unconstrained comparison with the best of $k = 2$ treatments. The disadvantage of comparison with the sample best formulation is it gives no inference when (4.41) does not hold, while unconstrained MCB may still be able to infer some treatment or treatments to be close to the best, and others not to be the best.

### 4.2.4 Unbalanced one-way model

Consider the unbalanced one-way model (4.10)

$$Y_{ia} = \mu_i + \epsilon_{ia}, \quad i = 1, \ldots, k, \quad a = 1, \ldots, n_i, \tag{4.42}$$

where $\mu_i$ is the effect of the $i$th treatment, and $\epsilon_{11}, \ldots, \epsilon_{kn_k}$ are i.i.d. normal with mean 0 and variance $\sigma^2$ unknown. We continue to use the notation

$$\hat{\mu}_i = \bar{Y}_i = \sum_{a=1}^{n_i} Y_{ia}/n_i,$$

$$\hat{\sigma}^2 = MSE = \sum_{i=1}^{k} \sum_{a=1}^{n_i} (Y_{ia} - \bar{Y}_i)^2 / \sum_{i=1}^{k} (n_i - 1)$$

for the sample means and the pooled sample variance.

### 4.2.4.1 Deduction from all-pairwise comparisons

As will be discussed in Section 5.2.1, when the sample sizes are unequal, the Tukey–Kramer confidence set is a set of (conservative) $100(1 - \alpha)\%$ simultaneous confidence intervals for all pairwise differences $\mu_i - \mu_j, i \neq j$, given by:

$$(\hat{\mu}_i - \hat{\mu}_j - |q^*|\hat{\sigma}\sqrt{n_i^{-1} + n_j^{-1}}, \hat{\mu}_i - \hat{\mu}_j + |q^*|\hat{\sigma}\sqrt{n_i^{-1} + n_j^{-1}}) \text{ for all } i \neq j. \tag{4.43}$$

Here $|q^*|$ is the solution to the equation

$$P\left\{ \frac{|Z_i - Z_j|}{\sqrt{2}\hat{\sigma}} \leq |q^*| \text{ for all } i > j \right\} = 1 - \alpha$$

in which $Z_1, \ldots, Z_k$ are i.i.d. standard normal random variables (making $|q^*|$ dependent on $\alpha, k$ and $\nu = \sum_{i=1}^{k}(n_i - 1)$ only). Unconstrained MCB confidence intervals can be deduced from these confidence intervals. However, the deduced confidence intervals for $\mu_i - \max_{j \neq i} \mu_j$ are no longer necessarily centered at $\hat{\mu}_i - \max_{j \neq i} \hat{\mu}_j$.

**Theorem 4.2.4** *For all* $\mu = (\mu_1, \ldots, \mu_k)$ *and* $\sigma^2$,

$$P_{\mu,\sigma^2}\{\min_{j \neq i}(\hat{\mu}_i - \hat{\mu}_j - |q^*|\hat{\sigma}\sqrt{n_i^{-1} + n_j^{-1}}) < \mu_i - \max_{j \neq i} \mu_j$$

$$< \min_{j \neq i}(\hat{\mu}_i - \hat{\mu}_j + |q^*|\hat{\sigma}\sqrt{n_i^{-1} + n_j^{-1}}) \text{ for } i = 1, \ldots, k\} \geq 1 - \alpha.$$

*Proof.* Consider the event

$$\begin{aligned}
E &= \{\hat{\mu}_i - \mu_i - (\hat{\mu}_j - \mu_j) < |q^*|\hat{\sigma}\sqrt{n_i^{-1} + n_j^{-1}} \text{ for all } j \neq i, \\
&\quad \text{ for } i = 1, \ldots, k\}.
\end{aligned}$$

$$\begin{aligned}
E &= \{\hat{\mu}_i - \hat{\mu}_j - |q^*|\hat{\sigma}\sqrt{n_i^{-1} + n_j^{-1}} < \mu_i - \mu_j \text{ for all } j \neq i, \\
&\quad \text{ for } i = 1, \ldots, k\} \\
&\subseteq \{\min_{m \neq i}(\hat{\mu}_i - \hat{\mu}_m - |q^*|\hat{\sigma}\sqrt{n_i^{-1} + n_m^{-1}}) < \mu_i - \mu_j \text{ for all } j \neq i, \\
&\quad \text{ for } i = 1, \ldots, k\}
\end{aligned}$$

$$= \{\min_{j \neq i}(\hat{\mu}_i - \hat{\mu}_j - |q^*|\hat{\sigma}\sqrt{n_i^{-1} + n_j^{-1}}) < \mu_i - \max_{j \neq i}\mu_j$$

$$\text{for } i = 1, \ldots, k\}$$

$$= E_1 \text{ (say)}.$$

$$E = \{\mu_i - \mu_j < \hat{\mu}_i - \hat{\mu}_j + |q^*|\hat{\sigma}\sqrt{n_i^{-1} + n_j^{-1}} \text{ for all } j \neq i,$$

$$\text{for } i = 1, \ldots, k\}$$

$$\subseteq \{\mu_i - \max_{m \neq i}\mu_m < \hat{\mu}_i - \hat{\mu}_j + |q^*|\hat{\sigma}\sqrt{n_i^{-1} + n_j^{-1}} \text{ for all } j \neq i,$$

$$\text{for } i = 1, \ldots, k\}$$

$$= \{\mu_i - \max_{j \neq i}\mu_j < \min_{j \neq i}(\hat{\mu}_i - \hat{\mu}_j + |q^*|\hat{\sigma}\sqrt{n_i^{-1} + n_j^{-1}})$$

$$\text{for } i = 1, \ldots, k\}$$

$$= E_2 \text{ (say)}.$$

Since $E \subseteq E_1 \cap E_2$ for all $\mu$ and $P\{E\} \geq 1 - \alpha$,

$$P_{\mu, \sigma^2}\{E_1 \cap E_2\} \geq 1 - \alpha$$

for all $\mu$ and $\sigma^2$.  $\square$

Again, the astute reader will notice that one can actually prove the following.

**Theorem 4.2.5** *For all* $\mu = (\mu_1, \ldots, \mu_k)$ *and* $\sigma^2$,

$$P_{\mu, \sigma^2}\{\min_{j \neq i}(\hat{\mu}_i - \hat{\mu}_j - |q^*|\hat{\sigma}\sqrt{n_i^{-1} + n_j^{-1}}) < \mu_i - \max_{j \neq i}\mu_j$$

$$< \min_{j \neq i}(\hat{\mu}_i - \hat{\mu}_j + |q^*|\hat{\sigma}\sqrt{n_i^{-1} + n_j^{-1}}) \text{ and}$$

$$\max_{j \neq i}(\hat{\mu}_i - \hat{\mu}_j - |q^*|\hat{\sigma}\sqrt{n_i^{-1} + n_j^{-1}}) < \mu_i - \min_{j \neq i}\mu_j$$

$$< \max_{j \neq i}(\hat{\mu}_i - \hat{\mu}_j + |q^*|\hat{\sigma}\sqrt{n_i^{-1} + n_j^{-1}}) \text{ for } i = 1, \ldots, k\} \geq 1 - \alpha.$$

That is, one can deduce multiple comparison with the best and multiple comparison with the worst inferences from the Tukey–Kramer all-pairwise comparison method simultaneously. (See Exercise 2.)

### 4.2.4.2 The Edwards–Hsu method

For each $i$, let $|d|^i$ be the critical value for Dunnett's method of two-sided MCC if treatment $i$ is the control. In other words, $|d|^i$ is the solution to

$$\int_0^\infty \int_{-\infty}^{+\infty} \prod_{j \neq i}[\Phi((\lambda_{ij}z + |d|^i s)/(1 - \lambda_{ij}^2)^{1/2})$$

$$-\Phi((\lambda_{ij}z - |d|^i s)/(1 - \lambda_{ij}^2)^{1/2})]d\Phi(z)\,\gamma(s)ds = 1 - \alpha$$

with $\lambda_{ij} = (1 + \frac{n_i}{n_j})^{-1/2}$, so that

$$P\{|\hat{\mu}_i - \mu_i - (\hat{\mu}_j - \mu_j)| < |d|^i \hat{\sigma} \sqrt{n_i^{-1} + n_j^{-1}} \text{ for all } j, j \neq i\} = 1 - \alpha.$$

Recall that the qmcc computer program with inputs $\lambda = (\lambda_{ij}, j \neq i)$, $\nu$ and $\alpha$ computes $|d|^i$. We will show that the intervals

$$[L_i, U_i], \quad i = 1, \ldots, k, \tag{4.44}$$

where

$$S = \{i : \min_{j \neq i}\{\hat{\mu}_i - \hat{\mu}_j + |d|^i \hat{\sigma} \sqrt{n_i^{-1} + n_j^{-1}}\} > 0\},$$

$$L_{ij} = \begin{cases} 0 \text{ if } i = j, \\ \hat{\mu}_i - \hat{\mu}_j - |d|^j \hat{\sigma} \sqrt{n_i^{-1} + n_j^{-1}} \text{ otherwise,} \end{cases}$$

$$L_i = \min_{j \in S} L_{ij},$$

$$U_{ij} = \begin{cases} 0 \text{ if } i = j, \\ -(\hat{\mu}_i - \hat{\mu}_j + |d|^j \hat{\sigma} \sqrt{n_i^{-1} + n_j^{-1}})^- \text{ otherwise,} \end{cases}$$

$$U_i = \max_{j \in S} U_{ij},$$

form a set of $100(1 - \alpha)\%$ simultaneous confidence intervals for

$$\mu_i - \max_{1 \leq j \leq k} \mu_j, \quad i = 1, \ldots, k.$$

Note that if $i \in S$, then $U_i = 0$. Further, if $S = \{M\}$, then $L_M = U_M = 0$ and the confidence intervals $[L_i, U_i], i \neq M$, are Dunnett's two-sided MCC confidence intervals with the $M$th treatment as the control.

**Theorem 4.2.6** *For all* $\mu = (\mu_i, \ldots, \mu_k)$ *and* $\sigma^2$,

$$P_{\mu,\sigma^2}\{\mu_i - \max_{1 \leq j \leq k} \mu_j \in [L_i, U_i] \text{ for } i = 1, \ldots, k\} \geq 1 - \alpha. \tag{4.45}$$

*Proof.* It will again be convenient to let $(k)$ denote the unknown index such that

$$\mu_{(k)} = \max_{1 \leq i \leq k} \mu_i.$$

Ties among $\mu_i$ can be broken in any fashion without affecting the validity of the derivation below.

Define the event $E$ as

$$E = \{\hat{\mu}_i - \mu_i - |d|^{(k)} \hat{\sigma} \sqrt{n_i^{-1} + n_{(k)}^{-1}} < \hat{\mu}_{(k)} - \mu_{(k)}$$

$$< \hat{\mu}_i - \mu_i + |d|^{(k)} \hat{\sigma} \sqrt{n_i^{-1} + n_{(k)}^{-1}} \text{ for all } i, i \neq (k)\}.$$

for all $i$,

$$E \subseteq \{\hat{\mu}_{(k)} > \hat{\mu}_i - |d|^{(k)}\hat{\sigma}\sqrt{n_i^{-1} + n_{(k)}^{-1}} \text{ for all } i, i \neq (k)\}$$
$$= \{(k) \in S\}.$$

*Derivation of upper confidence bounds*

$$E \subseteq \{(k) \in S \text{ and } \mu_{(k)} - \mu_{(k)} = 0 \text{ and}$$
$$\mu_i - \mu_{(k)} \leq -(\hat{\mu}_i - \hat{\mu}_{(k)} + |d|^{(k)}\hat{\sigma}\sqrt{n_i^{-1} + n_{(k)}^{-1}})^- \text{ for all } i, i \neq (k)\}$$
$$\subseteq \{\mu_i - \mu_{(k)} \leq U_i \text{ for } i = 1, \ldots, k\}$$
$$= E_1 \text{ (say).}$$

*Derivation of lower confidence bounds*

$$E \subseteq \{(k) \in S \text{ and } 0 = \mu_{(k)} - \mu_{(k)} \text{ and}$$
$$\hat{\mu}_i - \hat{\mu}_{(k)} - |d|^{(k)}\hat{\sigma}\sqrt{n_i^{-1} + n_{(k)}^{-1}} < \mu_i - \mu_{(k)} \text{ for all } i, i \neq (k)\}$$
$$\subseteq \{L_i \leq \mu_i - \mu_{(k)} \text{ for } i = 1, \ldots, k\}$$
$$= E_2 \text{ (say).}$$

We have shown $E \subseteq E_1 \cap E_2$. Thus,

$$1 - \alpha = P\{E\}$$
$$\leq P_{\mu,\sigma^2}\{E_1 \cap E_2\}$$
$$= P_{\mu,\sigma^2}\{L_i \leq \mu_i - \max_{1 \leq j \leq k} \mu_j \leq U_i \text{ for all } i\}. \quad \square$$

Starting with the same pivotal event $E$ as in Theorem 4.2.6, it is possible to derive unconstrained simultaneous confidence intervals for $\mu_i - \max_{j \neq i} \mu_j, i = 1, \ldots, k$. However, the expressions for the confidence bounds get complicated, and their derivations are left as an exercise for fanatics. (See Exercise 3(b).) Instead, we indicate in the following example how confidence intervals for $\mu_i - \max_{j \neq i} \mu_j, i = 1, \ldots, k$, can be given after confidence intervals for $\mu_i - \max_{1 \leq j \leq k} \mu_j, i = 1, \ldots, k$, have been computed.

### 4.2.5 Example: SAT scores (continued)

For the sample size pattern $n_1 = 103, n_2 = 31, n_3 = 122$, error degrees of freedom $\nu = 253$ (large enough to be taken as $\infty$ by some computer packages and the programs described in Appendix D), and $\alpha = 0.10$, we have

$$|q|^* = 2.052,$$
$$|d|^1 = 1.933,$$
$$|d|^2 = 1.852,$$
$$|d|^3 = 1.937.$$

$$|d|^3 \quad = 1.937.$$

(Had the computer codes not taken $\nu$ as $\infty$, the critical values would have been slightly larger.) The pooled sample standard deviation $\hat{\sigma} = 82.52$.

*Deduction from the Tukey–Kramer method*  The upper confidence bounds deduced from the Tukey–Kramer method are

$$\min\{619 - 629 + 2.052(82.52)\sqrt{1/103 + 1/31},$$
$$619 - 575 + 2.052(82.52)\sqrt{1/103 + 1/122}\} \quad = \quad 24.69$$

$$\min\{629 - 619 + 2.052(82.52)\sqrt{1/31 + 1/103},$$
$$629 - 575 + 2.052(82.52)\sqrt{1/31 + 1/122}\} \quad = \quad 44.69$$

$$\min\{575 - 619 + 2.052(82.52)\sqrt{1/122 + 1/103},$$
$$575 - 629 + 2.052(82.52)\sqrt{1/122 + 1/31}\} \quad = \quad -21.34,$$

while the lower confidence bounds deduced from the Tukey–Kramer method are

$$\min\{619 - 629 - 2.052(82.52)\sqrt{1/103 + 1/31},$$
$$619 - 575 - 2.052(82.52)\sqrt{1/103 + 1/122}\} \quad = \quad -44.69$$

$$\min\{629 - 619 - 2.052(82.52)\sqrt{1/31 + 1/103},$$
$$629 - 575 - 2.052(82.52)\sqrt{1/31 + 1/122}\} \quad = \quad -24.69$$

$$\min\{575 - 619 - 2.052(82.52)\sqrt{1/122 + 1/103},$$
$$575 - 629 - 2.052(82.52)\sqrt{1/122 + 1/31}\} \quad = \quad -88.06.$$

One can thus deduce from the Tukey–Kramer method

$$-44.69 \quad < \quad \mu_1 - \max_{j \neq 1} \mu_j \quad < \quad 24.69$$

$$-24.69 \quad < \quad \mu_2 - \max_{j \neq 2} \mu_j \quad < \quad 44.69$$

$$-88.06 \quad < \quad \mu_3 - \max_{j \neq 3} \mu_j \quad < \quad -21.34.$$

*The Edwards–Hsu method*  For the Edwards–Hsu method,

$$\min\{619 - 629 + 1.933(82.52)\sqrt{1/103 + 1/31},$$
$$619 - 575 + 1.933(82.52)\sqrt{1/103 + 1/122}\} \quad = \quad 22.68$$

$$\min\{629 - 619 + 1.852(82.52)\sqrt{1/31 + 1/103},$$
$$629 - 575 + 1.852(82.52)\sqrt{1/31 + 1/122}\} \quad = \quad 41.31$$

$$\min\{575 - 619 + 1.937(82.52)\sqrt{1/122 + 1/103},$$
$$575 - 629 + 1.937(82.52)\sqrt{1/122 + 1/31}\} \quad = \quad -22.61,$$

therefore $S = \{1, 2\}$. Thus, the upper confidence bounds are

$$U_1 = \max\{0, -(619 - 629 + 1.852(82.52)\sqrt{1/103 + 1/31})^-\} = \qquad 0$$

$$U_2 = \max\{0, -(629 - 619 + 1.933(82.52)\sqrt{1/31 + 1/103})^-\} = \qquad 0$$

$$U_3 = \quad \max\{-(575 - 619 + 1.933(82.52)\sqrt{1/122 + 1/103})^-, \\ -(575 - 629 + 1.852(82.52)\sqrt{1/122 + 1/31})^-\} = -22.65,$$

while the lower confidence bounds are

$$L_1 \quad = \quad \min\{0, 619 - 629 - 1.852(82.52)\sqrt{1/103 + 1/31}\} \quad = \quad -41.31$$

$$L_2 \quad = \quad \min\{0, 629 - 619 - 1.933(82.52)\sqrt{1/31 + 1/103}\} \quad = \quad -22.68$$

$$L_3 \quad = \quad \min\{575 - 619 - 1.933(82.52)\sqrt{1/122 + 1/103}, \\ 575 - 629 - 1.852(82.52)\sqrt{1/122 + 1/31}\} \quad = \quad -84.74.$$

The Edwards–Hsu method thus asserts

$$-41.31 \quad \leq \quad \mu_1 - \max_{1 \leq j \leq 3} \mu_j \quad \leq \quad 0$$

$$-22.68 \quad \leq \quad \mu_2 - \max_{1 \leq j \leq 3} \mu_j \quad \leq \quad 0$$

$$-84.74 \quad \leq \quad \mu_3 - \max_{1 \leq j \leq 3} \mu_j \quad \leq \quad -22.65.$$

Using the fact that if $\mu_i - \max_{1 \leq j \leq k} \mu_j$ is negative then $\mu_i - \max_{1 \leq j \leq k} \mu_j = \mu_i - \max_{j \neq i} \mu_j$, we can restate these confidence intervals as:

$$-41.31 \quad \leq \quad \mu_1 - \max_{j \neq 1} \mu_j \quad \leq \quad 22.68$$

$$-22.68 \quad \leq \quad \mu_2 - \max_{j \neq 2} \mu_j \quad \leq \quad 41.31$$

$$-84.74 \quad \leq \quad \mu_3 - \max_{j \neq 3} \mu_j \quad \leq \quad -22.65.$$

## 4.3 Nonparametric methods

In general, constrained multiple comparison with the best methods can be constructed based on one-sided multiple comparison with a control methods, while unconstrained multiple comparison with the best methods can either be constructed based on two-sided multiple comparison with a control methods, or deduced from all-pairwise comparison methods. Hsu (1981; 1984a) constructed nonparametric constrained MCB methods using one-sided MCC methods based on pairwise ranking, while Edwards and Hsu (1983) constructed nonparametric unconstrained MCB methods using two-sided MCC methods based on pairwise ranking (cf. Section 3.3.1).

Since joint ranking does not lead to valid nonparametric multiple comparison with a control methods (cf. Section 3.3.2), it does not lead to valid nonparametric multiple comparison with the best methods either. In fact, Rizvi and Woodworth (1970) showed that joint ranking does not lead to valid nonparametric subset selection methods. Recalling that subset selection inference is implied by constrained multiple comparison with the best inference, which in turn can be constructed from one-sided multiple comparison with a control methods, the result of Rizvi and Woodworth (1970) presaged the result of Fligner (1984) discussed in Section 3.3.2.

## 4.4 Exercises

1. Prove Theorem 4.2.2.

2. Prove Theorem 4.2.5.

3.(a) Consider the balanced one-way model (4.3). Let

$$S = \{i : \hat{\mu}_i - \hat{\mu}_j + |d|\hat{\sigma}\sqrt{2/n} > 0 \text{ for all } j, j \neq i\},$$

$$L_i^* = \begin{cases} \min_{j\neq i}\{\hat{\mu}_i - \hat{\mu}_j - |d|\hat{\sigma}\sqrt{2/n}\} & \text{if } S = \{i\}, \\ \min_{j\in S, j\neq i}\{\hat{\mu}_i - \hat{\mu}_j - |d|\hat{\sigma}\sqrt{2/n}\} & \text{otherwise}, \end{cases}$$

$$U_i^* = \begin{cases} \min_{j\neq i}\{\hat{\mu}_i - \hat{\mu}_j + |d|\hat{\sigma}\sqrt{2/n}\} & \text{if } i \in S \\ -(\max_{j\in S}\{\hat{\mu}_i - \hat{\mu}_j + |d|\hat{\sigma}\sqrt{2/n}\})^- & \text{if } i \notin S. \end{cases}$$

Show that the intervals

$$[L_i^*, U_i^*], \quad i = 1, \ldots, k,$$

form a set of $100(1-\alpha)\%$ simultaneous confidence intervals for

$$\mu_i - \max_{j\neq i}\mu_j, \quad i = 1, \ldots, k.$$

(b) Consider the unbalanced one-way model (4.10). Let

$$S = \{i : \min_{j\neq i}\{\hat{\mu}_i - \hat{\mu}_j + |d|^i\hat{\sigma}\sqrt{n_i^{-1} + n_j^{-1}}\} > 0\},$$

$$L_i^* = \begin{cases} (\min_{j\neq i}\{\hat{\mu}_i - \hat{\mu}_j - |d|^i\hat{\sigma}\sqrt{n_i^{-1} + n_j^{-1}}\})^+ & \text{if } S = \{i\}, \\ \min_{j\in S, j\neq i}\{\hat{\mu}_i - \hat{\mu}_j - |d|^j\hat{\sigma}\sqrt{n_i^{-1} + n_j^{-1}}\} & \text{otherwise}, \end{cases}$$

$$U_i^* = \begin{cases} \min_{j\neq i}\{\hat{\mu}_i - \hat{\mu}_j + |d|^i\hat{\sigma}\sqrt{n_i^{-1} + n_j^{-1}}\} & \text{if } i \in S \\ -(\max_{j\in S}\{\hat{\mu}_i - \hat{\mu}_j + |d|^j\hat{\sigma}\sqrt{n_i^{-1} + n_j^{-1}}\})^- & \text{if } i \notin S. \end{cases}$$

Show that the intervals

$$[L_i^*, U_i^*], \quad i = 1, \ldots, k$$

form a set of $100(1-\alpha)\%$ simultaneous confidence intervals for

$$\mu_i - \max_{j\neq i}\mu_j, \quad i = 1, \ldots, k.$$

4. *Deducing additional MCB-type inference from MCA inference: balanced design.* Consider the balanced one-way model (4.3). For each $i$, let

$$\mu_{(1)}^{(i)} \leq \cdots \leq \mu_{(k-1)}^{(i)}$$

denote the ordered $\{\mu_j, j \neq i\}$, let

$$\hat{\mu}_{[1]}^{(i)} \leq \cdots \leq \hat{\mu}_{[k-1]}^{(i)}$$

denote the ordered $\{\hat{\mu}_j, j \neq i\}$, and let $|q^*|$ be defined as in (4.34). Prove that, for all $\mu, \sigma^2$,

$$P_{\mu,\sigma^2}\{\hat{\mu}_i - \hat{\mu}_{[j]}^{(i)} - |q^*|\hat{\sigma}\sqrt{2/n} < \mu_i - \mu_{(j)}^{(i)} < \hat{\mu}_i - \hat{\mu}_{[j]}^{(i)} + |q^*|\hat{\sigma}\sqrt{2/n}$$
$$\text{for } j = 1, \ldots, k-1, i = 1, \ldots, k\} \geq 1 - \alpha.$$

Thus, a set of $100(1 - \alpha)\%$ simultaneous confidence intervals for $\mu_i - \mu_{(j)}^{(i)}, j = 1, \ldots, k-1, i = 1, \ldots, k$, is given by

$$\hat{\mu}_i - \hat{\mu}_{[j]}^{(i)} - |q^*|\hat{\sigma}\sqrt{2/n} < \mu_i - \mu_{(j)}^{(i)} < \hat{\mu}_i - \hat{\mu}_{[j]}^{(i)} + |q^*|\hat{\sigma}\sqrt{2/n}, \quad (4.46)$$

a result closely related to the results in Lam (1986).

Note that the simultaneous confidence intervals (4.36) constitute the subset of the confidence intervals (4.46) corresponding to $j = k - 1$; in short, Theorem 4.2.1 can be considerably strengthened.

Hint: For any constants $a, b$,

$$\{(x, y) : x < a, y < b\}$$
$$\subseteq \{(x, y) : \min\{x, y\} < \min\{a, b\}, \max\{x, y\} < \max\{a, b\}\}.$$

5. *Deducing additional MCB-type inference from MCA inference: unbalanced design.* For each $i$, let

$$\mu_{(1)}^{(i)} \leq \cdots \leq \mu_{(k-1)}^{(i)}$$

denote the ordered $\{\mu_j, j \neq i\}$. Show that if

$$L_{ij} < \mu_i - \mu_j < U_{ij} \text{ for all } i \neq j$$

then

$$L_{[j]}^{(i)} < \mu_i - \mu_{(j)}^{(i)} < U_{[j]}^{(i)} \text{ for all } i \neq j,$$

where

$$L_{[1]}^{(i)} \leq \cdots \leq L_{[k-1]}^{(i)}$$

denote the ordered $\{L_{ij}, j \neq i\}$, and

$$U_{[1]}^{(i)} \leq \cdots \leq U_{[k-1]}^{(i)}$$

denote the ordered $\{U_{ij}, j \neq i\}$; in short, Theorem 4.2.4 can be strengthened.

6. Apply constrained MCB to the FVC data (Table 2.4), using $\alpha = 0.01$. The results, with and without the non-inhaling smokers (NI) group, should be as in Table 4.3 and Table 4.4, respectively.

Table 4.3 *Constrained MCB confidence intervals for female FVC*

| Label | Group | $D_i^-$ | $D_i^+$ |
|-------|-------|---------|---------|
| NS | non-smokers | −0.007 | 0.247 |
| PS | passive smokers | −0.247 | 0.007 |
| LS | light smokers | −0.327 | 0.000 |
| MS | moderate smokers | −0.677 | 0.000 |
| HS | heavy smokers | −0.927 | 0.000 |

Table 4.4 *Constrained MCB confidence intervals for female FVC*

| Label | Group | $D_i^-$ | $D_i^+$ |
|-------|-------|---------|---------|
| NS | non-smokers | −0.026 | 0.243 |
| PS | passive smokers | −0.243 | 0.003 |
| NI | non-inhaling smokers | −0.355 | 0.026 |
| LS | light smokers | −0.323 | 0.000 |
| MS | moderate smokers | −0.673 | 0.000 |
| HS | heavy smokers | −0.923 | 0.000 |

# CHAPTER 5

# All-pairwise comparisons

For all-pairwise multiple comparisons (MCA), the parameters of primary interest are $\mu_i - \mu_j$ for all $i \neq j$, the $k(k-1)/2$ pairwise differences of treatment means.

## 5.1 Balanced one-way model

Suppose under the $i$th treatment a random sample $Y_{i1}, Y_{i2}, \ldots, Y_{in}$ of size $n$ is taken, where between the treatments the random samples are independent. Then under the usual normality and equality of variances assumptions, we have the one-way model

$$Y_{ia} = \mu_i + \epsilon_{ia}, \quad i = 1, \ldots, k, \quad a = 1, \ldots, n, \tag{5.1}$$

where $\mu_i$ is the effect of the $i$th treatment, $i = 1, \ldots, k$, and $\epsilon_{11}, \ldots, \epsilon_{kn}$ are i.i.d. normal with mean 0 and variance $\sigma^2$ unknown. We use the following notation

$$\hat{\mu}_i = \bar{Y}_i = \sum_{a=1}^{n} Y_{ia}/n,$$

$$\hat{\sigma}^2 = MSE = \sum_{i=1}^{k} \sum_{a=1}^{n} (Y_{ia} - \bar{Y}_i)^2 / [k(n-1)]$$

for the sample means and the pooled sample variance, and let $\nu = k(n-1)$, the degrees of freedom associated with $\hat{\sigma}^2$.

### 5.1.1 Tukey's method

Tukey's (1953) method provides the following simultaneous confidence intervals for all-pairwise differences:

$$\mu_i - \mu_j \in \hat{\mu}_i - \hat{\mu}_j \pm |q^*| \hat{\sigma} \sqrt{2/n} \text{ for all } i \neq j, \tag{5.2}$$

where $|q^*|$ is the solution to the equation

$$P\left\{ \frac{|\hat{\mu}_i - \mu_i - (\hat{\mu}_j - \mu_j)|}{\hat{\sigma}\sqrt{2/n}} \leq |q^*| \text{ for all } i > j \right\} = 1 - \alpha.$$

Numerically, $|q^*|$ is the solution of the equation

$$k \int_0^\infty \int_{-\infty}^{+\infty} [\Phi(z) - \Phi(z - \sqrt{2}|q^*|s)]^{k-1} d\Phi(z)\, \gamma(s) ds = 1 - \alpha. \qquad (5.3)$$

In (5.3), again $\Phi$ is the standard normal distribution function, and $\gamma$ is the density of $\hat{\sigma}/\sigma$. The qmca computer program described in Appendix D, with inputs $k, \nu$ and $\alpha$, computes $|q^*|$. Tables of $|q^*|$ can also be found in Appendix E, for example. Note that $\sqrt{2}|q^*|$ equals the critical value traditionally denoted by $q$, the upper $\alpha$th quantile of the Studentized range distribution with $k$ treatments and $\nu$ degrees of freedom. As a confidence set for $(\mu_1, \ldots, \mu_k)$, the set

$$T = \left\{ (\mu_1, \ldots, \mu_k) : \mu_i - \mu_j \in \hat{\mu}_i - \hat{\mu}_j \pm |q^*|\hat{\sigma}\sqrt{2/n} \text{ for all } i \neq j \right\}$$

is translation invariant, in the sense that if $(\mu_1, \ldots, \mu_k) \in T$ then $(\mu_1 + \delta, \ldots, \mu_k + \delta) \in T$ as well. In fact, $T$ is an infinite cylinder in $\Re^k$, centered at the vector $\{(\hat{\mu}_1 + \delta, \ldots, \hat{\mu}_k + \delta) : \delta \in \Re\}$, parallel to the vector $\{(\delta, \ldots, \delta) : \delta \in \Re\}$. Figure 5.1 depicts, for the case of $k = 3$ with $(\hat{\mu}_1, \hat{\mu}_2, \hat{\mu}_3) = (0, 1, 2)$, the hexagonal boundary of a section of Tukey's confidence set for $\mu_1, \mu_2, \mu_3$. For Maple codes that generate this depiction, which can be rotated, see Exercise 3.

**Theorem 5.1.1** *For all $\mu, \sigma^2$,*

$$P\left\{ \mu_i - \mu_j \in \hat{\mu}_i - \hat{\mu}_j \pm |q^*|\hat{\sigma}\sqrt{2/n} \text{ for all } i \neq j \right\} = 1 - \alpha.$$

*Proof.*

$$P\left\{ \mu_i - \mu_j \in \hat{\mu}_i - \hat{\mu}_j \pm |q^*|\hat{\sigma}\sqrt{2/n} \text{ for all } i \neq j \right\}$$

$$= \sum_{i=1}^k P\{ -|q^*|\hat{\sigma}\sqrt{2/n} < \hat{\mu}_i - \hat{\mu}_j - (\mu_i - \mu_j) < |q^*|\hat{\sigma}\sqrt{2/n}$$
$$\text{for all } j, j \neq i, \text{ and } \hat{\mu}_i - \mu_i = \max_{j=1,\ldots,k} (\hat{\mu}_j - \mu_j) \}$$

$$= \sum_{i=1}^k P\{ 0 < \hat{\mu}_i - \hat{\mu}_j - (\mu_i - \mu_j) < |q^*|\hat{\sigma}\sqrt{2/n} \text{ for all } j, j \neq i \}$$

$$= \sum_{i=1}^k P\{ 0 < \sqrt{n}(\hat{\mu}_i - \mu_i)/\sigma - \sqrt{n}(\hat{\mu}_j - \mu_j)/\sigma < \sqrt{2}|q^*|\hat{\sigma}/\sigma$$
$$\text{for all } j, j \neq i \}$$

$$= k \cdot P\{ 0 < \sqrt{n}(\hat{\mu}_1 - \mu_1)/\sigma - \sqrt{n}(\hat{\mu}_j - \mu_j)/\sigma < \sqrt{2}|q^*|\hat{\sigma}/\sigma$$
$$\text{for } j = 2, \ldots, k \}$$

$$= k \int_0^\infty \int_{-\infty}^{+\infty} [\Phi(z) - \Phi(z - \sqrt{2}|q^*|s)]^{k-1} d\Phi(z)\, \gamma(s) ds$$

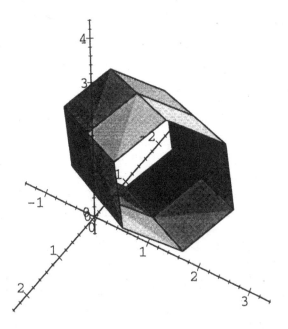

Figure 5.1 *Sample section of Tukey confidence set for* $k = 3$

$$= \quad 1 - \alpha. \quad \square$$

The expression of the coverage probability as a double integral allows the quantile $|q^*|$ to be efficiently computed. The critical value $|q^*|$ depends on $\alpha$, $k$ and $\nu$, the degrees of freedom associated with $\hat{\sigma}^2/\sigma^2$, but not the sample size $n$ explicitly. It will be convenient later on to denote $|q^*|$ by $|q^*|_{\alpha,k,\nu}$.

### 5.1.2 Bofinger's confident directions method

If, in a given situation, one takes the position that equalities among the $\mu_i$'s are impossible, then confident directions inference sharper than deduction from Tukey's simultaneous confidence intervals can be achieved using the following constrained $100(1 - \alpha)\%$ simultaneous confidence intervals for

$\mu_i - \mu_j$, given in Bofinger (1985) in a slightly different form:

$$\mu_i - \mu_j \in [-(\hat{\mu}_i - \hat{\mu}_j - q^*\hat{\sigma}\sqrt{2/n})^-, (\hat{\mu}_i - \hat{\mu}_j + q^*\hat{\sigma}\sqrt{2/n})^+] \text{ for all } i \neq j,$$
(5.4)

where

$$-x^- = \begin{cases} x & \text{if } x < 0 \\ 0 & \text{otherwise,} \end{cases}$$

$$x^+ = \begin{cases} x & \text{if } x > 0 \\ 0 & \text{otherwise,} \end{cases}$$

as in Section 4.1, and $q^*$ is the solution to the equation

$$P\left\{ \frac{\hat{\mu}_i - \mu_i - (\hat{\mu}_j - \mu_j)}{\hat{\sigma}\sqrt{2/n}} < q^* \text{ for all } i > j \right\} = 1 - \alpha,$$
(5.5)

i.e., $q^*$ is the critical value such that the event

$$E_1 = \{\mu_i - \mu_j > \hat{\mu}_i - \hat{\mu}_j - q^*\hat{\sigma}\sqrt{2/n} \text{ for all } i > j\}$$
(5.6)

occurs with probability $1 - \alpha$.

While the validity of the confidence intervals (5.4) does not depend on whether equalities among the $\mu_i$'s are possible, their interpretation does. If equalities among the treatment means are impossible and a larger treatment effect is better, then a lower confidence limit for $\mu_i - \mu_j$ equaling zero implies the $i$th treatment is better than the $j$th treatment, while an upper confidence limit for $\mu_i - \mu_j$ equaling zero implies the $j$th treatment is better than the $i$th treatment. If equalities among the treatment means are possible and a larger treatment effect is better, then a lower confidence limit for $\mu_i - \mu_j$ equaling zero implies the $i$th treatment is at least as good as the $j$th treatment, while an upper confidence limit for $\mu_i - \mu_j$ equaling zero implies the $i$th treatment is no better than the $j$th treatment.

Note that the critical value $q^*$ is always smaller than the critical value $|q^*|$ for Tukey's method, because $|q^*|$ is the constant such that the event

$$E = \{\mu_i - \mu_j \in \hat{\mu}_i - \hat{\mu}_j \pm |q^*|\hat{\sigma}\sqrt{2/n} \text{ for all } i \neq j\}$$

occurs with probability $1 - \alpha$, and $E \subset E_1$. Thus, assuming exact equalities among the $\mu_i$'s are impossible, then directional assertions are made more often by Bofinger's confidence intervals (5.4) than by Tukey's confidence intervals (5.2). That is, for the same data set and at the same error rate $\alpha$, Bofinger's method will make the same directional assertions as Tukey's method does, maybe more. However, having declared a treatment to be better than another treatment, in contrast to Tukey's method, Bofinger's method does not give a lower bound on *how much* the difference is.

**Theorem 5.1.2 (Bofinger 1985)** *For all $\mu, \sigma^2$,*

$$P_{\mu,\sigma^2}\{\mu_i - \mu_j \in [-(\hat{\mu}_i - \hat{\mu}_j - q^*\hat{\sigma}\sqrt{2/n})^-, (\hat{\mu}_i - \hat{\mu}_j + q^*\hat{\sigma}\sqrt{2/n})^+]$$

$$for \ all \ i \neq j\}$$
$$\geq \quad 1 - \alpha.$$

*Proof.* Let $\pi(1), \ldots, \pi(k)$ be a permutation of $1, \ldots, k$ such that $\mu_{\pi(i)} \leq \mu_{\pi(j)}$ for all $i > j$. Then, given that the event $E_1$ occurs with probability $1 - \alpha$, the event

$$E_2 \quad = \quad \{\mu_{\pi(i)} - \mu_{\pi(j)} > \hat{\mu}_{\pi(i)} - \hat{\mu}_{\pi(j)} - q^* \hat{\sigma} \sqrt{2/n} \text{ for all } i > j\}$$

also occurs with probability $1 - \alpha$. Now observe that

$$E_2 \quad \subseteq \quad \{\mu_i - \mu_j \in [-(\hat{\mu}_i - \hat{\mu}_j - q^* \hat{\sigma} \sqrt{2/n})^-, (\hat{\mu}_i - \hat{\mu}_j + q^* \hat{\sigma} \sqrt{2/n})^+]$$
$$for \ all \ i \neq j\},$$

because if $\mu_i - \mu_j > 0$, then only the upper confidence limit applies, while if $\mu_i - \mu_j < 0$, then only the lower confidence limit applies. $\qquad \square$

Exact computation of the critical value $q^*$ is non-trivial in general. Bofinger (1985) provided some conservative approximations to $q^*$. However, it turns out that the same critical value is subsequently used in the one-sided pairwise comparison method of Hayter (1990), discussed in Section 5.1.3, who gave exact tables of $q^*$ for $k \leq 9$. More recently, Hayter and Liu (1995) gave exact values of $q^*$ for $k \leq 20$.

*Remark.* Bofinger (1985) starts out with the assumption that exact equalities among the $\mu_i$'s are impossible, in which case the confidence intervals (5.4) can be given as open intervals instead of closed intervals.

### 5.1.3 Hayter's one-sided comparisons

Suppose it is suspected that

$$\mu_1 \leq \mu_2 \leq \cdots \leq \mu_k; \tag{5.7}$$

then one might be primarily interested in lower confidence bounds on $\mu_i - \mu_j$ for all $i > j$, which can be regarded as a version of one-sided all-pairwise comparisons. If $\mu_1, \ldots, \mu_k$ are mean responses corresponding to increasing doses in a dose-response study, for example, then (5.7) might be a reasonable guess. Of course, if one suspects

$$\mu_1 \geq \mu_2 \geq \cdots \geq \mu_k,$$

then lower confidence bounds on $\mu_i - \mu_j$ for all $i < j$ might be of primary interest.

Hayter (1990) derived the following simultaneous lower confidence bounds on $\mu_i - \mu_j$ for all $i > j$ :

$$\mu_i - \mu_j > \hat{\mu}_i - \hat{\mu}_j - q^* \hat{\sigma} \sqrt{2/n} \text{ for all } i > j, \tag{5.8}$$

where $q^*$ is the same critical value as in the simultaneous confidence intervals (5.4) of Bofinger (1985), i.e., $q^*$ is the solution to equation (5.5).

Hayter (1990) gave tables of $q^*$ for $k \leq 9$, and Hayter and Liu (1995) gave exact values of $q^*$ for $k \leq 20$. The confidence bounds (5.8) are valid confidence bounds for $\mu_i - \mu_j$ for all $i > j$ regardless of whether (5.7) is true or not. By forgoing upper confidence bounds and thus being able to use a smaller critical value than Tukey's method, if (5.7) turns out to be true, Hayter's method will likely make more directional assertions than Tukey's method (while retaining the ability to give lower bounds on *how much* the differences are).

**Theorem 5.1.3 (Hayter 1990)** *For all $\mu, \sigma^2$,*

$$P\{\mu_i - \mu_j > \hat{\mu}_i - \hat{\mu}_j - q^* \hat{\sigma} \sqrt{2/n} \text{ for all } i > j\} = 1 - \alpha.$$

*Proof.* Recall that $q^*$ is defined to be the critical value such that event

$$E_1 = \{\mu_i - \mu_j > \hat{\mu}_i - \hat{\mu}_j - q^* \hat{\sigma} \sqrt{2/n} \text{ for all } i > j\}$$

occurs with probability $1 - \alpha$. Noting that $E_1$ corresponds to correct coverage of the simultaneous confidence intervals (5.8), the proof follows. □

It is easy to verify that

$$\sum_{i=1}^{k} c_i \mu_i = \sum_{i=1}^{k-1} \left( -\sum_{j=1}^{i} c_j \right) (\mu_{i+1} - \mu_i)$$

for any contrast, because $c_k = -\sum_{i=1}^{k-1} c_i$. Therefore, the simultaneous confidence bounds (5.8) readily extend to simutanous confidence bounds for all so-called *non-negative* contrasts, which are contrasts $\sum_{i=1}^{k} c_i \mu_i$ with $\sum_{i=1}^{k} c_i = 0$ and $\sum_{j=1}^{i} c_j \leq 0$ for $i = 1, \ldots, k - 1$. Note that the likelihood ratio test for

$$H_0 : \mu_1 = \cdots = \mu_k$$

against the ordered alternative

$$H_a : \mu_1 \leq \cdots \leq \mu_k \text{ (with at least one strict inequality)},$$

discussed extensively in Barlow, Bartholomew, Bremner, Brunk (1972) and Robertson, Wright, and Dykstra (1988), pivots to give simultaneous confidence bounds on the so-called *monotone* contrasts, which are contrasts $\sum_{i=1}^{k} c_i \mu_i$ with $\sum_{i=1}^{k} c_i = 0$ and $c_i \geq c_j \ \forall \ i > j$ (see Williams 1977 and Marcus 1978). Monotone contrasts form a (not easily interpreted) proper subset of non-negative contrasts. For example, $\mu_i - \mu_j$, $i > j$, are non-negative contrasts but not monotone contrasts. In order to obtain confidence bounds on non-negative contrasts from a likelihood ratio test for $H_0 : \mu_1 = \cdots = \mu_k$ against an ordered alternative, the alternative needs to be a transformed version of $H_a : \mu_1 \leq \cdots \leq \mu_k$ (with at least one strict inequality). For $k = 3$, for example, the alternative needs to be

$$H_a : \frac{\mu_2 + \mu_3}{2} \geq \mu_1 \text{ and } \mu_3 \geq \frac{\mu_1 + \mu_2}{2} \text{ (with at least one strict inequality)}.$$

See Marcus (1978) for details.

### 5.1.4 Multiple range tests

If only confident inequalities are desired, then sharper inference than deduction from Tukey's method may be possible using a multiple range test.

Multiple range tests for all-pairwise comparisons, as described by Keuls (1952), Tukey (1953) (see Tukey 1994, Part F, pp. 251–275), Duncan (1955), Lehmann and Shaffer (1977), Finner (1990b), and many others, proceed as follows. Let $c_2, \ldots, c_k$ denote the critical values for comparing sets of $2, \ldots, k$ means, respectively. Let $[1], \ldots, [k]$ denote the random indices such that

$$\hat{\mu}_{[1]} \leq \cdots \leq \hat{\mu}_{[k]}.$$

(Since $\hat{\mu}_1, \ldots, \hat{\mu}_k$ are continuous random variables, ties occur among them with probability zero.)

As a first step, compare the $k$ range $\hat{\mu}_{[k]} - \hat{\mu}_{[1]}$ with $c_k \hat{\sigma} \sqrt{2/n}$. If

$$\hat{\mu}_{[k]} - \hat{\mu}_{[1]} \leq c_k \hat{\sigma} \sqrt{2/n}$$

then stop; else assert

$$\mu_{[1]} \neq \mu_{[k]}$$

and the two $(k-1)$ ranges $\hat{\mu}_{[k-1]} - \hat{\mu}_{[1]}$ and $\hat{\mu}_{[k]} - \hat{\mu}_{[2]}$ are compared with $c_{k-1} \hat{\sigma} \sqrt{2/n}$. If both ranges are less than or equal to $c_{k-1} \hat{\sigma} \sqrt{2/n}$, then stop. Otherwise, assert

$$\mu_{[1]} \neq \mu_{[k-1]}$$

if

$$\hat{\mu}_{[k-1]} - \hat{\mu}_{[1]} > c_{k-1} \hat{\sigma} \sqrt{2/n}$$

and/or assert

$$\mu_{[2]} \neq \mu_{[k]}$$

if

$$\hat{\mu}_{[k]} - \hat{\mu}_{[2]} > c_{k-1} \hat{\sigma} \sqrt{2/n},$$

and the three or appropriate one $(k-2)$ range(s) are compared with $c_{k-2} \hat{\sigma} \sqrt{2/n}$; and so on. Once a range has been found to be less than or equal to its scaled critical value, its subranges are no longer tested.

The execution of the multiple range test described above can be performed by the familiar 'underlining' procedure, as follows.

List the ordered $\hat{\mu}_i$ horizontally. If

$$\frac{\hat{\mu}_{[k]} - \hat{\mu}_{[1]}}{\hat{\sigma} \sqrt{2/n}} \leq c_k,$$

then draw a line underneath from $\hat{\mu}_{[1]}$ to $\hat{\mu}_{[k]}$. If

$$\frac{\hat{\mu}_{[k]} - \hat{\mu}_{[1]}}{\hat{\sigma} \sqrt{2/n}} > c_k,$$

then do not draw a line underneath from $\hat{\mu}_{[1]}$ to $\hat{\mu}_{[k]}$. For any set of ordered sample means $\hat{\mu}_{[i]} < \cdots < \hat{\mu}_{[i+m-1]}$ which have not been connected by an unbroken line, if

$$\frac{\hat{\mu}_{[i+m-1]} - \hat{\mu}_{[i]}}{\hat{\sigma}\sqrt{2/n}} \leq c_m,$$

then draw a line underneath from $\hat{\mu}_{[i]}$ to $\hat{\mu}_{[i+m-1]}$. If

$$\frac{\hat{\mu}_{[i+m-1]} - \hat{\mu}_{[i]}}{\hat{\sigma}\sqrt{2/n}} > c_m,$$

then do not draw a line underneath from $\hat{\mu}_{[i]}$ to $\hat{\mu}_{[i+m-1]}$. When the process is completed, treatment means whose estimates are not connected by an unbroken line beneath them are asserted to be unequal.

Different multiple range tests differ on the choice of the critical values $c_2, \ldots, c_k$ or, equivalently, the probabilities

$$\alpha_m = P_{\mu_1 = \cdots = \mu_m} \left\{ \max_{1 \leq i,j \leq m} \frac{\hat{\mu}_i - \hat{\mu}_j}{\hat{\sigma}\sqrt{2/n}} > c_m \right\}, \quad m = 2, \ldots, k.$$

When the critical values are monotone

$$c_2 < \cdots < c_k,$$

which is intuitively a reasonable requirement because the critical value for comparing a larger number of treatments should be larger than the critical value for comparing a smaller number of treatments, Lehmann and Shaffer (1977) gave an exact expression for the probability of at least one incorrect assertion when the error degrees of freedom $\nu$ is infinity.

**Theorem 5.1.4 (Lehmann and Shaffer 1977)** *If* $c_1 \leq \cdots \leq c_k$ *and* $\nu = \infty$, *then*

$$\sup_{\mu_1, \ldots, \mu_k} P_\mu\{\text{at least one incorrect assertion}\}$$

$$= \max_{g=1,\ldots,k-1} \max_{p_1 + \cdots + p_g = k} \left(1 - \prod_{i=1}^{g} (1 - \alpha_{p_i})\right)$$

*where* $p_1, \ldots, p_g$ *are positive integers and* $\alpha_1 = 0$.

Using Corollary A.1.1, they also showed that the same expression is an upper bound on the probability of at least one incorrect assertion when the error degrees of freedom $\nu$ is finite.

**Corollary 5.1.1 (Lehmann and Shaffer 1977)** *If* $c_1 \leq \cdots \leq c_k$, *then*

$$\sup_{\mu_1, \ldots, \mu_k, \sigma^2} P_{\mu,\sigma^2}\{\text{at least one incorrect assertion}\}$$

$$\leq \max_{g=1,\ldots,k-1} \max_{p_1 + \cdots + p_g = k} \left(1 - \prod_{i=1}^{g} (1 - \alpha_{p_i})\right)$$

*where $p_1, \ldots, p_g$ are positive integers and $\alpha_1 = 0$.*

While we will use these elegant results, stated without proof, to show directly certain multiple range tests to be confident directions methods, an alternative (less direct) proof is given in Section 5.1.6.

### 5.1.4.1 Newman–Keuls multiple range test

Recalling that the stepdown MCC method of Naik, Marcus, Peritz and Gabriel, discussed in Section 3.1.1.2, uses critical values associated with size-$\alpha$ tests for comparing sets of new treatments of different sizes with a control, one might contemplate setting

$$\alpha_m = \alpha$$

in all-pairwise comparisons as well. This leads to the multiple range test of 'Student' (1927), Newman (1939), and Keuls (1952), commonly referred to as the Newman–Keuls multiple range test. The choice $\alpha_m = \alpha$ leads to critical values $c_m = |q^*|_{\alpha,m,\nu}$ which are clearly increasing in $m$. Therefore, Theorem 5.1.4 applies. This theorem, or the following direct computation, readily shows that the Newman–Keuls multiple range test is not a confident inequalities method when $\nu = \infty$, unless $k \leq 3$.

Suppose $\mu_1 = \mu_2 \ll \cdots \ll \mu_{2\lfloor k/2 \rfloor - 1} = \mu_{2\lfloor k/2 \rfloor}$ with $k \geq 4$, where $\ll$ means 'much smaller than' and $\lfloor x \rfloor$ is the greatest integer which is less than or equal to $x$. Then all ranges except $\hat{\mu}_{i+1} - \hat{\mu}_i$, $i = 1, 3, \ldots, 2\lfloor k/2 \rfloor - 1$ will exceed their scaled critical values with virtual certainty. Therefore,

$$
\begin{aligned}
&P_{\mu,\sigma^2}\{\text{at least one incorrect assertion}\} \\
=\ &P_{\mu,\sigma^2}\{\text{assert } \mu_1 \neq \mu_2 \text{ or } \cdots \text{ or } \mu_{2\lfloor k/2 \rfloor - 1} \neq \mu_{2\lfloor k/2 \rfloor}\} \\
\overset{\nu \to \infty}{\longrightarrow}\ &1 - (1 - \alpha)^{\lfloor k/2 \rfloor} \\
>\ &\alpha
\end{aligned}
$$

unless $k \leq 3$.

For example, suppose $k = 4$, $\alpha = 0.05$, and $\mu_1 = \mu_2 \ll \mu_3 = \mu_4$. Then all ranges except $\hat{\mu}_2 - \hat{\mu}_1$ and $\hat{\mu}_4 - \hat{\mu}_3$ will exceed their scaled critical values with virtual certainty. Thus,

$$
\begin{aligned}
&P_{\mu,\sigma^2}\{\text{at least one incorrect assertion}\} \\
=\ &P_{\mu,\sigma^2}\{\text{assert } \mu_1 \neq \mu_2 \text{ or } \mu_3 \neq \mu_4\} \\
\overset{\nu \to \infty}{\longrightarrow}\ &1 - (0.95)^2 \\
=\ &0.0975.
\end{aligned}
$$

Therefore, the Newman–Keuls multiple range test is not a confident inequalities method and cannot be recommended.

## 5.1.4.2 The Methods of Ryan, Einot and Gabriel, and Welsch

To obtain a confident inequalities multiple range test, if one chooses $\alpha_m$, $m = 2, \ldots, k$, to satisfy

$$\max_{g=1,\ldots,k-1} \max_{p_1+\cdots+p_g=k} 1 - \prod_{i=1}^{g}(1 - \alpha_{p_i}) \leq \alpha, \tag{5.9}$$

then, provided the corresponding critical values $c_m$ are increasing in $m$, Corollary 5.1.1 guarantees

$$P_{\mu,\sigma^2}\{\text{at least one incorrect assertion}\} \leq \alpha.$$

Choosing

$$\alpha_m = 1 - (1 - \alpha)^{m/k} \tag{5.10}$$

satisfies (5.9) because then

$$\max_{g=1,\ldots,k-1} \max_{p_1+\cdots+p_g=k} 1 - \prod_{i=1}^{g}(1 - \alpha_{p_i})$$

$$= \max_{g=1,\ldots,k-1} \max_{p_1+\cdots+p_g=k} 1 - \prod_{i=1}^{g}(1 - \alpha)^{p_i/k}$$

$$= \max_{g=1,\ldots,k-1} \max_{p_1+\cdots+p_g=k} 1 - (1 - \alpha)^{(p_1+\cdots+p_g)/k}$$

$$= 1 - (1 - \alpha)^{k/k}$$

$$= \alpha.$$

This choice of $\alpha_m$ was suggested by Einot and Gabriel (1975). Finner (1990b) proved that, for this choice of $\alpha_m$, the corresponding critical value $c_m$ is increasing in $m$ when $\nu = \infty$. He also observed that, for $\nu < \infty$, the critical value $c_m = |q^*|_{1-(1-\alpha)^{m/k},m,\nu}$ is usually increasing in $m$ except when the error degrees of freedom $\nu$ is relatively small.

A popular modification of $\alpha_{k-1}$, dating back to Tukey (1953) (see Tukey 1994, p. 268), and utilized by Welsch (1977), Ramsey (1978), Begun and Gabriel (1981), among others, is to set $\alpha_{k-1} = \alpha$ which, when combined with (5.10), gives

$$\alpha_m = \begin{cases} 1 - (1 - \alpha)^{m/k} & \text{if } m = 2, \ldots, k-2, \\ \alpha & \text{if } m = k-1, k. \end{cases} \tag{5.11}$$

This choice of $\alpha_m$ also satisfies (5.9), because $\alpha_1 = 0$, meaning that if $k-1$ treatments are equal to each other and one lone treatment is different, then no inequality assertion concerning that lone treatment can be incorrect. However, with this choice of $\alpha_m$, there is no guarantee that $c_{k-2} < c_{k-1}$ even when $\nu = \infty$. Therefore, it is usually suggested that the critical value $c_{k-1}$ be replaced by $c_{k-2}$ whenever the latter is larger (e.g., Finner 1990b).

*Historical remark.* Using the Bonferroni inequality and two-sample $t$ statistics instead of range statistics, Ryan (1960) suggested

$$\alpha_m = \begin{cases} (m/k)\alpha & \text{if } m = 2, \ldots, k-2, \\ \alpha & \text{if } m = k-1, k. \end{cases}$$

which is slightly more conservative than (5.10). Using range statistics, Welsch (1977) independently made the same suggestion and noted further that the critical values need to be monotone.

The REGWQ statement of the MEANS option in PROC GLM of Version 6.09 of the SAS system implements the multiple range test with $\alpha_m$ set according to (5.11), without checking for the monotonicity of the critical value $c_m$. Therefore, as discussed in more detail in Section 5.1.6, this implementation is guaranteed to result in a confident inequalities method only when the critical values are monotonically non-decreasing in $m$ for the particular data set being analyzed (see Finner 1990b, p. 193). Before the implementation is fixed, you can assert the inequalities reported (only) after manually verifying from the listing that the critical values $c_m$ are monotonically non-decreasing in $m$.

There remains the important question whether this multiple range test is a confident directions method. For the case of $k = 3$, the technique in Finner (1990a) can in principle be adapted to show that it is. For $k > 3$, as yet there is neither a proof that the multiple range test is a confident directions method nor a counter-example that it is not.

### 5.1.4.3 Duncan's multiple range test

Duncan's (1955) multiple range test sets

$$\alpha_m = 1 - (1 - \alpha)^{m-1} \tag{5.12}$$

which, in contrast to the Newman–Keuls multiple range test and the methods of Ryan, Einot and Gabriel, and Welsch, leads to testing the $k$ range $\hat{\mu}_{[k]} - \hat{\mu}_{[1]}$ at the first step at a level *higher* than $\alpha$ (for $k > 2$). If you examine table of critical values of Duncan's method (e.g., Harter 1960), you will be struck by the observation that, for finite $\nu$, the critical value $c_m$ typically first increases but then remains constant as $m$ increases. Actually, for finite $\nu$, the choice (5.12) makes $c_m$ *decrease* beyond certain $m$, which seems counter-intuitive (again, the critical value for comparing a larger number of treatments should be larger than the critical value for comparing a smaller number of treatments). The tabulated critical values have been modified to ensure monotonicity. Thus modified, Duncan's multiple range test controls the so-called per comparison error rate, since its $\alpha$ level for each $H_{\{i,j\}} : \mu_i = \mu_j$ equals $\alpha$. This is implemented as the DUNCAN statement of the MEANS option in PROC GLM of the SAS system. But since its $\alpha_m$'s are at least as high as the $\alpha_m$'s of the Newman–Keuls multiple range test (higher for $k > 2$), which is not a confident inequalities method, Duncan's multiple range test is not a confident inequalities method and cannot be recommended either.

In the words of Tukey (1991), Duncan's multiple range test was a 'distraction' in the history of multiple comparisons, amounting to 'talking 5% while using more than 5% simultaneous.'

### 5.1.5 Scheffé's method

Recall the standard result that, under the null hypothesis

$$H_0 : \mu_1 = \cdots = \mu_k$$

the statistic

$$\frac{\sum_{i=1}^{k} n(\hat{\mu}_i - \hat{\bar{\mu}})^2 / (k - 1)}{\hat{\sigma}^2},$$

where

$$\hat{\bar{\mu}} = \frac{\hat{\mu}_1 + \cdots + \hat{\mu}_k}{k},$$

has an $F$ distribution with $k - 1$ numerator and $\nu$ denominator degrees of freedom. Since the means of $\hat{\mu}_1 - \mu_1, \ldots, \hat{\mu}_k - \mu_k$, as well as the mean of $\hat{\bar{\mu}} - \bar{\mu}$, where

$$\bar{\mu} = \frac{\mu_1 + \cdots + \mu_k}{k},$$

are zero,

$$\frac{\sum_{i=1}^{k} n(\hat{\mu}_i - \mu_i - (\hat{\bar{\mu}} - \bar{\mu}))^2 / (k - 1)}{\hat{\sigma}^2}$$

also has the $F$ distribution with $k-1$ numerator and $\nu$ denominator degrees of freedom for all $\boldsymbol{\mu} = (\mu_1, \ldots, \mu_k)$ and $\sigma^2$. Therefore

$$P\left\{ \sum_{i=1}^{k} (\sqrt{n}(\hat{\mu}_i - \hat{\bar{\mu}}) - \sqrt{n}(\mu_i - \bar{\mu}))^2 \le (k - 1)\hat{\sigma}^2 F_{\alpha, k-1, \nu} \right\} = 1 - \alpha,$$

where $F_{\alpha, k-1, \nu}$ is the upper $\alpha$ quantile of the $F$ distribution with $k - 1$ and $\nu$ degrees of freedom. Thus the set

$$S = \left\{ (\mu_1, \ldots, \mu_k) : \sum_{i=1}^{k} (\sqrt{n}(\hat{\mu}_i - \hat{\bar{\mu}}) - \sqrt{n}(\mu_i - \bar{\mu}))^2 \le (k - 1)\hat{\sigma}^2 F_{\alpha, k-1, \nu} \right\}$$

is a $100(1-\alpha)\%$ confidence set for $\mu_1, \ldots, \mu_k$. This confidence set is translation invariant, in the sense that if $(\mu_1, \ldots, \mu_k) \in S$ then $(\mu_1 + \delta, \ldots, \mu_k + \delta) \in S$ as well. In fact, $S$ is an infinite cylinder in $\Re^k$, centered at the vector $\{(\hat{\mu}_1 + \delta, \ldots, \hat{\mu}_k + \delta) : \delta \in \Re\}$, parallel to the vector $(1, \ldots, 1)$, with a spherical cross-section. Figure 5.2 depicts, for the case of $k = 3$ with $(\hat{\mu}_1, \hat{\mu}_2, \hat{\mu}_3) = (0, 1, 2)$, the circular boundary of a section of the confidence set $S$ for $\mu_1, \mu_2, \mu_3$. For Maple codes that generate this depiction,

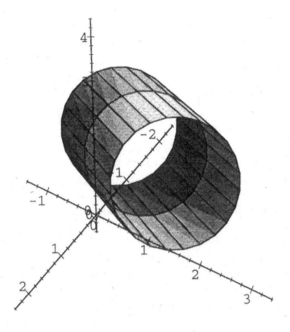

Figure 5.2 *Sample section of Scheffé confidence set for $k = 3$*

which can be rotated, see Exercise 4. The projection of $S$ on a plane, *any* plane, perpendicular to $(1, \ldots, 1)$ (i.e., a cross-section of $S$) gives Scheffé's $100(1-\alpha)\%$ simultaneous confidence intervals for all contrasts of $\mu_1, \ldots, \mu_k$: For all $c_1 + \cdots + c_k = 0$,

$$\sum_{i=1}^{k} c_i \mu_i \in \sum_{i=1}^{k} c_i \hat{\mu}_i \pm \sqrt{(k-1) F_{\alpha, k-1, \nu}} \, \hat{\sigma} (\sum_{i=1}^{k} c_i^2 / n)^{1/2}.$$

The derivation of these simultaneous confidence intervals from the projection is no more difficult for the unbalanced one-way model than it is for the balanced one-way model, and is therefore deferred until Section 5.2.3 (see Theorem 5.2.3). The reason for discussing Scheffé's method at this point is that it is easier to compare Scheffé's method and Tukey's method in the present, simpler, setting.

One can obviously deduce pairwise comparisons from Scheffé's confidence set by specializing to $c = c_{ij}$, the $k$-dimensional vector with 1 in the $i$th coordinate, $-1$ in the $j$th coordinate, and 0 in all other coordinates, to obtain a set of (conservative) $100(1-\alpha)\%$ simultaneous confidence intervals for pairwise differences $\mu_i - \mu_j$ :

$$\mu_i - \mu_j \in \hat{\mu}_i - \hat{\mu}_j \pm \sqrt{(k-1)F_{\alpha,k-1,\nu}}\,\hat{\sigma}\sqrt{2/n}, \text{ for all } i,j, \ i \neq j.$$

The rejection of the $F$-test for $H_0 : \mu_1 = \cdots = \mu_k$ corresponds to the exclusion of the vector $\{(\delta,\ldots,\delta) : \delta \in \Re\}$ by Scheffé's confidence set for $\mu_1,\ldots,\mu_k$. Thus, it is true that the rejection of the $F$-test implies there exists a contrast $c_1\mu_1 + \cdots + c_k\mu_k$ for which Scheffé's confidence interval does not contain zero. In fact, if we let $c^* = (\hat{\mu}_1 - \hat{\bar{\mu}}, \ldots, \hat{\mu}_k - \hat{\bar{\mu}})$, then $c_1^*\mu_1 + \cdots + c_k^*\mu_k$ is such a contrast. Further, if Scheffé's confidence interval for $c_1^*\mu_1 + \cdots + c_k^*\mu_k$ does not contain zero, then its confidence intervals for any contrast sufficiently 'close' to $c_1^*\mu_1 + \cdots + c_k^*\mu_k$ will not contain zero either. However, there is no guarantee that, when the $F$-test rejects, the collection of contrasts for which Scheffé's confidence intervals do not contain zero contains a pairwise difference $\mu_i - \mu_j$ for some $i \neq j$. Geometrically, this situation occurs when the infinite cylinder $S$, Scheffé's confidence set, does not contain $\{(\delta,\ldots,\delta) : \delta \in \Re\}$ but does intersect the hyperplanes $\mu_i = \mu_j$ for all $i \neq j$. Figure 5.3 is a perspective view of such a situation from a viewpoint which parallels the vector $\{(\delta,\ldots,\delta) : \delta \in \Re\}$, in which case the hyperplane $\mu_1 = \mu_2$ coincides with the $\mu_3$ axis, the hyperplane $\mu_2 = \mu_3$ coincides with the $\mu_1$ axis, and the $\mu_3 = \mu_1$ hyperplane coincides with the $\mu_2$ axis. For Maple codes that generate this view, which can be rotated, see Exercise 6. For examples of data that lead to this situation, see Exercise 5.

Given that, for pairwise comparisons in the balanced one-way model, Tukey's method is exact while Scheffé's method is conservative, the latter is only recommended when contrasts that are not pairwise differences are of primary interest.

### 5.1.6 A general stepdown testing scheme, with application to multiple F-tests

Tukey's method, as well as the methods of Ryan, Einot and Gabriel, and Welsch, are based on range statistics. As confident inequalities methods, the latter are typically sharper than deduction from Tukey's method. Thus, one might contemplate whether confident inequalities methods sharper than deduction from Scheffé's method can be constructed using $F$ statistics. Toward that end, consider the following general stepdown testing scheme.

*A general stepdown testing scheme* For any $I \subseteq K = \{1,\ldots,k\}$, let $H_I$ denote the hypothesis

$$H_I : \mu_i = \mu_j \text{ for all } i,j \in I,$$

and let $|I|$ denote the number of elements in $I$.

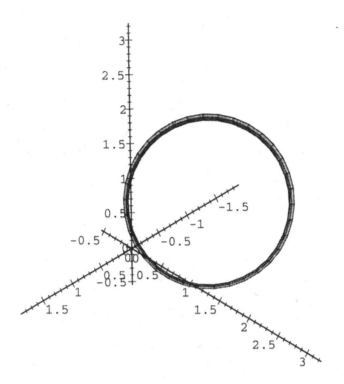

Figure 5.3 *F-test rejects but no significant pairwise difference*

| Step 1 |
|:---:|

Test, at $\alpha = \alpha_k$, $H_K : \mu_1 = \cdots = \mu_k$.
If   $H_K$   is accepted
    then   accept $H_I$ for all $I \subset K$;
           stop
    else   assert $\mu_i \neq \mu_j$ for some $i, j \in K$;
           go to step 2.

| Step 2 |
|:---:|

Test, at $\alpha = \alpha_{k-1}$, each $H_J$ with $|J| = k - 1$ that has not been accepted.
If   $H_J$   is accepted
    then   accept $H_I$ for all $I \subset J$;
    else   assert $\mu_i \neq \mu_j$ for some $i, j \in J$
If   all $H_J$   are accepted
    then   stop
    else   go to step 3.

$$\boxed{\text{Generic step}}$$

Test, at $\alpha = \alpha_{|J|}$, each $H_J$ that has not been accepted at an earlier step.

    If   $H_J$   is accepted

        then   accept $H_I$ for all $I \subset J$

        else   assert $\mu_i \neq \mu_j$ for some $i, j \in J$.

Continue testing until no $H_J$ remains to be tested. Finally, assert $\mu_i \neq \mu_j$ if all $H_J$ such that $i, j \in J$ are rejected.

The following theorem shows confident inequalities methods can be constructed by applying range statistics or $F$ statistics to the stepdown testing scheme and setting $\alpha_m$ in accordance with (5.11).

**Theorem 5.1.5** *Suppose the test statistic $T_I$ and critical value $b_I$ for each hypothesis $H_I$ satisfy*

$$P_{H_I}\{T_I > b_I\} \leq \begin{cases} 1 - (1-\alpha)^{|I|/k} & \text{if } |I| = 2, \ldots, k-2, \\ \alpha & \text{if } |I| = k-1, k, \end{cases} \qquad (5.13)$$

*and whenever $I_1, \ldots, I_g$ are disjoint subsets of $\{1, \ldots, k\}$, the test statistics for $H_{I_1}, \ldots, H_{I_g}$ satisfy the condition of Corollary A.1.1. Then for the stepdown method*

$$P_{\mu,\sigma^2}\{\text{at least one incorrect assertion}\} \leq \alpha. \qquad (5.14)$$

*Proof.* Suppose an $H_I$ with $|I| = k$ or $k-1$ is true. Then only it, and hypotheses implied by it, are true. Since it is tested at level $\alpha$, (5.14) holds.

Now let $I_1, \ldots, I_g$ ($g \geq 2$) be the disjoint subsets of $\{1, \ldots, k\}$ of size at most $k-2$ each such that only $H_{I_1}, \ldots, H_{I_g}$, and hypotheses implied by them, are true. Then since each $H_{I_i}$, if tested, is tested at level $(1-\alpha)^{|I_i|/k}$,

$$\begin{aligned} &P_{\mu,\sigma^2}\{\text{at least one incorrect assertion}\} \\ \leq\ & 1 - P_{H_{I_1}, \ldots, H_{I_g}}\{T_{I_1} \leq b_{I_1} \text{ and } \cdots \text{ and } T_{I_g} \leq b_{I_g}\} \\ \leq\ & 1 - \prod_{i=1}^{g} P_{H_{I_i}}\{T_{I_i} \leq b_{I_i}\} \\ \leq\ & 1 - \prod_{i=1}^{g} (1-\alpha)^{|I_i|/k} \\ =\ & 1 - (1-\alpha)^{(|I_1| + \cdots + |I_g|)/k} \\ \leq\ & 1 - (1-\alpha) \\ =\ & \alpha, \end{aligned}$$

where the second inequality follows from Corollary A.1.1. $\square$

Consider the range test which rejects $H_I$ when

$$\max_{i,j \in I} \frac{\hat{\mu}_i - \hat{\mu}_j}{\hat{\sigma}\sqrt{2/n}} > c_{|I|}.$$

The Studentized range statistics for testing $H_{I_1}, \ldots, H_{I_g}$ are independent conditional on $\hat{\sigma}$ if $I_1, \ldots, I_g$ are disjoint, so the condition of Corollary A.1.1 is satisfied. If one lets

$$
c_m = \begin{cases} |q^*|_{1-(1-\alpha)^{m/k}, m, \nu} & \text{if } m = 2, \ldots, k-2, \\ |q^*|_{\alpha, m, \nu} & \text{if } m = k-1, k, \end{cases}
$$

in accordance with (5.11) and executes the stepdown method using range tests, then a confident inequalities method results. Now, provided $c_{|I|}$ is increasing in $|I|$, if $H_{\{[i], \ldots, [i+m-1]\}}$ (with consecutive indices $[i], \ldots, [i+m-1]$) is rejected by the stepdown method, then any $H_J$ such that $\{[i], [i+m-1]\} \subseteq J \subset \{[i], \ldots, [i+m-1]\}$ will be rejected by the stepdown method as well, since the value of the range statistic for any such $H_J$ is the same as for $H_{\{[i], \ldots, [i+m-1]\}}$ and is compared to a smaller scaled critical value. Therefore, in this case, the assertion $\mu_{[i]} \neq \mu_{[i+m-1]}$ will be made and the form of the stepdown method simplifies to the underlining scheme of multiple range tests described at the beginning of Section 5.1.4. (This provides a proof that the Ryan/Einot–Gabriel/Welsch multiple range test with increasing critical values is a confident inequalities method.) Indeed, it would be rather frustrating to reject $H_{\{[i], \ldots, [i+m-1]\}}$ but not be able to assert as different the two treatments furtherest apart in terms of sample means among those with indices in $\{[i], \ldots, [i+m-1]\}$. Recall, however, setting $\alpha_m$ in accordance with (5.11) does not always generate $c_m$ increasing in $m$. To understand the pitfall of applying non-monotone critical values to the underlining scheme described at the beginning of Section 5.1.4, consider a data set with $k = 10$, $n_1 = \cdots = n_{10} = 2$, $\hat{\mu}_1 = \cdots = \hat{\mu}_8 = 0$, $\hat{\mu}_9 = 3.88$, $\hat{\mu}_{10} = 4$, $\hat{\sigma} = 1$, and suppose $\alpha = 0.05$. Then

$$
\begin{aligned}
c_{10} &= |q^*|_{0.05, 10, 10} &&= 3.959 \\
c_9 &= |q^*|_{0.05, 9, 10} &&= 3.861 \\
c_8 &= |q^*|_{1-0.95^{(8/10)}, 8, 10} &&= 3.899.
\end{aligned}
$$

Noting $\hat{\sigma}\sqrt{2/n} = 1$ and

$$
\begin{aligned}
\hat{\mu}_{10} - \hat{\mu}_1 &= 4 &&> &&3.959, \\
\hat{\mu}_9 - \hat{\mu}_1 &= 3.88 &&> &&3.861,
\end{aligned}
$$

the underlining scheme described at the beginning of Section 5.1.4 asserts

$$
\begin{aligned}
\mu_1 &\neq \mu_{10}, \\
\mu_1 &\neq \mu_9
\end{aligned}
$$

(as does the implementation of REGWQ under the MEANS option in PROC GLM of Version 6.09 of the SAS system). However, since

$$
\hat{\mu}_9 - \hat{\mu}_1 = 3.88 < 3.899,
$$

the hypothesis

$$H_0 : \mu_1 = \cdots = \mu_7 = \mu_9$$

is accepted by the stepdown testing scheme, and it is a mistake to assert $\mu_1 \neq \mu_9$. (You can use the artificial data set provided in Exercise 8, which has summary statistics matching those above, to test this aspect of any computer implementation of the multiple range test.) When the critical values $c_m$ for the Studentized range statistics are not monotone, if one uses the modified critical values $c'_m = \max_{2 \leq i \leq m} c_i, m = 2, \ldots, k$, instead, which are monotone and of course guarantee (5.13) also, then the stepdown testing scheme can be implemented using the simple underlining scheme described at the beginning of Section 5.1.4.

Now consider the $F$ test which rejects $H_I$ when

$$\frac{\sum_{i \in I} n(\hat{\mu}_i - \hat{\mu}_I)^2/(|I| - 1)}{\hat{\sigma}^2} > c_{|I|}$$

where

$$\hat{\mu}_I = \frac{\sum_{i \in I} \sum_{a=1}^{n} Y_{ia}}{\sum_{i \in I} n}.$$

The $F$ statistics for testing $H_{I_1}, \ldots, H_{I_g}$ are independent conditional on $\hat{\sigma}^2$ if $I_1, \ldots, I_g$ are disjoint, so again the condition of Corollary A.1.1 is satisfied. If we set

$$c_m = \begin{cases} F_{1-(1-\alpha)^{m/k}, m-1, \nu} & \text{if } m = 2, \ldots, k-2, \\ F_{\alpha, m-1, \nu} & \text{if } m = k-1, k. \end{cases}$$

where $F_{\gamma, m-1, \nu}$ is the upper $\gamma$ quantile of the $F$ distribution with $m - 1$ numerator and $\nu = k(n - 1)$ denominator degrees of freedom, and assert $\mu_i \neq \mu_j$ if and only if all $H_I$ such that $i, j \in I$ are rejected by the stepdown method, then the resulting confident inequalities method can be called a multiple $F$ test (in analogy with multiple range tests). In contrast to the multiple range test, however, no short-cut such as testing only hypotheses of the form

$$H_{\{[i], \ldots, [i+m-1]\}} : \mu_{[i]} = \cdots = \mu_{[i+m-1]}$$

(with consecutive indices $[i], \ldots, [i + m - 1]$) is possible. To understand the importance of testing all subset hypotheses prescribed by the stepdown testing scheme, consider the following example with $k = 4, n_1 = n_2 = n_3 = n_4 = 2, \hat{\mu}_1 = \hat{\mu}_2 = -1.575, \hat{\mu}_3 = \hat{\mu}_4 = 1.575, \hat{\sigma} = 1$, and suppose $\alpha = 0.05$. Then

$$c_4 = F_{0.05,3,4} \quad = 6.591,$$
$$c_3 = F_{0.05,2,4} \quad = 6.944.$$

Noting

$$\frac{\sum_{i \in \{1,2,3,4\}} 2(\hat{\mu}_i - \hat{\mu}_{\{1,2,3,4\}})^2/(4-1)}{1} = 6.615 \quad > \quad 6.591,$$

the assertion $\mu_i \neq \mu_j$ for some $1 \leq i < j \leq 4$ can be made. However, since

$$\frac{\sum_{i \in \{1,3,4\}} 2(\hat{\mu}_i - \hat{\bar{\mu}}_{\{1,3,4\}})^2 / (3 - 1)}{1} = 6.615 \ < \ 6.944,$$

it would be a mistake to assert $\mu_1 \neq \mu_4$ (which the implementation of REGWF under the MEANS option in PROC GLM of Version 6.09 of the SAS system does). (I was led to this problem with REGWF in SAS PROC GLM by Helmut Finner.) (You can use the artificial data set provided in Exercise 9, which has summary statistics matching those above, to test this aspect of any computer implementation of the multiple $F$ test.) This example also shows that, in contrast to a multiple range test (with monotone critical values), a multiple $F$ test does not have the property that if $H_{\{[i],\ldots,[i+m-1]\}}$ is rejected then any $H_J$ such that $\{[i], [i+m-1]\} \subseteq J \subset \{[i], \ldots, [i+m-1]\}$ will be rejected; that is, the rejection of $H_{\{[i],\ldots,[i+m-1]\}}$ by a multiple $F$ test does not necessarily allow one to assert as different the two treatments furthest apart in terms of sample means among those with indices in $\{[i], \ldots, [i + m - 1]\}$.

### 5.1.7 The closed testing method, with applications to closed multiple range tests

A powerful, generally applicable, technique of constructing tests which control the probability of at least one incorrect assertion is the so-called 'closure' technique, discussed by Marcus, Peritz and Gabriel (1976), for example.

*The closed testing method* Let $\{H_i, i = 1, \ldots, L\}$ be a finite family of hypotheses. Form the closure of this family by taking all non-empty intersections $H_V = \cap_{i \in V} H_i$ for $V \subseteq \{1, \ldots, L\}$. Suppose a level-$\alpha$ test is available for each hypothesis $H_V$. Then the closed testing method asserts 'not $H_V$' if and only if every $H_U$ with $U \supseteq V$ is rejected by its associated level-$\alpha$ test.

**Theorem 5.1.6** *For the closed testing method,*

$$P\{at\ least\ one\ incorrect\ assertion\} \leq \alpha.$$

*Proof.* If there is no true null hypothesis, then the theorem is trivially true. So let $\{H_i, i \in V\}$ be the non-empty collection of true null hypotheses. Then

$$
\begin{aligned}
&P_{\mu,\sigma^2}\{\text{at least one incorrect assertion}\} \\
&= \quad P\{\text{assert 'not } H_i\text{' for some } i \in V\} \\
&\leq \quad P\{H_V \text{ is rejected}\} \\
&\leq \quad \alpha. \qquad \square
\end{aligned}
$$

If the hypotheses being tested are hypotheses of equalities, then Theorem 5.1.6 shows the closed testing method is a confident inequalities method.

Whether the probability of at least one incorrect assertion is still controlled when one makes directional assertions upon rejection of the hypotheses of equality is not yet known.

For all-pairwise comparisons, one starts with the family of hypotheses $\{H_{ij} : \mu_i = \mu_j, i < j\}$ (so $L = k(k-1)/2$) and then forms the closure of these hypotheses. For example, suppose $k = 4$ then one starts with

$$
\begin{aligned}
H_{12} &: \mu_1 = \mu_2, \\
H_{13} &: \mu_1 = \mu_3, \\
H_{14} &: \mu_1 = \mu_4, \\
H_{23} &: \mu_2 = \mu_3, \\
H_{24} &: \mu_2 = \mu_4, \\
H_{34} &: \mu_3 = \mu_4.
\end{aligned}
\tag{5.15}
$$

By forming closure, the additional hypotheses

$$
\begin{aligned}
H_{12,34} &: \mu_1 = \mu_2, \ \mu_3 = \mu_4, \\
H_{13,24} &: \mu_1 = \mu_3, \ \mu_2 = \mu_4, \\
H_{14,23} &: \mu_1 = \mu_4, \ \mu_2 = \mu_3,
\end{aligned}
\tag{5.16}
$$

$$
\begin{aligned}
H_{123} &: \mu_1 = \mu_2 = \mu_3, \\
H_{124} &: \mu_1 = \mu_2 = \mu_4, \\
H_{134} &: \mu_1 = \mu_3 = \mu_4, \\
H_{234} &: \mu_2 = \mu_3 = \mu_4,
\end{aligned}
\tag{5.17}
$$

$$
H_{1234} : \mu_1 = \mu_2 = \mu_3 = \mu_4,
\tag{5.18}
$$

are obtained. One asserts $\mu_2 \neq \mu_3$, for example, if $H_{1234}, H_{123}, H_{234}, H_{14,23}$, and $H_{23}$ are rejected.

Each of the hypotheses in (5.15), (5.17) and (5.18) is tested at level $\alpha$. To test a hypothesis in (5.16), $H_{12,34}$ say, Peritz (1970) would test each of the subhypotheses $H'_{12} : \mu_1 = \mu_2$ and $H'_{34} : \mu_3 = \mu_4$ at level $1 - (1 - \alpha)^{1/2}$ in accordance with (5.11) whether range statistics or $F$ statistics are used, as described in Begun and Gabriel (1981). An alternative, proposed by Finner (1987) and Royen (1989), is to reject a hypothesis when the maximum of the range statistics of its subhypotheses is too large.

Whether range statistics or $F$ statistics are used, Peritz's closed testing method is more powerful than the corresponding Ryan/Einot–Gabriel/ Welsch multiple range test or multiple $F$ test. That is, for the same data and with the same guarantee on the probability of at least one incorrect assertion, the closed testing method will make the same assertions of pairwise inequalities as the corresponding Ryan/Einot–Gabriel/Welsch method, sometimes more. For example, suppose $k = 4$,

$$\hat{\mu}_1 < \hat{\mu}_2 < \hat{\mu}_3 < \hat{\mu}_4,$$

and the hypotheses $H_{1234}, H_{123}, H_{234}, H_{14,23}$ are rejected (at level $\alpha$). Then whereas the Ryan/Einot–Gabriel/Welsch methods will assert '$\mu_2 \neq \mu_3$' if the subhypothesis $H'_{23} : \mu_2 = \mu_3$ is rejected at level $1 - (1 - \alpha)^{1/2}$, the closed testing method asserts the same if that subhypothesis is rejected at level $\alpha$.

### 5.1.8 Fisher's least significant difference methods

Fisher's protected least significant difference (LSD) method (Fisher 1935) is a two-step method. It starts with a size-$\alpha$ $F$-test for

$$H_{\{1,\ldots,k\}} : \mu_1 = \cdots = \mu_k.$$

If $H_{\{1,\ldots,k\}}$ is accepted, then the method stops and no assertion is made. If $H_{\{1,\ldots,k\}}$ is rejected, then the directional version of this method asserts $\mu_i > \mu_j$ for every $i \neq j$ such that

$$\frac{\hat{\mu}_i - \hat{\mu}_j}{\hat{\sigma}\sqrt{2/n}} > t_{\alpha/2,\nu}, \tag{5.19}$$

while the non-directional version of the method asserts $\mu_i \neq \mu_j$ for every $i \neq j$ such that

$$\frac{|\hat{\mu}_i - \hat{\mu}_j|}{\hat{\sigma}\sqrt{2/n}} > t_{\alpha/2,\nu}. \tag{5.20}$$

The protected LSD method controls the probability of at least one incorrect assertion when $\mu_1 = \cdots = \mu_k$, as it makes an assertion only if the size-$\alpha$ $F$-test for $H_{\{1,\ldots,k\}}$ rejects. For $k = 3$, Finner (1990a) showed that the directional version controls its probability of at least one incorrect assertion at level $\alpha$. So, for $k = 3$, the protected LSD is a confident directions method. However, Hayter (1986) showed that (for the balanced one-way model under consideration here) the maximum probability of at least one incorrect assertion of the non-directional version is $P\{Q > \sqrt{2}t_{\alpha/2,\nu}\}$, where $Q$ has the Studentized range distribution with $k - 1$ treatments and $\nu$ degrees of freedom. This error rate, attained when $\mu_1 = \cdots = \mu_{k-1} \ll \mu_k$ and the degrees of freedom $\nu$ of $\hat{\sigma}^2$ is infinity, exceeds $\alpha$ when $k > 3$ and increases rapidly as $k$ increases. For example, with $\alpha = 0.05$, the error rate exceeds 50% by the time $k$ reaches 9. Thus, for $k > 3$, the protected LSD is not a confident inequalities method and is not recommended. A simple remedy, proposed in Hayter (1986), is to use the critical value $|q^*|_{\alpha,k-1,\nu}$ instead of $t_{\alpha/2,\nu}$ in (5.20).

Fisher's unprotected LSD method (Fisher 1935) asserts $\mu_i \neq \mu_j$ for every

$i \neq j$ such that

$$\frac{|\hat{\mu}_i - \hat{\mu}_j|}{\hat{\sigma}\sqrt{2/n}} > t_{\alpha/2,\nu}, \tag{5.21}$$

i.e., no adjustment is made for the multiplicity of the pairwise comparisons. Clearly, the unprotected LSD is not a confident inequalities method and is only suitable when the rate of simultaneous correctness of the assertions is of no interest.

### 5.1.9 Example: insect traps

Recall that Wilson and Shade (1967) reported on the relative attractiveness of colors to insects, and a hypothetical data set patterned after their data was used to illustrate MCB methods in the last chapter. Their data actually came from a randomized complete block experiment. To illustrate all-pairwise comparisons for the balanced one-way model, we take a subset of their data, on the number of cereal leaf beetles trapped when six boards of each of five colors were placed in a field of oats, and pretend the data comes from a one-way experiment. Summary statistics of the number of beetles trapped are given in Table 5.1.

Table 5.1 *Summary statistics of number of beetles trapped*

| Color | Treatment label | Sample size | Sample mean |
|---|---|---|---|
| Yellow | 1 | 6 | 49.2 |
| Orange | 2 | 6 | 34.9 |
| Red | 3 | 6 | 26.4 |
| Blue | 4 | 6 | 17.2 |
| White | 5 | 6 | 15.0 |

The $\hat{\sigma} = 7.956$ from the original experiment has 91 degrees of freedom. However, for the purpose of illustration, we will pretend that $\hat{\sigma}$ has $\nu = 5(6 - 1) = 25$ degrees of freedom as it would from a one-way experiment. We take $\alpha = 0.05$.

*Tukey's method*  For five treatments ($k = 5$) and 25 error degrees of freedom ($\nu = 25$), we have $|q^*| = 2.937$. Thus, with a common sample size of six ($n = 6$),

$$|q^*|\hat{\sigma}\sqrt{2/n} = 13.491.$$

Therefore 95% Tukey's confidence intervals are

$$
\begin{array}{rcccl}
0.81 & < & \mu_1 - \mu_2 & < & 27.79 \\
9.31 & < & \mu_1 - \mu_3 & < & 36.29 \\
18.51 & < & \mu_1 - \mu_4 & < & 45.49 \\
20.71 & < & \mu_1 - \mu_5 & < & 47.69 \\
-4.99 & < & \mu_2 - \mu_3 & < & 21.99 \\
4.21 & < & \mu_2 - \mu_4 & < & 31.19 \\
6.41 & < & \mu_2 - \mu_5 & < & 33.39 \\
-4.29 & < & \mu_3 - \mu_4 & < & 22.69 \\
-2.09 & < & \mu_3 - \mu_5 & < & 24.89 \\
-11.29 & < & \mu_4 - \mu_5 & < & 15.69.
\end{array}
$$

Thus, in terms of 0-1 directional inference, Tukey's method asserts 'Yellow is more attractive than all other colors; orange is more attractive than blue and white.'

*Bofinger's confident directions method*   With five treatments ($k = 5$) and 25 error degrees of freedom ($\nu = 25$), according to Table 3 in Hayter (1990), $q^* = 2.671$. Thus, with a common sample size of six ($n = 6$),

$$q^* \hat{\sigma} \sqrt{2/n} = 12.269.$$

Therefore Bofinger's 95% simultaneous constrained confidence intervals are

$$
\begin{array}{rcccl}
0 & \leq & \mu_1 - \mu_2 & \leq & 26.57 \\
0 & \leq & \mu_1 - \mu_3 & \leq & 35.07 \\
0 & \leq & \mu_1 - \mu_4 & \leq & 44.27 \\
0 & \leq & \mu_1 - \mu_5 & \leq & 46.47 \\
-3.77 & \leq & \mu_2 - \mu_3 & \leq & 20.77 \\
0 & \leq & \mu_2 - \mu_4 & \leq & 29.97 \\
0 & \leq & \mu_2 - \mu_5 & \leq & 32.17 \\
-3.07 & \leq & \mu_3 - \mu_4 & \leq & 21.47 \\
-0.87 & \leq & \mu_3 - \mu_5 & \leq & 23.67 \\
-10.07 & \leq & \mu_4 - \mu_5 & \leq & 14.47.
\end{array}
$$

Thus, assuming that beetles are not color-blind and are unlikely to find two different colors exactly equally attractive, Bofinger's method makes the same directional assertions as Tukey's method.

*Hayter's one-sided comparisons*   To illustrate Hayter's one-sided comparisons, let us pretend that there is reason, *a priori*, for lower confidence bounds on $\mu_i - \mu_j$, $i < j$, to be of primary interest. (For a more realistic example, see Exercise 1.)

Hayter's method and Bofinger's method share the same critical value, which is $q^* = 2.671$ for this data. So again

$$q^* \hat{\sigma} \sqrt{2/n} = 12.269.$$

Therefore Hayter's 95% simultaneous lower confidence bounds are

$$
\begin{array}{rcl}
2.03 & < & \mu_1 - \mu_2 \\
10.53 & < & \mu_1 - \mu_3 \\
19.73 & < & \mu_1 - \mu_4 \\
21.93 & < & \mu_1 - \mu_5 \\
-3.77 & < & \mu_2 - \mu_3 \\
5.43 & < & \mu_2 - \mu_4 \\
7.63 & < & \mu_2 - \mu_5 \\
-3.07 & < & \mu_3 - \mu_4 \\
-0.87 & < & \mu_3 - \mu_5 \\
-10.07 & < & \mu_4 - \mu_5.
\end{array}
$$

Thus, while in terms of 0-1 directional inference Hayter's method makes the same assertions as Tukey's method, it gives sharper bounds on *how much* more attractive yellow is than the other colors, and *how much* more attractive orange is than blue and white.

*Multiple range tests*   The Newman–Keuls multiple range test sets

$$
\alpha_m = \alpha, \quad m = 2, \ldots, k,
$$

which leads to

| $m$ | 2 | 3 | 4 | 5 |
|---|---|---|---|---|
| $c_m$ | 2.060 | 2.491 | 2.751 | 2.937 |
| $c_m \hat{\sigma} \sqrt{2/n}$ | 9.460 | 11.442 | 12.635 | 13.491 |

The multiple range test of Ryan, Einot and Gabriel, and Welsch sets

$$
\alpha_m = \begin{cases} 1 - (1 - \alpha)^{m/k} & \text{if } m = 2, \ldots, k - 2, \\ \alpha & \text{if } m = k - 1, k, \end{cases}
$$

which leads to

| $m$ | 2 | 3 | 4 | 5 |
|---|---|---|---|---|
| $c_m$ | 2.478 | 2.721 | 2.751 | 2.937 |
| $c_m \hat{\sigma} \sqrt{2/n}$ | 11.384 | 12.500 | 12.635 | 13.491 |

Note $c_m$ is increasing in $m$ in this case.

Working through the details of the underlining scheme, we find that both the Newman–Keuls multiple range test and the Ryan/Einot–Gabriel/Welsch multiple range test assert 'Yellow is different from all other colors; orange is different from blue and white,' the same inequalities as can be deduced from Tukey's method.

Duncan's multiple range test sets

$$
\alpha_m = 1 - (1 - \alpha)^{m-1}, m = 2, \ldots, k,
$$

which leads to

| $m$ | 2 | 3 | 4 | 5 |
|---|---|---|---|---|
| $c_m$ | 2.060 | 2.163 | 2.230 | 2.278 |
| $c_m\hat{\sigma}\sqrt{2/n}$ | 9.460 | 9.938 | 10.245 | 10.462 |

Working through the details of the underlining scheme, we find that Duncan's multiple range test asserts 'Yellow is different from all other color; orange is different from blue and white;' the same inequalities as can be deduced from Tukey's method, and additionally 'Red is different from white.'

*Scheffé's method* For five treatments ($k = 5$) and 25 error degrees of freedom ($\nu = 25$), we have $F_{\alpha,k-1,\nu} = 2.759$. Thus, with a common sample size of six ($n = 6$),

$$\sqrt{(k-1)F_{\alpha,k-1,\nu}}\,\hat{\sigma}\sqrt{2/n} = 15.259.$$

Therefore 95% Scheffé's confidence intervals are

$$
\begin{aligned}
-0.96 &< \mu_1 - \mu_2 < 29.56\\
7.54 &< \mu_1 - \mu_3 < 38.06\\
16.74 &< \mu_1 - \mu_4 < 47.26\\
18.94 &< \mu_1 - \mu_5 < 49.46\\
-6.76 &< \mu_2 - \mu_3 < 23.76\\
2.44 &< \mu_2 - \mu_4 < 32.96\\
4.64 &< \mu_2 - \mu_5 < 35.16\\
-6.06 &< \mu_3 - \mu_4 < 24.46\\
-3.86 &< \mu_3 - \mu_5 < 26.66\\
-13.06 &< \mu_4 - \mu_5 < 17.46.
\end{aligned}
$$

Notice these confidence intervals are wider than the corresponding Tukey confidence intervals. In terms of 0-1 directional inference, Scheffé's method asserts 'Yellow is more attractive than red, blue, and white; orange is more attractive than blue and white,' which Tukey's method asserts as well, while failing to assert 'Yellow is better than orange,' which Tukey's method does.

*Multiple F-tests* The multiple $F$ test of Ryan, Einot and Gabriel, and Welsch sets

$$\alpha_m = \begin{cases} 1 - (1-\alpha)^{m/k} & \text{if } m = 2,\ldots,k-2,\\ \alpha & \text{if } m = k-1,k, \end{cases}$$

which leads to

| $m$ | 2 | 3 | 4 | 5 |
|---|---|---|---|---|
| $c_m$ | 6.142 | 4.034 | 2.991 | 2.759 |

Working through the details of the stepdown $F$-tests, we find the Ryan/Einot–Gabriel/Welsch multiple $F$ test asserts 'Yellow is different from all other colors; orange is different from blue and white,' the same inequalities as can be deduced from Tukey's method.

*The closed testing method of Peritz* A fact associated with Peritz's closed testing method (based on range statistics) which makes its execution easier is that it will assert a pair of treatments are unequal if the Ryan/Einot–Gabriel/Welsch multiple range test so asserts, and it will not assert a pair of treatments are unequal unless the Newman–Keuls multiple range test so asserts. (Treatments asserted to be unequal by the Newman–Keuls multiple range test but not by the Ryan/Einot–Gabriel/Welsch multiple range test may or may not be asserted to be unequal by Peritz's method.)

For the traps data set, since the Newman–Keuls multiple range test and the Ryan/Einot–Gabriel/Welsch multiple range test make exactly the same assertions, so does Peritz's method (based on range statistics).

*Fisher's protected and unprotected LSD methods* For five treatments ($k = 5$) and 25 error degrees of freedom ($\nu = 25$), we have $F_{\alpha,k-1,\nu} = 2.759$ and $t_{\alpha/2,\nu} = 2.060$. Thus, with a common sample size of six ($n = 6$),

$$t_{\alpha/2,\nu}\hat{\sigma}\sqrt{2/n} = 9.460.$$

The $F$-test statistic turns out to be 18.57, so $H_0 : \mu_1 = \cdots = \mu_k$ is rejected. Working through the details, we find that both the protected and the unprotected LSD assert 'Yellow is different from all other colors; orange is different from blue and white; red is different from white,' the same as Duncan's multiple range test.

## 5.2 Unbalanced one-way model

Suppose now the sample sizes are unequal, and under the $i$th treatment a random sample $Y_{i1}, Y_{i2}, \ldots, Y_{in_i}$ is taken, where between the treatments the random samples are independent. Then under the usual normality and equality of variance assumptions, we have the one-way model (1.1)

$$Y_{ia} = \mu_i + \epsilon_{ia}, \quad i = 1, \ldots, k, \quad a = 1, \ldots, n_i, \tag{5.22}$$

where $\epsilon_{11}, \ldots, \epsilon_{kn_k}$ are i.i.d. normal with mean 0 and variance $\sigma^2$ unknown. We use the notation

$$\hat{\mu}_i \quad = \bar{Y}_i \quad = \sum_{a=1}^{n_i} Y_{ia}/n_i,$$

$$\hat{\sigma}^2 \quad = MSE \quad = \sum_{i=1}^{k}\sum_{a=1}^{n_i}(Y_{ia} - \bar{Y}_i)^2 / \sum_{i=1}^{k}(n_i - 1)$$

for the sample means and the pooled sample variance.

### 5.2.1 The Tukey–Kramer method

For the unequal sample sizes case, a natural generalization of (5.2) is

$$\mu_i - \mu_j \in \hat{\mu}_i - \hat{\mu}_j \pm |q^e|\hat{\sigma}\sqrt{n_i^{-1} + n_j^{-1}} \text{ for all } i \neq j, \qquad (5.23)$$

where $|q^e|$ is the critical value such that the coverage probability is exactly $1 - \alpha$:

$$P\{\mu_i - \mu_j \in \hat{\mu}_i - \hat{\mu}_j \pm |q^e|\hat{\sigma}\sqrt{n_i^{-1} + n_j^{-1}} \text{ for all } i \neq j\} = 1 - \alpha. \quad (5.24)$$

In this case, $|q^e|$ depends on the sample size configuration, not just $\alpha, k$ and $\nu$. As the coverage probability is a continuous increasing function of $|q^e|$, equation (5.24) has a unique solution in $|q^e|$. However, when the sample sizes are unequal, expressing the coverage probability as a double integral or a sum of double integrals for efficient numerical computation is no longer possible. (Exercise 10 asks you to try.)

For $k \leq 4$, one can compute the coverage probability by numerical integration (cf. Spurrier and Isham 1985; Uusipaikka 1985). In particular, for $k = 3$, Spurrier and Isham (1985) provide tables for $|q^e|$ for $n_1 + n_2 + n_3 < 30$, as well as an accurate approximation to $|q^e|$ for $n_1 + n_2 + n_3 \geq 30$. But for $k > 4$ the exact solution $|q^e|$ is difficult to compute.

Thus, for the unequal sample size case, Tukey (1953) and Kramer (1956) proposed what we now call the Tukey–Kramer approximate simultaneous confidence intervals for all-pairwise differences:

$$\mu_i - \mu_j \in \hat{\mu}_i - \hat{\mu}_j \pm |q^*|\hat{\sigma}\sqrt{n_i^{-1} + n_j^{-1}} \text{ for all } i \neq j. \qquad (5.25)$$

In other words, substitute $|q^*|$, the solution to the equation

$$P\left\{\frac{|Z_i - Z_j|}{\sqrt{2}\hat{\sigma}} \leq |q^*| \text{ for all } i > j\right\} = 1 - \alpha$$

in which $Z_1, \ldots, Z_k$ are i.i.d. standard normal random variables (making $|q^*|$ dependent on $\alpha, k$ and $\nu = \sum_{i=1}^{k}(n_i - 1)$ only) for $|q^e|$. Tukey (1953) stated 'the approximation ... is apparently in the conservative direction' (see Tukey 1994, p. 49). There was some early skepticism (e.g. Miller 1981, p.45). Over time, however, simulation evidence supporting Tukey's conjecture mounted, the most extensive of such studies being Dunnett (1980). Thus, interest in proving Tukey's conjecture was renewed. In particular, Brown (1979) proved Tukey's conjecture for the cases of $k = 3, 4$ and 5. In a breakthrough paper, Hayter (1984) showed that the Tukey–Kramer approximation is conservative:

**Theorem 5.2.1 (Hayter 1984)** *For all* $n_1, \ldots, n_k$,

$$P\{\mu_i - \mu_j \in \hat{\mu}_i - \hat{\mu}_j \pm |q^*|\hat{\sigma}\sqrt{n_i^{-1} + n_j^{-1}} \text{ for all } i \neq j\} \geq 1 - \alpha$$

*with equality if and only if* $n_1 = \cdots = n_k$.

*Proof.* This is one proof not given or even indicated in this book, as I know of no way of simplifying the rather intricate proof of Hayter (1984). The reader is referred to the original article for details.  $\square$

Prior to the publication of Hayter (1984), Spjøtvoll and Stoline (1973) showed that, if one replaces the upper $\alpha$ quantile $|q^*|$ in (5.25) by the upper $\alpha$ quantile $|q^a|$ of the Studentized *augmented* range, then the simultaneous coverage probability of the intervals is at least $1 - \alpha$. As $|q^a| > |q^*|$ always, in view of Hayter's proof, one should use the critical value $|q^*|$.

### 5.2.1.1 Graphical representation of the Tukey–Kramer method

Tukey (1992) states:

> multiple comparisons are somewhat complicated – as a result, graphical pre-
> sentation of multiple comparison results is even more important than graphical
> representation of simpler results.

A good representation of the Tukey–Kramer method should let the user achieve both the significant difference and the practical equivalence infer-ential tasks described in Section 2.3. In addition, a good representation should accomplish the following perceptual tasks.

Intrinsic dependencies among the $k(k-1)/2$ confidence intervals should be reflected in any representation of the Tukey–Kramer method. Specifi-cally, the point estimates of the elementary contrasts $\mu_i - \mu_j$ are clearly additive:

$$\hat{\mu}_i - \hat{\mu}_k = (\hat{\mu}_i - \hat{\mu}_j) + (\hat{\mu}_j - \hat{\mu}_k). \tag{5.26}$$

Therefore, a graphical representation of the Tukey–Kramer method should center the confidence interval for $\mu_i - \mu_k$ at the 'sum,' in some sense, of the centers of the confidence intervals for $\mu_i - \mu_j$ and $\mu_j - \mu_k$. Further, there is an inherent transitivity in all-pairwise multiple comparisons, as stated in the following theorem.

**Theorem 5.2.2** *Let* $\sqrt{v_j^i}$ *denote* $\sqrt{n_i^{-1} + n_j^{-1}}$. *If*

$$\hat{\mu}_i - \hat{\mu}_j > c\sigma\sqrt{v_j^i}$$

*and*

$$\hat{\mu}_j - \hat{\mu}_k > c\sigma\sqrt{v_k^j}$$

*then*

$$\hat{\mu}_i - \hat{\mu}_k > c\sigma\sqrt{v_k^i}.$$

*Proof.* Computing the variances of both sides of (5.26), we obtain

$$\sigma^2 v_k^i = \sigma^2 v_j^i + \sigma^2 v_k^j + 2Cov(\hat{\mu}_i - \hat{\mu}_j, \hat{\mu}_j - \hat{\mu}_k).$$

Noting that

$$Cov(\hat{\mu}_i - \hat{\mu}_j, \hat{\mu}_j - \hat{\mu}_k) \leq \sigma^2\sqrt{v_j^i v_k^j},$$

the result follows. □

Thus, a good representation of the Tukey–Kramer method should convey visually this transitivity in significant difference inference, in the sense that, if the Tukey–Kramer method declares $\mu_i > \mu_j$ and $\mu_j > \mu_k$, then it necessarily declares $\mu_i > \mu_k$ as well.

Consider the following representations of the Tukey–Kramer method.

*Underlining* Perhaps the oldest representation of the Tukey–Kramer method is the underlining representation described in Section 5.1.4. Among statistical packages, the RANGES option of the ONEWAY command in SPSS includes this representation; the MEANS option of PROC GLM in SAS uses this representation for balanced designs by default.

In terms of significant difference inference, it is impossible to implement the underlining representation consistently in some unequal sample size situations. For example, suppose

$$\hat{\mu}_1 < \hat{\mu}_2 < \hat{\mu}_3$$

but the sample sizes are such that the estimated standard deviation of $\hat{\mu}_3 - \hat{\mu}_2$ is smaller than the estimated standard deviation of $\hat{\mu}_3 - \hat{\mu}_1$ (which can occur, for example, when $n_1 < n_3 < n_2$). Then the Tukey–Kramer method may find $\mu_2$ and $\mu_3$ significantly different, but $\mu_1$ and $\mu_3$ not significantly different, and this is impossible to represent by underlining.

Underlining is incapable of representing practical equivalence, as it does not allow the user to distinguish between a confidence interval for $\mu_i - \mu_j$ that contains zero and is tight around zero, signifying practical equivalence, and a wide interval that contains zero, signifying insufficient sample size relative to noise in the data.

Thus the usefulness of the underlining representation is limited to 0-1 significant difference inference for balanced designs.

*Line-by-line displays* One can of course list numerically the values of the upper and lower bounds of all $k(k-1)/2$ confidence intervals for the pairwise differences $\mu_i - \mu_j$, $i < j$, given by the Tukey–Kramer method.

For example, the MEANS option of PROC GLM in the SAS system lists, line by line, the numerical values of the upper and lower bounds of all $k(k-1)$ confidence intervals for $\mu_i - \mu_j$, $i \neq j$ (by default for unbalanced designs, or if the CLDIFF option is invoked when the design is balanced). See Figure 5.4.

A variation is to list the numerical values of the upper and lower bounds of the $k(k-1)/2$ confidence intervals for $\mu_i - \mu_j$, $i < j$, in the lower left of a $k \times k$ matrix, as implemented in the TUKEY subcommand of the ONEWAY command in MINITAB. See Figure 5.5.

Tukey (1992) suggested the same representation except that the matrix is rotated by 45° and, instead of the confidence limits, the values of $\hat{\mu}_i - \hat{\mu}_j$

```
                      General Linear Models Procedure

           Tukey's Studentized Range (HSD) Test for variable: COUNT

           NOTE: This test controls the type I experimentwise error rate.

                 Alpha= 0.05  Confidence= 0.95  df= 25  MSE= 63.3
                      Critical Value of Studentized Range= 4.153
                      Minimum Significant Difference= 13.49

          Comparisons significant at the 0.05 level are indicated by '***'.
```

|          | Simultaneous | Difference | Simultaneous |     |
|----------|--------------|------------|--------------|-----|
|          | Lower        |            | Upper        |     |
| COLOR    | Confidence   | Between    | Confidence   |     |
| Comparison | Limit      | Means      | Limit        |     |
| 1  - 2   | 0.810        | 14.300     | 27.790       | *** |
| 1  - 3   | 9.310        | 22.800     | 36.290       | *** |
| 1  - 4   | 18.510       | 32.000     | 45.490       | *** |
| 1  - 5   | 20.710       | 34.200     | 47.690       | *** |
| 2  - 1   | -27.790      | -14.300    | -0.810       | *** |
| 2  - 3   | -4.990       | 8.500      | 21.990       |     |
| 2  - 4   | 4.210        | 17.700     | 31.190       | *** |
| 2  - 5   | 6.410        | 19.900     | 33.390       | *** |
| 3  - 1   | -36.290      | -22.800    | -9.310       | *** |
| 3  - 2   | -21.990      | -8.500     | 4.990        |     |
| 3  - 4   | -4.290       | 9.200      | 22.690       |     |
| 3  - 5   | -2.090       | 11.400     | 24.890       |     |
| 4  - 1   | -45.490      | -32.000    | -18.510      | *** |
| 4  - 2   | -31.190      | -17.700    | -4.210       | *** |
| 4  - 3   | -22.690      | -9.200     | 4.290        |     |
| 4  - 5   | -11.290      | 2.200      | 15.690       |     |
| 5  - 1   | -47.690      | -34.200    | -20.710      | *** |
| 5  - 2   | -33.390      | -19.900    | -6.410       | *** |
| 5  - 3   | -24.890      | -11.400    | 2.090        |     |
| 5  - 4   | -15.690      | -2.200     | 11.290       |     |

Figure 5.4 *Sample SAS representation of the Tukey–Kramer method*

are displayed in the cells, with a larger font size indicating a significant difference at the 5% level.

Program P7D in BMDP draws, line by line and on a one-dimensional scale, all $k(k-1)/2$ confidence intervals for $\mu_i - \mu_j$, $i < j$. See Figure 5.6.

However, line-by-line displays do not readily convey to the user the additivity of the point estimates of contrasts (5.26), or the transitivity in significant difference inference (Theorem 5.2.2).

Tukey's pairwise comparisons

```
     Family error rate = 0.0500
Individual error rate = 0.00706
```

Critical value = 4.15

Intervals for (column level mean) - (row level mean)

|   | 1 | 2 | 3 | 4 |
|---|---|---|---|---|
| 2 | 0.82<br>27.78 | | | |
| 3 | 9.32<br>36.28 | -4.98<br>21.98 | | |
| 4 | 18.52<br>45.48 | 4.22<br>31.18 | -4.28<br>22.68 | |
| 5 | 20.72<br>47.68 | 6.42<br>33.38 | -2.08<br>24.88 | -11.28<br>15.68 |

Figure 5.5 *Sample MINITAB representation of the Tukey–Kramer method*

TUKEY STUDENTIZED RANGE METHOD

95% CONFIDENCE INTERVALS

| GROUP<br>NO. LABEL | GROUP<br>NO. LABEL | MEAN<br>DIFF |
|---|---|---|
| 1 *1 | 2 *2 | 14.30 |
| 1 *1 | 3 *3 | 22.80 |
| 1 *1 | 4 *4 | 32.00 |
| 1 *1 | 5 *5 | 34.20 |
| 2 *2 | 3 *3 | 8.50 |
| 2 *2 | 4 *4 | 17.70 |
| 2 *2 | 5 *5 | 19.90 |
| 3 *3 | 4 *4 | 9.20 |
| 3 *3 | 5 *5 | 11.40 |
| 4 *4 | 5 *5 | 2.20 |

```
----+---------+---------+---------+-----
-37.50   -12.50    12.50     37.50
```

Figure 5.6 *Sample BMDP representation of the Tukey–Kramer method*

*Error bars*   A very popular graphical representation of the Tukey–Kramer method is obtained by attaching *error bars* to a scatter plot of estimated treatment effects versus treatment labels. The lengths of the error bars are adjusted so that the population means of a pair of treatments can be inferred to be different if their bars do not overlap. A variation of this representation, designed to infer significant differences between medians instead of means, is the 'notched boxplot' representation, suggested by McGill, Tukey and Larson (1978). With this variation, notches are added to side-by-side boxplots so that the population medians of a pair of treatments can be inferred to be different if their notches do not overlap. As implemented in DataDesk, these notches are drawn as grey error bars. See Plate 1 (with grey changed to blue). Tukey (1993) has since suggested a modification, with more emphasis on the notches and less on the boxes.

In terms of 0-1 significant difference inference, it is impossible to implement the error bar representation consistently in some unequal sample size situations. This is because, when the sample sizes are unequal, there is no guarantee that one can determine the half-lengths $w_1, \ldots, w_k$ of the error bars in such a way that

$$|\hat{\mu}_i - \hat{\mu}_j| > |q^*|\hat{\sigma}\sqrt{n_i^{-1} + n_j^{-1}}$$

if and only if

$$|\hat{\mu}_i - \hat{\mu}_j| > w_i + w_j$$

for all $i \neq j$. The following example was given in Hochberg, Weiss and Hart (1982). Suppose $k = 4$, $n_1 = n_4 = 1$, $n_2 = n_3 = \infty$, $\hat{\mu}_1 = 0$, $\hat{\mu}_2 = 0.7$, $\hat{\mu}_3 = 0.9$, $\hat{\mu}_4 = 1.6$, and $|q^*|\hat{\sigma}\sqrt{2} = 1$. Then the only significant differences are between treatments 1 and 4, and between treatments 2 and 3, which clearly does not allow an error bar representation.

Techniques have been proposed by Gabriel (1978), Andrews, Snee, and Sarner (1980) and Hochberg, Weiss and Hart (1982) for choosing $w_1, \ldots, w_k$ so as to make the error bar representation as faithful as possible to Tukey's 0-1 significant difference inference. However, a graphical representation totally faithful to the Tukey–Kramer inference is perhaps preferable.

It is also questionable whether the error bar representation is capable of confidence interval inference, of either the significant difference or the practical equivalence type. To deduce the upper confidence bound on $\mu_i - \mu_j$, the user has to perceive the vertical distance between the top of the error bar of the $i$th treatment and the bottom of the error bar of the $j$th treatment. Even with bars not far apart, as illustrated in Cleveland (1985, p. 276), it is not easy for the eye–brain combination to perceive accurately derived vertical distances.

*Comparison circles*   Problems with the error bar representation in the unequal sample size situation arise because the 'yardsticks' $\sqrt{n_i^{-1} + n_j^{-1}}$

are *nonlinear* functions of $n_1^{-1}, \ldots, n_k^{-1}$, while the error bar representation forces the yardsticks to be *linear* functions of $w_1, \ldots, w_k$.

A clever representation which exploits Pythagoras' theorem is based on the *comparison circles* proposed by Sall (1992a). The treatment effect estimates are plotted on a vertical scale, and, for each treatment $i$, a circle is drawn, centered at $\hat{\mu}_i$ and with radius $|q^*|\hat{\sigma}\sqrt{n_i^{-1}}$. By Pythagoras' theorem, a pair of treatments can be inferred to be different if either their respective circles do not overlap, or if they overlap and the outer angle formed by the tangent lines through the point at which the circles intersect is greater than $90°$.

This representation will always faithfully represent the 0-1 significant difference inference given by the Tukey–Kramer method. It is true that, as James (1992) commented, the angle of intersection may be difficult to judge in some instances. However, as pointed out in Sall (1992b), the implementation of the comparison circles in JMP very effectively overcomes this problem by utilizing the dimension of color (gray scale on a monochrome display). When the user 'clicks' on the circle corresponding to a treatment, the colors (shades of gray on a monochrome monitor) of the other circles change to indicate which other treatments can be inferred to be different. See Plate 2. Thus the comparison circles effectively represent 0-1 significant difference inference.

However, no feature of the comparison circles allows one to perform confidence interval inference, of either the significant difference type or the practical equivalence type. The optional Means Comparisons text window on the Fit Y by X platform of JMP could have been used to list the numerical values of the confidence bounds, but was not.

*Mean-mean scatter plot* Depicting the Tukey–Kramer confidence set as an object in the $k$-dimensional space of $\mu_1, \ldots, \mu_k$ as in Figure 5.1 for the case of $k = 3$, or as an object in the $(k-1)$-dimensional space of contrasts of $\mu_1, \ldots, \mu_k$, is difficult when $k > 4$.

The *mean-mean scatter plot* representation of Hsu and Peruggia (1994) can be motivated by an analogy with simple linear regression, where the observed vector $Y$ and the the predictor vector $X$ can be thought of as either a pair of points in $n$-dimensional Euclidean space, or as $n$ points in the familiar two-dimensional scatter plot space of $y$ versus $x$. Similarly, instead of thinking of the Tukey–Kramer confidence set as an object in the $k$-dimensional space of $\mu_1, \ldots, \mu_k$ bounded by the $k(k-1)/2$ pairs of planes

$$\mu_i - \mu_j = \hat{\mu}_i - \hat{\mu}_j \pm |q^*|\hat{\sigma}\sqrt{n_i^{-1} + n_j^{-1}}$$

for all $i \neq j$, one can visualize it as $k(k-1)/2$ line segments in the two-dimensional scatter plot space of $\mu$ versus $\mu$.

More specifically, consider the two-dimensional Euclidean space of $\mu_i \times \mu_j$

for a particular pair of treatment means $\mu_i$ and $\mu_j$. Then the 45° line through the origin represents the set of points satisfying $\mu_i = \mu_j$, and the directional (perpendicular) distance of any point $(\mu_i, \mu_j)$ from this line is equal to $(\mu_i - \mu_j)/\sqrt{2}$.

Therefore, in this space, a line segment centered at $(\hat{\mu}_i, \hat{\mu}_j)$ of slope $-1$ and half-width $|q^*|\hat{\sigma}\sqrt{n_i^{-1} + n_j^{-1}}/\sqrt{2}$ represents the Tukey–Kramer confidence interval for $\mu_i - \mu_j$. To perform 0-1 significant difference inference, one simply checks whether the line segment crosses the 45° line. By aligning the end points of the line segment against a scale which contracts the $\mu_i$, $\mu_j$ scales by a factor of $\sqrt{2}$ and is placed along the $-45°$ line, one can accomplish 0-1 practical equivalence inference as well as confidence interval inference, of either the significant difference or practical equivalence type.

All $k(k-1)/2$ confidence intervals given by the Tukey–Kramer method can be completely faithfully represented by plotting, for each pair $(\mu_i, \mu_j)$, the appropriate line segment centered at $(\hat{\mu}_i, \hat{\mu}_j)$. To help identify treatments, one can draw the lines $x = \hat{\mu}_i$ and $y = \hat{\mu}_i$ for each sample mean $\hat{\mu}_i$. To avoid overlapping, one can draw only the intervals for $\mu_i - \mu_j$, with $\hat{\mu}_i > \hat{\mu}_j$, i.e. only intervals below the 45° line. See Plate 3. To help gauge the numerical values of the confidence limits, one can optionally draw and label an axis contracted by a factor of $\sqrt{2}$ along the $-45°$ line.

Intrinsic dependencies among the confidence intervals are conveyed visually by the mean-mean scatter plot, as the three arrows in Figure 5.7 illustrate for the case in which $\hat{\mu}_i > \hat{\mu}_j > \hat{\mu}_k$. The additivity of the contrast estimators (5.26) is represented visually in a fashion similar to the familiar graphical representation of vector addition $\vec{a} + \vec{b} = \vec{c}$. The transitivity of significant difference inference (Theorem 5.2.2) is conveyed by the fact that if the confidence intervals for $\mu_i - \mu_j$ and $\mu_j - \mu_k$ are both entirely below the 45° line, then the confidence interval for $\mu_i - \mu_k$ must be as well.

Hsu and Peruggia (1994) described an event-driven implementation using XLISP-STAT (Tierney 1990) on a workstation running the X11 windowing system and on a Macintosh. As illustrated in Plate 4, this representation rotates the scatter plot by 45° counterclockwise. The top horizontal axis is the scale against which the confidence intervals should be aligned, and two treatments are significantly different if their confidence interval does not intersect the vertical line which bisects the two axes. Color coding of the treatments, indicated by an array of color buttons and treatment labels, helps to identify the treatments pair to which each confidence interval corresponds. The convention is that, if $R$ and $L$ denote the treatments corresponding to the colors of the right and left tips of an interval respectively, then the confidence interval is for $\mu_R - \mu_L$. Among the options available, the one illustrated in Plate 4 is clicking on the button corresponding to the $i$th treatment highlights as well as displays the numerical values of the confidence intervals for $\mu_i - \mu_j$ for all $j$, $j \neq i$.

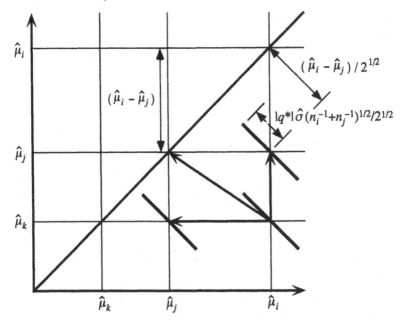

Figure 5.7 *Mean-mean scatter plot*

Other graphical representations are possible. Sampson (1980) proposed representing significant differences by chords connecting treatment labels placed equidistantly on a circle. Dumouchel (1988) describes an implementation which graphically elicits a prior from the user for a Bayesian model, and then lets the user assess the differences between the treatments using side-by-side error bars.

### 5.2.2 Hayter's method

Hayter (1992) proposed the following $100(1-\alpha)\%$ simultaneous confidence intervals for one-sided MCA inference:

$$\mu_i - \mu_j > \hat{\mu}_i - \hat{\mu}_j - q^e \hat{\sigma} \sqrt{n_i^{-1} + n_j^{-1}} \text{ for all } i > j, \qquad (5.27)$$

where $q^e$ is the solution to the equation

$$P \left\{ \frac{\hat{\mu}_i - \mu_i - (\hat{\mu}_j - \mu_j)}{\hat{\sigma} \sqrt{n_i^{-1} + n_j^{-1}}} \leq q^e \text{ for all } i > j \right\} = 1 - \alpha.$$

For $k = 3$ and $\alpha = 0.10, 0.05, 0.01$, Hayter (1992) gives values of $q^e$ for $3 \leq n_1, n_2, n_3 \leq 10$. But for $k > 3$, (deterministic) computation of $q^e$ is generally difficult for unbalanced designs. One might wonder whether the analog of the Tukey–Kramer approximation, namely, substituting the solution $q^*$ to the equation

$$P\left\{\frac{Z_i - Z_j}{\sqrt{2}\hat{\sigma}} \leq q^* \text{ for all } i > j\right\} = 1 - \alpha$$

(where $Z_1, \ldots, Z_k$ are i.i.d. standard normal random variables) for $q^e$ results in conservative methods. Unfortunately, the answer is 'no.' To see this, consider the case of $k = 3, \nu = \infty$, and $\sigma^2$. Then $\hat{\mu}_2 - \hat{\mu}_1, \hat{\mu}_3 - \hat{\mu}_1, \hat{\mu}_3 - \hat{\mu}_2$ have a (singular) multivariate normal distribution with correlations

$$Corr(\hat{\mu}_2 - \hat{\mu}_1, \hat{\mu}_3 - \hat{\mu}_1) = \frac{1}{\sqrt{1 + \frac{n_1}{n_2}}\sqrt{1 + \frac{n_1}{n_3}}},$$

$$Corr(\hat{\mu}_2 - \hat{\mu}_1, \hat{\mu}_3 - \hat{\mu}_2) = \frac{-1}{\sqrt{1 + \frac{n_2}{n_1}}\sqrt{1 + \frac{n_2}{n_3}}},$$

$$Corr(\hat{\mu}_3 - \hat{\mu}_1, \hat{\mu}_3 - \hat{\mu}_2) = \frac{1}{\sqrt{1 + \frac{n_3}{n_1}}\sqrt{1 + \frac{n_3}{n_2}}}.$$

Thus, for fixed $n_1$ and $n_3$, the coverage probability of (5.27) is an increasing function of $n_2$ by Slepian's inequality (A.3.1). That is, in contrast to two-sided inference, the minimum coverage of probability of (5.27) cannot occur at $n_1 = n_2 = n_3$.

### 5.2.3 Scheffé's method

Recall the standard result that, under the null hypothesis

$$H_0 : \mu_1 = \cdots = \mu_k$$

the statistic

$$\frac{\sum_{i=1}^{k} n_i(\hat{\mu}_i - \hat{\bar{\mu}})^2/(k-1)}{\hat{\sigma}^2},$$

where

$$\hat{\bar{\mu}} = \frac{n_1\hat{\mu}_1 + \cdots + n_k\hat{\mu}_k}{n_1 + \cdots + n_k},$$

has an $F$ distribution with $k - 1$ numerator and $\nu$ denominator degrees of freedom. Thus, for all $\mu = (\mu_1, \ldots, \mu_k)$ and $\sigma^2$,

$$\frac{\sum_{i=1}^{k} n_i(\hat{\mu}_i - \mu_i - (\hat{\bar{\mu}} - \bar{\mu}))^2/(k-1)}{\hat{\sigma}^2},$$

where

$$\bar{\mu} = \frac{n_1\mu_1 + \cdots + n_k\mu_k}{n_1 + \cdots + n_k},$$

has the same distribution and

$$P\left\{\sum_{i=1}^{k}(\sqrt{n_i}(\hat{\mu}_i - \hat{\bar{\mu}}) - \sqrt{n_i}(\mu_i - \bar{\mu}))^2 \le (k-1)\hat{\sigma}^2 F_{\alpha,k-1,\nu}\right\} = 1-\alpha,$$

where $F_{\alpha,k-1,\nu}$ is the upper $\alpha$ quantile of the $F$ distribution with $k-1$ and $\nu$ degrees of freedom. Therefore the set

$$S = \{(\mu_1,\ldots,\mu_k) : \sum_{i=1}^{k}(\sqrt{n_i}(\hat{\mu}_i - \hat{\bar{\mu}}) - \sqrt{n_i}(\mu_i - \bar{\mu}))^2 \le (k-1)\hat{\sigma}^2 F_{\alpha,k-1,\nu}\}$$

is a $100(1-\alpha)\%$ confidence set for $\mu_1,\ldots,\mu_k$. The set $S$ is translation invariant, in the sense that if $(\mu_1,\ldots,\mu_k) \in S$ then $(\mu_1 + \delta,\ldots,\mu_k + \delta) \in S$ as well. Thus, as $S$ is an infinite cylinder in $\Re^k$ parallel to the vector $(1,\ldots,1)$, it is completely described by its cross-section on a plane, any plane, perpendicular to $(1,\ldots,1)$. The projection of $S$ on such a plane forms Scheffé's $100(1-\alpha)\%$ simultaneous confidence intervals for all contrasts of $\mu_1,\ldots,\mu_k$:

$$\sum_{i=1}^{k}c_i\mu_i \in \sum_{i=1}^{k}c_i\hat{\mu}_i \pm \sqrt{(k-1)F_{\alpha,k-1,\nu}}\,\hat{\sigma}(\sum_{i=1}^{k}c_i^2/n_i)^{1/2} \text{ for all } \sum_{i=1}^{k}c_i = 0.$$

**Theorem 5.2.3 (Scheffé 1953)**

$$P\{\sum_{i=1}^{k}c_i\mu_i \in \sum_{i=1}^{k}c_i\hat{\mu}_i \pm \sqrt{(k-1)F_{\alpha,k-1,\nu}}\,\hat{\sigma}(\sum_{i=1}^{k}c_i^2/n_i)^{1/2}$$

$$\text{for all } (c_1,\ldots,c_k) \in \mathcal{C}\}$$

$$= 1-\alpha$$

*where* $\mathcal{C} = \{(c_1,\ldots,c_k) : c_1 + \cdots + c_k = 0\}$.

*Proof.* While a more mathematically detailed proof can be found in Miller (1981), for example, our humble Lemma B.1.1 in Appendix B makes possible a geometric proof.

Letting $z_i = \sqrt{n_i}(\mu_i - \bar{\mu})$, we see that $z = (z_1,\ldots,z_k)$ satisfying

$$\sum_{i=1}^{k}(z_i - \sqrt{n_i}(\hat{\mu}_i - \hat{\bar{\mu}}))^2 \le (k-1)F_{\alpha,k-1,\nu}\hat{\sigma}^2$$

constitutes the interior of a $k$-dimensional sphere centered at $(\sqrt{n_1}(\hat{\mu}_1 - \hat{\bar{\mu}}),\ldots,\sqrt{n_k}(\hat{\mu}_k - \hat{\bar{\mu}}))$ with radius $\sqrt{(k-1)F_{\alpha,k-1,\nu}}\,\hat{\sigma}$. Therefore, by applying the *if* portion of Lemma B.1.1 in Appendix B to

$$c = \left(\frac{c_1}{\sqrt{n_1}},\ldots,\frac{c_k}{\sqrt{n_k}}\right)$$

with $c_1 + \cdots + c_k = 0$, we see that if $(\mu_1, \ldots, \mu_k) \in S$, then

$$\sum_{i=1}^{k} \frac{c_i}{\sqrt{n_i}} \sqrt{n_i}(\mu_i - \bar{\mu}) \in$$

$$\sum_{i=1}^{k} \frac{c_i}{\sqrt{n_i}} \sqrt{n_i}(\hat{\mu}_i - \hat{\bar{\mu}}) \pm \sqrt{(k-1)F_{\alpha,k-1,\nu}} \hat{\sigma}(\sum_{i=1}^{k} c_i^2/n_i)^{1/2},$$

which simplifies to

$$\sum_{i=1}^{k} c_i \mu_i \in \sum_{i=1}^{k} c_i \hat{\mu}_i \pm \sqrt{(k-1)F_{\alpha,k-1,\nu}} \hat{\sigma}(\sum_{i=1}^{k} c_i^2/n_i)^{1/2}. \qquad (5.28)$$

Now suppose $(\mu_1, \ldots, \mu_k) \notin S$ then

$$\sum_{i=1}^{k} (\sqrt{n_i}(\hat{\mu}_i - \hat{\bar{\mu}}) - \sqrt{n_i}(\mu_i - \bar{\mu}))^2 > (k-1)\hat{\sigma}^2 F_{\alpha,k-1,\nu}.$$

Letting $c = (n_1(\mu_1 - \bar{\mu} - (\hat{\mu}_1 - \hat{\bar{\mu}})), \ldots, n_k(\mu_k - \bar{\mu} - (\hat{\mu}_k - \hat{\bar{\mu}}))) \in \mathcal{C}$, we see that

$$\sum_{i=1}^{k} c_i(\mu_i - \hat{\mu}_i)$$

$$= \sum_{i=1}^{k} c_i(\mu_i - \bar{\mu} - (\hat{\mu}_i - \hat{\bar{\mu}}))$$

$$= \|(\sqrt{n_1}(\mu_1 - \bar{\mu} - (\hat{\mu}_1 - \hat{\bar{\mu}})), \ldots, \sqrt{n_k}(\mu_k - \bar{\mu} - (\hat{\mu}_k - \hat{\bar{\mu}})))\|^2$$

$$> \sqrt{(k-1)\hat{\sigma}^2 F_{\alpha,k-1,\nu}}(\sum_{i=1}^{k} c_i^2/n_i)^{1/2},$$

that is, (5.28) is not satisfied.    $\square$

One can obviously deduce simultaneous confidence intervals for $\mu_i - \mu_j$, $i \neq j$, from Scheffé's confidence set by applying Theorem 5.2.3 to $c = c_{ij}$, the $k$-dimensional vector with 1 in the $i$th coordinate, $-1$ in the $j$th coordinate, and 0 in all other coordinates, to obtain the following.

**Corollary 5.2.1**

$$P\{\mu_i - \mu_j \in \hat{\mu}_i - \hat{\mu}_j \pm \sqrt{(k-1)F_{\alpha,k-1,\nu}} \hat{\sigma}\sqrt{n_i^{-1} + n_j^{-1}},$$
$$\text{for all } i,j, \ i \neq j\} > 1 - \alpha.$$

However, usually for pairwise comparisons Scheffé's method gives much wider confidence intervals than the Tukey–Kramer method, and is thus not recommended.

### 5.2.4 Stepdown tests

When the sample sizes are unequal, consider the range test which rejects

$$H_I : \mu_i = \mu_j \text{ for all } i,j \in I$$

for any $I \subseteq \{1,\ldots,k\}$ when

$$\max_{i,j \in I} \frac{\hat{\mu}_i - \hat{\mu}_j}{\hat{\sigma}\sqrt{n_i^{-1} + n_j^{-1}}} > c_{|I|}.$$

If one lets

$$c_m = \begin{cases} |q^*|_{1-(1-\alpha)^{m/k},m,\nu} & \text{if } m = 2,\ldots,k-2, \\ |q^*|_{\alpha,m,\nu} & \text{if } m = k-1,k. \end{cases}$$

then Theorem 5.2.1 implies (5.11) is satisfied. So if the stepdown testing scheme in Section 5.1.6 is executed, then by Theorem 5.1.5 a confident inequalities method results. However, the short-cut underlining execution scheme described in Section 5.1.4 is no longer possible. This is because the largest statistic among

$$\frac{\hat{\mu}_i - \hat{\mu}_j}{\hat{\sigma}\sqrt{n_i^{-1} + n_j^{-1}}}$$

for $i,j \in I$ may not come from the $i,j$ pair with the largest $\hat{\mu}_i, i \in I$, and smallest $\hat{\mu}_j, j \in I$, when the sample sizes are unequal. One can only assert $\mu_i \neq \mu_j$ if all $H_I$ such that $i,j \in I$ are rejected by the stepdown method.

Now consider the $F$-test which rejects $H_I$ when

$$\frac{\sum_{i \in I} n_i (\hat{\mu}_i - \hat{\bar{\mu}}_I)^2 / (|I| - 1)}{\hat{\sigma}^2} > c_{|I|}$$

where

$$\hat{\bar{\mu}}_I = \frac{\sum_{i \in I} \sum_{a=1}^{n_i} Y_{ia}}{\sum_{i \in I} n_i}.$$

If we set

$$c_m = \begin{cases} F_{1-(1-\alpha)^{m/k},m-1,\nu} & \text{if } m = 2,\ldots,k-2, \\ F_{\alpha,m-1,\nu} & \text{if } m = k-1,k, \end{cases}$$

where $F_{\gamma,m-1,\nu}$ is the upper $\gamma$ quantile of the $F$ distribution with $m-1$ numerator and $\nu = k(n-1)$ denominator degrees of freedom, and assert $\mu_i \neq \mu_j$ if and only if all $H_I$ such that $i,j \in I$ are rejected by the stepdown method, then Theorem 5.1.5 continues to apply and a confident inequalities method results.

### 5.2.5 The Miller–Winer method

For the unbalanced one-way model, the so-called Miller–Winer method uses the harmonic mean

$$\tilde{n}_I = (\sum_{i \in I} \frac{1}{n_i}/|I|)^{-1}$$

of the sample sizes of the treatments in $I \subseteq \{1, \ldots, k\}$ as the common sample size $n_I$ for the treatments in $I$ to be compared and the chosen method for the balanced one-way model is executed. This leads to invalid statistical inference in general. In particular, when the Miller–Winer method is applied to Tukey's method, the (marginal) probability that the assertions associated with $\mu_i - \mu_j, i, j \in J$, are correct is

$$P\left\{|\hat{\mu}_i - \hat{\mu}_j - (\mu_i - \mu_j)| < |q^*|\hat{\sigma}\sqrt{2/\tilde{n}} \text{ for all } i, j \in J\right\} \qquad (5.29)$$

$$= P\left\{|\hat{\mu}_i - \hat{\mu}_j - (\mu_i - \mu_j)|/\hat{\sigma} < |q^*|\sqrt{2/\tilde{n}} \text{ for all } i, j \in J\right\}$$

$$= P\left\{\frac{|\hat{\mu}_i - \hat{\mu}_j - (\mu_i - \mu_j)|}{\hat{\sigma}\sqrt{1/n_i + 1/n_j}} < |q^*|\sqrt{\frac{2/\tilde{n}}{1/n_i + 1/n_j}} \text{ for all } i, j \in J\right\}.$$

When the sample sizes $n_i$ and $n_j$ are relatively small, the factor

$$\sqrt{\frac{2/\tilde{n}}{1/n_i + 1/n_j}} \qquad (5.30)$$

may be so substantially less than one that the probability (5.29) becomes less than $1 - \alpha$. For example, suppose $k = 20$ and the sample sizes are $n_1 = \cdots = n_4 = 2, n_5 = \cdots = n_{20} = 20$. Then $\tilde{n} = 1/0.14$ and (5.30) equals $\sqrt{2(0.14)}$ for $1 \le i < j \le 4$. With $\alpha = 0.05$, using the qmca program with $\nu = 308$, we find $|q^*| = 3.544$ and the probability (5.29) with $J = \{1, 2, 3, 4\}$ turns out to be equal to 0.761, which is much less than 0.95. The probability that all the assertions associated with $\mu_i - \mu_j, i > j$, are simultaneously correct is of course lower still.

Thus, the Miller–Winer method (which is applied to the Tukey–Kramer method by the HARMONIC ALL option in the ONEWAY command of SPSS-X (SPSS 1988) as well as to the method of Ryan, Einot–Gabriel, and Welsch by the REGWQ and REGWF options of the MEANS statement of PROC GLM of the SAS system, for example) should therefore be avoided.

### 5.2.6 Fisher's least significant difference methods

For the unbalanced one-way model, Fisher's protected least significant difference method (Fisher 1935) remains as described in Section 5.1.8, except that the denominators in (5.19), (5.20), and (5.21) become $\sqrt{n_i^{-1} + n_j^{-1}}$ instead of $\sqrt{2/n}$. Finner's (1990a) proof that the directional version controls its probability of at least one incorrect assertion for $k = 3$ is applicable to the unbalanced one-way model (applicable to the general linear model, in fact). So, for $k = 3$, the protected LSD is a confident directions method. However, as described in Section 5.1.8, neither the protected LSD nor the unprotected LSD is a confident inequalities method for $k > 3$, and neither is recommended.

### 5.2.7 Methods based on probabilistic inequalities

For two-sided all-pairwise comparisons, let

$$E_{ij} = \{|\hat{\mu}_i - \hat{\mu}_j - (\mu_i - \mu_j)|/\hat{\sigma}\sqrt{n_i^{-1} + n_j^{-1}} < |q^e|\}, \tag{5.31}$$

while for Hayter's one-sided all-pairwise comparisons let

$$E_{ij} = \{(\hat{\mu}_i - \hat{\mu}_j - (\mu_i - \mu_j))/\hat{\sigma}\sqrt{n_i^{-1} + n_j^{-1}} < q^e\}. \tag{5.32}$$

The Bonferroni inequality (A.3) states

$$P(\bigcup_{i<j} E_{ij}^c) \le \sum_{i<j} P(E_{ij}^c).$$

If each $E_{ij}^c$ is such that

$$P(E_{ij}^c) = \alpha/[k(k-1)/2],$$

then

$$P(\bigcup_{i<j} E_{ij}^c) \le \alpha.$$

Thus, a conservative approximation to $|q^e|$ is

$$|q^*|_{\text{Bonferroni}} = t_{\frac{\alpha}{k(k-1)},\nu}, \tag{5.33}$$

which is implemented as the BON option of the MEANS statement of PROC GLM of the SAS system, while a conservative approximation to $q^e$ is

$$q^*_{\text{Bonferroni}} = t_{\frac{2\alpha}{k(k-1)},\nu}. \tag{5.34}$$

Šidák's inequality (Theorem A.4.1) states

$$P(\bigcap_{i<j} E_{ij} \mid \hat{\sigma}) \ge \prod_{i<j} P(E_{ij} \mid \hat{\sigma}).$$

Thus, a conservative approximation to $|q^e|$, proposed by Hochberg (1974), is

$$|q^*|_{\text{GT2}} = |m|_{\alpha,k(k-1)/2,\nu},$$

which is implemented as the GT2 option in the MEANS statement of PROC GLM of the SAS system (as it was referred to as the GT2 method in Hochberg 1974b).

Now $P(E_{ij}|\hat{\sigma}), i < j$, are monotone in $\hat{\sigma}$ in the same direction, so one can further apply Corollary A.1.1 to get

$$P(\bigcap_{i<j} E_{ij}) \geq E_{\hat{\sigma}}[\prod_{i<j} P(E_{ij} \mid \hat{\sigma})] \geq \prod_{i<j} E_{\hat{\sigma}} P(E_{ij}|\hat{\sigma}) = \prod_{i<j} P(E_{ij}).$$

If each $E_{ij}$ is such that

$$P(E_{ij}) = (1 - \alpha)^{1/[k(k-1)/2]},$$

then

$$P(\bigcap_{i<j} E_{ij}) \geq 1 - \alpha.$$

Thus, a conservative approximation to $|q^e|$ is

$$|q^*|_{\text{Šidák}} = t_{\frac{1-(1-\alpha)^{2/[k(k-1)]}}{2},\nu},$$

which is implemented as the SIDAK option of the MEANS statement of PROC GLM of the SAS system, and a conservative approximation to $q^e$ is

$$q^*_{\text{Slepian}} = t_{1-(1-\alpha)^{2/[k(k-1)]},\nu}.$$

From the discussion in Section 1.3.3, we know that the approximation $|q^*|_{\text{GT2}}$ is very slightly better than $|q^*|_{\text{Šidák}}$ (unless the degrees of freedom $\nu$ is very small). From the discussion in Section 1.3.5, we know that, in turn, the approximation $|q^*|_{\text{Šidák}}$ is very slightly better than $|q^*|_{\text{Bonferroni}}$ and the approximation $q^*_{\text{Slepian}}$ is very slightly better than $q^*_{\text{Bonferroni}}$. Relative to $|q^*|$ and $q^e$, these (first-order) approximations are not very accurate, and are not recommended.

One could obtain better (less conservative) approximations by using the (second-order) Hunter–Worsley inequality (A.6). But even these approximations are quite conservative. To illustrate, we take the balanced one-way model (5.1) and plot the lower bound on the confidence level given by the Hunter–Worsley inequality when one uses the exact critical value $|q^*|$ for Tukey's two-sided confidence intervals (Figure 5.8) or $q^*$ for either Hayter's one-sided confidence intervals or Bofinger's constrained confidence intervals (Figure 5.9).

### 5.2.8 Example: SAT scores

We use the data in Campbell and McCabe (1984) to illustrate all-pairwise comparisons for the unbalanced one-way model. Recall from Section 4.1.5 that they reported on the Scholastic Aptitude Test (SAT) mathematics scores of computer science, engineering and other majors. The sample sizes and sample means were given in Table 4.2, and the pooled sample standard deviation was $\hat{\sigma} = 82.52$. Again we use $\alpha = 0.10$. (Note, however, in contrast

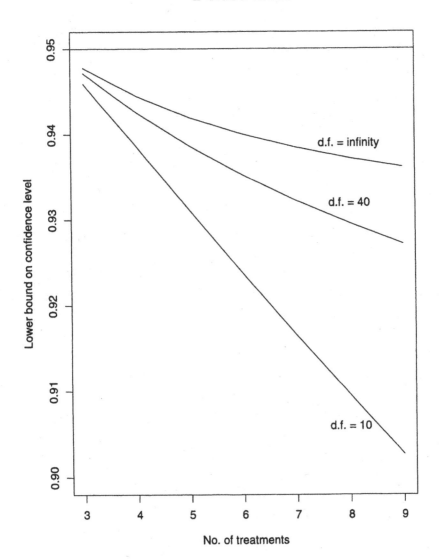

Figure 5.8 *Hunter–Worsley bounds on confidence level (actual level = 0.95)*

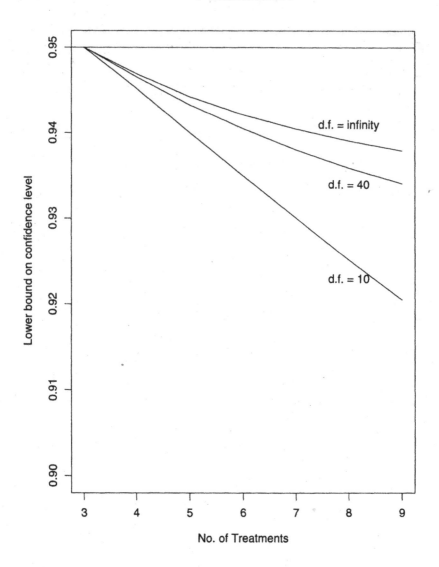

Figure 5.9 *Hunter–Worsley bounds on confidence level (actual level = 0.95)*

to the calculations in Section 4.1.5 and 4.2.5, the calculations below are based on computer codes that do not consider $\nu = 253$ large enough to be taken as $\infty$.)

*Exact method* For $\alpha = 0.10$, three treatments ($k = 3$), sample size pattern $n_1 = 103, n_2 = 31, n_3 = 122$, and 253 error degrees of freedom ($\nu = 253$), we have $|q^e| = 2.047$, which can be computed using the algorithm in Soong and Hsu (1995). (Approximation based on Spurrier and Isham 1985 gives a value of 2.049.) Therefore exact 90% confidence intervals for pairwise differences are

$$-44.60 \;<\; \mu_1 - \mu_2 \;<\; 24.60$$
$$21.40 \;<\; \mu_1 - \mu_3 \;<\; 66.60$$
$$20.02 \;<\; \mu_2 - \mu_3 \;<\; 87.98.$$

Thus, in terms of 0-1 directional inference, the exact method asserts 'both computer science and engineering majors have higher mean SAT mathematics scores than other majors.'

*Tukey–Kramer method* For $\alpha = 0.10$, three treatments ($k = 3$), and 253 error degrees of freedom ($\nu = 253$), we have $|q^*| = 2.062$. Therefore 90% Tukey–Kramer confidence intervals are

$$-44.85 \;<\; \mu_1 - \mu_2 \;<\; 24.85$$
$$21.24 \;<\; \mu_1 - \mu_3 \;<\; 66.76$$
$$19.78 \;<\; \mu_2 - \mu_3 \;<\; 88.22.$$

These confidence intervals are slightly wider than the exact confidence intervals. In terms of 0-1 directional inference, the Tukey–Kramer method makes the same assertions as the exact method.

*Stepdown tests* One way to obtain confident directions methods from the general stepdown testing scheme in Section 5.1.6 is to set

$$\alpha_m = \begin{cases} 1 - (1 - \alpha)^{m/k} & \text{if } m = 2, \ldots, k - 2, \\ \alpha & \text{if } m = k - 1, k, \end{cases}$$

which leads to $\alpha_2 = \alpha_3 = \alpha$, and

| $m$ | 2 | 3 |
|-----|-------|-------|
| $c_m$ | 1.651 | 2.062 |

for range statistics (note $c_m$ is increasing in $m$) and

| $m$ | 2 | 3 |
|-----|-------|-------|
| $c_m$ | 2.725 | 2.324 |

for $F$ statistics. Working through the details of the stepdown methods, we find that both methods make the same assertions of inequality as can be deduced from the Tukey–Kramer method.

*Scheffé's Method* For $\alpha = 0.10$, three treatments ($k = 3$), and 253 error degrees of freedom ($\nu = 253$), we have $F_{\alpha,k-1,\nu} = 2.324$. Therefore 90%

Scheffé confidence intervals are

$$
\begin{array}{ccccc}
-46.44 & < & \mu_1 - \mu_2 & < & 26.44 \\
20.20 & < & \mu_1 - \mu_3 & < & 67.80 \\
18.22 & < & \mu_2 - \mu_3 & < & 89.78.
\end{array}
$$

Notice while these confidence intervals are wider than the corresponding Tukey–Kramer confidence intervals, in terms of 0-1 directional inference, Scheffé's method makes the same assertions as the Tukey–Kramer method.

*Fisher's protected and unprotected LSD methods* For $\alpha = 0.10$, three treatments ($k = 3$), and 253 error degrees of freedom ($\nu = 253$), we have $F_{\alpha,k-1,\nu} = 2.324$ and $t_{\alpha/2,\nu} = 1.651$. The $F$-test statistic turns out to be 10.23, so $H_0 : \mu_1 = \cdots = \mu_k$ is rejected. Since $k = 3$, the directional protected LSD is a confident directions method. It makes the same directional assertions as can be deduced from the Tukey–Kramer method. The unprotected LSD makes the same assertions of inequality as can be deduced from the Tukey–Kramer method.

*Methods based on probabilistic inequalities* For $\alpha = 0.10$, three treatments ($k = 3$), and 253 error degrees of freedom ($\nu = 253$), we have

$$
\begin{array}{rcl}
|q^*|_{\text{GT2}} & = & 2.125, \\
|q^*|_{\text{Šidák}} & = & 2.126, \\
|q^*|_{\text{Bonferroni}} & = & 2.140.
\end{array}
$$

Therefore 90% Hochberg's (GT2) confidence intervals are

$$
\begin{array}{ccccc}
-45.92 & < & \mu_1 - \mu_2 & < & 25.92, \\
20.54 & < & \mu_1 - \mu_3 & < & 67.46, \\
18.74 & < & \mu_2 - \mu_3 & < & 89.26,
\end{array}
$$

while 90% confidence intervals based on Šidák's inequality are

$$
\begin{array}{ccccc}
-45.93 & < & \mu_1 - \mu_2 & < & 25.93, \\
20.53 & < & \mu_1 - \mu_3 & < & 67.47, \\
18.72 & < & \mu_2 - \mu_3 & < & 89.28,
\end{array}
$$

and 90% confidence intervals based on the Bonferroni inequality are

$$
\begin{array}{ccccc}
-46.17 & < & \mu_1 - \mu_2 & < & 26.17, \\
20.37 & < & \mu_1 - \mu_3 & < & 67.63, \\
18.49 & < & \mu_2 - \mu_3 & < & 89.51.
\end{array}
$$

They are almost the same, all wider than the Tukey–Kramer confidence intervals. In terms of directional inference, they lead to the same conclusions as the Tukey–Kramer method.

### 5.3 Nonparametric methods

Suppose the one-way model (5.22) holds except that the error distribution may not be normal:

$$Y_{ia} = \mu_i + \epsilon_{ia}, \quad i = 1, \ldots, k, \quad a = 1, \ldots, n_i, \tag{5.35}$$

where $\epsilon_{11}, \ldots, \epsilon_{kn_k}$ are i.i.d. with distribution $F$, which is absolutely continuous but otherwise unknown. Again one can use the fact that, under

$$H_0 : \mu_1 = \cdots = \mu_k,$$

all rankings of

$$Y_{11}, \ldots, Y_{1n_1}, Y_{21}, \ldots, Y_{2n_2}, \ldots, Y_{k1}, \ldots, Y_{kn_k}$$

are equally likely to construct statistical methods that hold their error rates at the nominal level for all possible $F$ and any sample size.

#### 5.3.1 Pairwise ranking methods

Let $R_{ia}^j(\delta_i)$ denote the rank of $Y_{ia} - \delta_i$ in the combined sample of size $n_i + n_j$

$$Y_{i1} - \delta_i, \ldots, Y_{in_i} - \delta_i, Y_{j1}, \ldots, Y_{jn_j}$$

and let

$$R_i^j(\delta_i) = \sum_{a=1}^{n_i} R_{ia}^j(\delta_i).$$

Let

$$D_{i[1]}^j \leq \cdots \leq D_{i[n_i n_j]}^j$$

denote the ordered $n_i n_j$ differences

$$Y_{ia} - Y_{jb}, 1 \leq a \leq n_i, \ 1 \leq b \leq n_j,$$

with the additional understanding that

$$D_{i[0]}^j = -\infty$$

and

$$D_{i[n_i n_j + 1]}^j = +\infty.$$

Using the well-known relationship between $R_i^j(\delta_i)$ and the $D_{i[a]}^j$'s stated in Lemma 3.3.1, level $1 - \alpha$ simultaneous MCA confidence intervals can be constructed as follows.

For two-sided MCA inference under the one-way model (5.35) with balanced sample sizes $n_1 = \cdots = n_k = n$, let $|r|$ be the smallest integer such that, under $H_0$,

$$P_{H_0}\{n(2n+1) - |r| \leq R_i^j(0) \leq |r| \text{ for all } i \neq j\} \geq 1 - \alpha;$$

then again $|r|$ can be computed without knowledge of $F$.

## Theorem 5.3.1

$$P\{D^j_{i[n(3n+1)/2-|r|]} \le \mu_i - \mu_j < D^j_{i[|r|-n(n+1)/2+1]} \text{ for all } i \ne j\} \ge 1 - \alpha.$$

*Proof.* Noting that changing the value of $\delta_i$ in $R^j_i(\delta_i)$ may change the value of $R^j_i(\delta_i)$, but not the value of any other $R^j_m(\delta_m), m \ne i$, we have

$$
\begin{aligned}
1 - \alpha \;\le\;& P_{H_0}\{n(2n+1) - |r| \le R^j_i(0) \le |r| \text{ for all } i \ne j\} \\
=\;& P\{n(2n+1) - |r| \le R^j_i(\mu_i - \mu_j) \le |r| \text{ for all } i \ne j\} \\
=\;& P\{D^j_{i[n(3n+1)/2-|r|]} \le \mu_i - \mu_j < D^j_{i[|r|-n(n+1)/2+1]} \text{ for all } i \ne j\}.
\end{aligned}
$$

$\square$

Steel (1960) and Dwass (1960) proposed making 0-1 decisions based on pairwise ranking. For $k = 3$, Steel (1960) gave exact values of $|r|$ for $n = 2, 3, 4$. Exact values of $|r|$ for $2 \le n \le 7$ can be obtained from Critchlow and Fligner (1991). For large $n$, the critical value $|r|$ can be approximated by

$$|r| \approx \frac{n(2n+1)}{2} + |q^*|_\infty \sqrt{\frac{n^2(2n+1)}{12}} + 0.5,$$

where $|q^*|_\infty$ is the critical value of Tukey's method for the same one-way model, but with degrees of freedom $\nu = \infty$.

For nonparametric directional MCA inference and one-sided MCA inference under the one-way location model (5.35) with balanced sample sizes $n_1 = \cdots = n_k = n$, let $r$ be the smallest integer such that, under $H_0$,

$$P_{H_0}\{R^j_i(0) \le r \text{ for all } i > j\} \ge 1 - \alpha.$$

Then since under $H_0$ all rankings of

$$Y_{11}, \ldots, Y_{1n}, Y_{21}, \ldots, Y_{2n}, \ldots, Y_{k1}, \ldots, Y_{kn}$$

are equally likely, $r$ can be computed without knowledge of $F$. Note that, by the symmetry of the distribution of $R^j_i(0), i > j$, under $H_0$, the critical value $r$ is also the smallest integer such that, under $H_0$,

$$P\{R^j_i(0) \ge n(2n+1) - r \text{ for all } i > j\} \ge 1 - \alpha.$$

## Theorem 5.3.2

$$P\{\mu_i - \mu_j < D^j_{i[r-n(n+1)/2+1]} \text{ for all } i > j\} \ge 1 - \alpha$$

*and*

$$P\{\mu_i - \mu_j \ge D^j_{i[n(3n+1)/2-r]} \text{ for all } i > j\} \ge 1 - \alpha.$$

*Proof.*

$$
\begin{aligned}
1 - \alpha \;\le\;& P_{H_0}\{R^j_i(0) \le r \text{ for all } i > j\} \\
=\;& P\{R^j_i(\mu_i - \mu_j) \le r \text{ for all } i > j\} \\
=\;& P\{\mu_i - \mu_j \ge D^j_{i[n(3n+1)/2-r]} \text{ for all } i > j\},
\end{aligned}
$$

$$1 - \alpha \leq P_{H_0}\{R_i^j(0) \geq n(2n+1) - r \text{ for all } i > j\}$$
$$= P\{R_i^j(\mu_i - \mu_j) \geq n(2n+1) - r \text{ for all } i > j\}$$
$$= P\{\mu_i - \mu_j < D_{i[r-n(n+1)/2+1]}^j \text{ for all } i > j\}. \quad \square$$

Exact values of $r$ for $3 \leq n \leq 7$ can be obtained from Hayter and Stone (1991). For large $n$, using the asymptotic multivariate normality of $R_1^2(0)$, $\ldots, R_{k-1}^k(0)$ under $H_0$, the critical value $r$ can be approximated by

$$r \approx \frac{n(2n+1)}{2} + q_\infty^* \sqrt{\frac{n^2(2n+1)}{12}} + 0.5,$$

where $q_\infty^*$ is the critical value of Bofinger's and Hayter's methods for the same one-way model, but with degrees of freedom $\nu = \infty$.

Again, if we let the asymptotic efficiency of confidence set $A$ relative to confidence set $B$ be defined as the reciprocal of the limiting ratio of sample sizes needed so that the two confidence sets have the same asymptotic probability of excluding the same sequence of false parameters $\mu_i^{(n)} - \mu_j^{(n)}, i > j$, approaching $(0, \ldots, 0)$ at the rate of $n^{1/2}$, arising from

$$\mu_i^{(n)} = \frac{\Delta_i}{\sqrt{n}}, \quad i = 1, \ldots, k,$$

where $\Delta_1, \ldots, \Delta_k$ are fixed positive constants, then (under some regularity conditions) the asymptotic relative efficiency (ARE) of the pairwise ranking methods relative to the corresponding sample means methods is

$$e(F) = 12\sigma^2 \left[\int_{-\infty}^{\infty} f^2(y)dy\right]^2, \tag{5.36}$$

where $f$ is the density of $F$ and $\sigma^2$ is the variance of a typical $Y_{ia}$ (cf. Koziol and Reid 1977). Recalling the discussion in Section 3.3, if the normality assumption in the location model (5.35) is in doubt, then the pairwise ranking methods are preferable to the sample means methods.

For two-sided MCA inference under the unbalanced one-way location model (5.35), in analogy to Theorem 5.3.1, it is easy to deduce from Lemma 3.3.1 that if $|r_{21}|, \ldots, |r_{k(k-1)}|$ are integers such that

$$P_{H_0}\{n_i(n_i + n_j + 1) - |r_{ij}| \leq R_i^j(0) \leq |r_{ij}| \text{ for all } i > j\} \geq 1 - \alpha, \tag{5.37}$$

then

$$P\{D_{i[n_i n_j + n_i(n_i+1)/2 - |r_{ij}|]}^j \leq \mu_i - \mu_j < D_{i[|r_{ij}| - n_i(n_i+1)/2+1]}^j$$
$$\text{for all } i > j\} \geq 1 - \alpha.$$

In general, there may be different combinations of $|r_{21}|, \ldots, |r_{k(k-1)}|$ satisfying (5.37). For $2 \leq n_1 \leq n_2 \leq n_3 \leq 7$, the smallest value of $w$ such that

$$P\{(n_i + n_j)(n_i + n_j + 1)/2 - w[n_i n_j(n_i + n_j + 1)/24]^{1/2} < R_i^j(0)$$
$$< w[n_i n_j(n_i + n_j + 1)/24]^{1/2} \text{ for all } i > j\} \geq 1 - \alpha$$

can be obtained from Critchlow and Fligner (1991). For large $n_1, \ldots, n_k$, one can use the asymptotic multivariate normality of $R_i^j(0), 1 \leq j < i \leq k$, under $H_0$ and approximate the $|r_{ij}|$'s by

$$|r_{ij}| \approx \frac{n_i(n_i + n_j + 1)}{2} + |q^*|_\infty \sqrt{\frac{n_i n_j(n_i + n_j + 1)}{12}} + 0.5,$$

where $|q^*|_\infty$ is the critical value of the Tukey–Kramer method for the same one-way model, but with degrees of freedom $\nu = \infty$.

For nonparametric directional MCA inference and one-sided MCA inference under the unbalanced one-way location model (5.35), in analogy to Theorem 5.3.2, it is easy to deduce from Lemma 3.3.1 that, if $r_{21}, \ldots, r_{k(k-1)}$ are integers satisfying

$$P\{R_i^j(0) \leq r_{ij} \text{ for all } i > j\} \geq 1 - \alpha, \tag{5.38}$$

then

$$P\{\mu_i - \mu_j < D_{i[r_{ij} - n_i(n_i+1)/2+1]}^j \text{ for all } i > j\} \geq 1 - \alpha$$

and

$$P\{\mu_i - \mu_j \geq D_{i[n_i n_k + n_i(n_i+1)/2 - r_{ij}]}^j \text{ for all } i > j\} \geq 1 - \alpha.$$

For $3 \leq n_1, n_2, n_3 \leq 7$, exact values of the smallest $c$ such that

$$P\{R_i^j(0) < n_i n_j/2 + c[n_i n_j(n_i + n_j + 1)/24]^{1/2} \text{ for all } i > j\} \geq 1 - \alpha,$$

can be obtained from Hayter and Stone (1991). For large $n_1, \ldots, n_k$, one can use the asymptotic multivariate normality of $R_i^j(0), 1 \leq j < i \leq k$ under $H_0$ to approximate the $r_{ij}$'s by

$$r_{ij} \approx \frac{n_i(n_i + n_j + 1)}{2} + q_\infty^* \sqrt{\frac{n_i n_j(n_i + n_j + 1)}{12}} + 0.5,$$

where $q_\infty^*$ is the critical value of sample means method for the same one-way model, but with degrees of freedom $\nu = \infty$.

### 5.3.2 Joint ranking methods

Under the one-way location model (5.35) with equal sample sizes $n_1 = \cdots = n_k = n$, let $\boldsymbol{\delta} = (\delta_1, \ldots, \delta_k)$ and let $R_{ia}(\boldsymbol{\delta})$ denote the rank of $Y_{ia} - \delta_i$ in the sample combining all $kn$ observations

$$Y_{11} - \delta_1, \ldots, Y_{1n} - \delta_1, \ldots, Y_{k1} - \delta_k, \ldots, Y_{kn} - \delta_k.$$

Let $R_i(\delta)$ denote the rank sum of the $i$th treatment (in this joint ranking of all $k$ treatments):

$$R_i(\delta) = \sum_{a=1}^{n} R_{ia}(\delta).$$

Suppose $|r|_J$ is the smallest integer such that

$$P_{H_0}\{|R_i - R_j| \leq |r|_J \text{ for all } i > j\} \geq 1 - \alpha \qquad (5.39)$$

under

$$H_0 : \mu_1 = \cdots = \mu_k,$$

where $R_i = R_i(0, \ldots, 0), i = 1, \ldots, k$; then since

$$(5.39) = P_{\boldsymbol{\mu}}\{|R_i(\boldsymbol{\mu}) - R_j(\boldsymbol{\mu})| \leq |r|_J \text{ for all } i > j\},$$

a $100(1 - \alpha)\%$ confidence set for $\mu_1, \ldots, \mu_k$ is

$$C = \{(\delta_1, \ldots, \delta_k) : |R_i(\delta) - R_j(\delta)| \leq |r|_J \text{ for all } i > j\}.$$

This set $C$ is translation invariant in the sense that if $(\delta_1, \ldots, \delta_k) \in C$, then $(\delta_1 + \delta, \ldots, \delta_k + \delta) \in C$ as well. Therefore, by examining a cross-section of $C$ perpendicular to $(1, \ldots, 1)$, simultaneous confidence intervals for $\mu_i - \mu_j, i \neq j$, can be deduced. Unfortunately, no general technique for computing this cross-section is known. The difficulty is that, in contrast to the pairwise ranking method, changes in the value of $\delta_i$ not only induce changes in $R_i(\delta) - R_j(\delta), j \neq i$, but changes in $R_m(\delta) - R_j(\delta), m \neq j$ as well.

A popular two-sided MCA method, attributed to Dunn (1964), asserts

$$\mu_i \neq \mu_j \text{ for all } i \neq j \text{ such that } |R_i - R_j| > |r|_J.$$

When $H_0$ is true, clearly

$$P_{H_0}\{\text{at least one incorrect assertion}\} \leq \alpha. \qquad (5.40)$$

However, as we indicate below, at least for some $k$,

$$\lim_{n \to \infty} \sup_{F, \boldsymbol{\mu}} P_{F, \boldsymbol{\mu}}\{\text{at least one incorrect assertion}\} > \alpha$$

at the usual $\alpha$ levels.

Suppose $\mu_1 = \delta, \mu_2 = \cdots = \mu_k = 0$, so that an assertion of $\mu_i \neq \mu_j$ for any $2 \leq j < i \leq k$ is incorrect. Oude Voshaar (1980) showed that, for this parameter configuration, asymptotically $(n^3 b)^{-1/2}(R_2 - R_k, \ldots, R_{k-1} - R_k)$ and $Z_2 - Z_k, \ldots, Z_{k-1} - Z_k$ have the same distribution, where $Z_2, \ldots, Z_k$ are i.i.d. standard normal random variables, and $b = b(F, \delta) = k^2/12 + (2Cov(F(X), F(X - \delta)) - 1/6)(k-1) + Var(F(X - \delta)) - 1/12$. Oude Voshaar (1980) further showed

$$\sup_{F, \delta} b(F, \delta) \geq (k^2 + k/2 + 5/16)/12.$$

Thus, noting (Hollander and Wolfe 1973, p. 124)

$$\lim_{n \to \infty} |r|_J = \sqrt{2}|q^*|_{k-1} n(k(kn+1)/12)^{1/2}$$

where $|q^*|_{k-1}$ is the critical value for the Tukey's MCA method for comparing $k-1$ treatments with infinite $MSE$ degrees of freedom ($\nu = \infty$), we have

$$\lim_{n \to \infty} \sup_{F,\mu} P_{F,\mu}\{\text{at least one incorrect assertion}\}$$

$$\geq \quad P\{|Z_i - Z_j| > \sqrt{2}|q^*|_{k-1} k/(k^2 + k/2 + 5/16)^{1/2}$$
$$\text{for some } 2 \leq j < i \leq k\}.$$

Table 5.2 tabulates this probability which, as can be seen, can be greater than $\alpha$. So this 'nonparametric' MCA method is not even a confident inequalities method, and is not recommended.

Table 5.2 *Lower bound on asymptotic error rate of two-sided joint ranking method*

|         |        |        | $k$    |        |        |        |
|---------|--------|--------|--------|--------|--------|--------|
| $\alpha$ | 3      | 5      | 7      | 10     | 15     | 20     |
| 0.10    | 0.0612 | 0.0909 | 0.0987 | 0.1025 | 0.1039 | 0.1041 |
| 0.05    | 0.0325 | 0.0478 | 0.0514 | 0.0529 | 0.0532 | 0.0530 |
| 0.01    | 0.0079 | 0.0109 | 0.0114 | 0.0114 | 0.0112 | 0.0111 |

## 5.4 Exercises

1. Consider the female forced vital capacity (FVC) data in Table 2.4. It is reasonable to expect pulmonary health not to improve as one progresses from non-smokers (NS) through passive smokers (PS) through light smokers (LS) through moderate smokers (MS) to heavy smokers (HS). Ignoring the non-inhaling smokers (NI) group, apply Hayter's one-sided MCA method to this data at $\alpha = 0.01$.

   Note that where non-inhaling smokers (NI) rank theoretically is not obvious. Indeed, in the same retrospective study of White and Froeb (1980), the forced mid-expiratory flow (FEF) of males data in Table 3.5 shows a different sample ranking for the non-inhaling smokers. Despite this, if non-inhaling smokers had a sample size of 200, then since Hayter confidence bounds are valid regardless of whether the assumed ranking is correct or not, the non-inhaling smokers could be included in the analysis.

2. Stevenson, Lee and Stigler (1986) first reported on the mathematics

achievement of Chinese, Japanese and American children. In an expanded study, Stevenson, Chen and Lee (1993) reported that the mean Japanese–American difference was 8.36 in 1980, 5.39 in 1984, and 6.92 in 1990. The mean Chinese–American difference was 5.83 in 1980, 7.59 in 1984, and 9.65 in 1990. They also reported that the mean Japanese–Chinese difference was 2.53 in 1980, $-2.19$ in 1984, and $-2.73$ in 1990 (as can be deduced). The sample standard deviations and sample sizes are given in Table 5.3.

Table 5.3 *Standard deviations and sample sizes of mathematics tests*

| Year | | United States | Taiwan | Japan |
|------|--|--------|--------|-------|
| 1980 | Standard deviation | 5.6 | 7.4 | 5.1 |
|      | Sample size | 288 | 286 | 280 |
| 1984 | Standard deviation | 5.3 | 5.5 | 5.2 |
|      | Sample size | 237 | 241 | 240 |
| 1990 | Standard deviation | 6.2 | 5.7 | 7.5 |
|      | Sample size | 238 | 241 | 239 |

Apply the Tukey–Kramer method to compare mean mathematics test scores among Chinese, Japanese and American children, separately for 1980, 1984 and 1990.

3. Use the following Maple commands to plot a sample Tukey confidence set for $k = 3$.

```
with(plots):
tubeplot(
{[t,t+1,t+2,t=-1..1,radius=1,tubepoints=7,numpoints=5],
[t,t+1,t+2,t=-1..1,radius=.03,tubepoints=5,numpoints=5]},
axes=NORMAL, scaling=CONSTRAINED, projection=0.5,
style=PATCHNOGRID);
```

4. Use the following Maple commands to plot a sample Scheffé confidence set for $k = 3$.

```
with(plots):
tubeplot(
{[t,t+1,t+2,t=-1..1,radius=1,tubepoints=35,numpoints=3],
[t,t+1,t+2,t=-1..1,radius=.03,tubepoints=5,numpoints=3]},
axes=NORMAL, scaling=CONSTRAINED, projection=0.5,
style=PATCHNOGRID);
```

5. Consider the balanced one-way model (5.1) with $k = 3$. Show that if $\hat{\mu}_1 = 0$ and $\hat{\mu}_2 = \hat{\mu}_3$, then

(a) the $F$-test for $H_0 : \mu_1 = \mu_2 = \mu_3$ rejects if $\hat{\mu}_2^2 = \hat{\mu}_3^2 > 3F_{\alpha,2,\nu}\hat{\sigma}^2/n$;

(b) Scheffé's confidence intervals for $\mu_1 - \mu_2$ and $\mu_1 - \mu_3$ contain zero if $\hat{\mu}_2^2 = \hat{\mu}_3^2 < 4F_{\alpha,2,\nu}\hat{\sigma}^2/n$ (of course, when $\hat{\mu}_2 = \hat{\mu}_3$, Scheffé's confidence interval for $\mu_2 - \mu_3$ contains zero);

(c) Scheffé's confidence interval for $\mu_1 - \frac{\mu_2+\mu_3}{2}$ does not contain zero if $\hat{\mu}_2^2 = \hat{\mu}_3^2 > 3F_{\alpha,2,\nu}\hat{\sigma}^2/n$.

6. Use the following Maple commands to plot a Scheffé confidence set corresponding to the situation where

(a) the $F$-test for $H_0 : \mu_1 = \mu_2 = \mu_3$ rejects;

(b) Scheffé's confidence intervals for $\mu_1 - \mu_2, \mu_1 - \mu_3$ and $\mu_2 - \mu_3$ all contain zero;

(c) Scheffé's confidence interval for $\mu_1 - \frac{\mu_2+\mu_3}{2}$ does not contain zero.

```
with(plots):
tubeplot(
{[t,t+1.35,t+1.35,t=-1..1,radius=1,tubepoints=35,
numpoints=3],
[t,t+1.35,t+1.35,t=-1..1,radius=.03,tubepoints=5,
numpoints=3]},
axes=NORMAL, scaling=CONSTRAINED, projection=1,
style=PATCHNOGRID);
```

7. Consider a data set with $k = 4$, $n_1 = \cdots = n_4 = 2$, $\hat{\mu}_1 = 0, \hat{\mu}_2 = 2, \hat{\mu}_3 = 3.55, \hat{\mu}_4 = 7.05$, $\hat{\sigma} = 1$, and suppose $\alpha = 0.05$. Then

$$
\begin{aligned}
c_4 &= |q^*|_{0.05,4,4} &= 4.071 \\
c_3 &= |q^*|_{0.05,3,4} &= 3.564 \\
c_2 &= |q^*|_{1-0.95^{(2/4)},2,4} &= 3.481,
\end{aligned}
$$

which are monotone. Noting $\hat{\sigma}\sqrt{2/n} = 1$ and

$$
\begin{aligned}
\hat{\mu}_4 - \hat{\mu}_1 &= 7.05 &> &\quad 4.071, \\
\hat{\mu}_4 - \hat{\mu}_2 &= 5.05 &> &\quad 3.564, \\
\hat{\mu}_3 - \hat{\mu}_1 &= 3.55 &< &\quad 3.564, \\
\hat{\mu}_4 - \hat{\mu}_3 &= 3.50 &> &\quad 3.481,
\end{aligned}
$$

the multiple range test asserts

$$
\begin{aligned}
\mu_1 &\neq \mu_4, \\
\mu_2 &\neq \mu_4
\end{aligned}
$$

and

$$\mu_3 \neq \mu_4$$

but not

$$\mu_1 \neq \mu_3,$$

even though
$$\hat{\mu}_3 - \hat{\mu}_1 = 3.55 > \hat{\mu}_4 - \hat{\mu}_3 = 3.50,$$
which may seem counter-intuitive. (You can execute the multiple range test by applying a computer implementation such as REGWQ under the MEANS option of PROC GLM of the SAS system to the artificial data set in Table 5.4, which has summary statistics matching those above.)

Table 5.4 *Data leading to counter-intuitive multiple range test assertions*

| Treatment | Observation |
|:---:|:---:|
| 1 | 0.7071068 |
| 1 | −0.7071068 |
| 2 | 2.7071068 |
| 2 | 1.2928932 |
| 3 | 4.2571068 |
| 3 | 2.8428932 |
| 4 | 7.7571068 |
| 4 | 6.3428932 |

8. Apply the multiple range test to the artificial data in Table 5.5. Verify that, with a prespecified
$$P\{\text{at least one incorrect assertion}\} \leq 0.05,$$
the only assertion that can be made is
$$\mu_1 \neq \mu_{10}.$$

9. Apply the multiple $F$ test to the artificial data in Table 5.6. Verify that, with a prespecified
$$P\{\text{at least one incorrect assertion}\} \leq 0.05,$$
the only assertion that can be made is
$$\mu_i \neq \mu_j \text{ for some } 1 \leq i < j \leq 4.$$

10. Consider the unbalanced one-way model (5.22). Write the distribution function of the range
$$\max_{i \neq j} \frac{\hat{\mu}_i - \hat{\mu}_j}{\hat{\sigma}}$$
of $\hat{\mu}_i/\hat{\sigma}, i = 1, \ldots, k$, as the sum of $k$ double integrals. Can you write the distribution function of
$$\max_{i \neq j} \frac{\hat{\mu}_i - \hat{\mu}_j}{\hat{\sigma}\sqrt{n_i^{-1} + n_j^{-1}}}$$
as a sum of double integrals?

Table 5.5 *Data useful for testing* REGWQ *implementation*

| Treatment | Observation |
|-----------|-------------|
| 1 | 0.7071068 |
| 1 | −0.7071068 |
| 2 | 0.7071068 |
| 2 | −0.7071068 |
| 3 | −0.7071068 |
| 3 | 0.7071068 |
| 4 | 0.7071068 |
| 4 | −0.7071068 |
| 5 | −0.7071068 |
| 5 | 0.7071068 |
| 6 | 0.7071068 |
| 6 | −0.7071068 |
| 7 | 0.7071068 |
| 7 | −0.7071068 |
| 8 | −0.7071068 |
| 8 | 0.7071068 |
| 9 | 4.5871068 |
| 9 | 3.1728932 |
| 10 | 3.2928932 |
| 10 | 4.7071068 |

Table 5.6 *Data useful for testing* REGWF *implementation*

| Treatment | Observation |
|-----------|-------------|
| 1 | −2.2821 |
| 1 | −0.8679 |
| 2 | −0.8679 |
| 2 | −2.2821 |
| 3 | 2.2821 |
| 3 | 0.8679 |
| 4 | 2.2821 |
| 4 | 0.8679 |

# Abuses and misconceptions in multiple comparisons

According to one survey, multiple comparison methods constitute the second most frequently applied group of statistical methods, second only to the $F$-test (Mead and Pike 1975). If they rank second in frequency of *use*, they rank perhaps first in frequency of *abuse*.

## 6.1 Not adjusting for multiplicity

One kind of abuse is to perform many comparisons without adjusting for multiplicity. This kind of abuse can be difficult to detect if only a select number of comparisons (typically those showing the most significant differences) are reported. Such an abuse allegedly occured in a Swedish study on the effect of electro-magnetic forces (EMF) emanating from high voltage power transmission lines on humans, as reported in the Public Broadcasting System (PBS) Frontline program titled Currents of Fear (PBS 1995).

## 6.2 Inflation of strength of inference

Another kind of abuse is to use a method capable of only low-strength inference to give high-strength inference. It is common to see the inference 'means with the same letter are not significantly different' given based on the Newman–Keuls multiple range test, Duncan's multiple range test, or Fisher's LSD (protected or unprotected). In this case, a test of homogeneity or an individual comparison method has been employed as a confident inequalities method.

In fact, the same methods have often been used to give the inference 'means followed by * are not significantly different from the best,' with the implication that those *not* followed by * are different from the best. In this case, a test of homogeneity or an individual comparison method has been employed as a confident directions method, for the assertions are *directional:* treatments different from the best are worse than the best.

*Proof by confusion* Justifications for such abuses have often been attempted using 'proof by confusion,' i.e., by appealing to concepts which are totally irrelevant. One favorite such device is the incorrect invocation of the concept of the experimentwise error rate, by showing that the error rate is controlled under the null hypothesis

$$H_0 : \mu_1 = \cdots = \mu_k. \tag{6.1}$$

Unfortunately this sanctifies precious little, as $H_0$ is rarely, if ever, true. In fact, Tukey (1992) states: 'The "null hypothesis" *never* holds.'* Even if you do not subscribe to this statement, the following example illustrates that an unscrupulous experimenter can make any desired assertion while controlling the error rate under the null hypothesis (6.1) at any level ($> 0$).

*The Yummy Cat Chow example*  Suppose it is desired to assert that Yummy Cat Chow is the best among ten brands. The brands are to be compared by blind tasting using a panel of expert cats. Consider the following procedure:

$$\text{Test } H_0 : \mu_1 = \cdots = \mu_k \text{ at level } \alpha.$$

If $H_0$ is $\begin{cases} \text{accepted then assert nothing;} \\ \text{rejected then assert Yummy Cat Chow is best.} \end{cases}$

The error rate of this procedure clearly is no more than $\alpha$ under the null hypothesis (6.1).

Prior to experimentation, let us introduce treatment 11: Tabasco sauce. Since treatment 11 is so different from the rest, the null hypothesis of homogeneity $H_0 : \mu_1 = \cdots = \mu_k$ is guaranteed to be rejected. Thus we are guaranteed the inference 'Yummy Cat Chow is the best,' regardless of data on the ten treatments of interest. (See Figure 6.1.) This example shows that a guarantee on the error rate which only applies when the null hypothesis $H_0 : \mu_1 = \cdots = \mu_k$ is true is not much of a guarantee. As Tukey (1992) states:

> A significant $F$ fails to tell us what is different – and seduces many innocent users to the entirely unfounded view that 'since the effects are (omnibus-wise) statistically significant, we can believe any and all appearances as truth.' What could be further from wisdom?*

Rothman (1990) argues against making multiplicity adjustment in multiple comparisons, in particular the adjustment of $p$-values computed under (6.1), which he calls the 'universal null hypothesis.' His arguments were:

- The universal null hypothesis is untenable.

- $p$-values, small or large, are not necessarily informative.

In this book, the universal null hypothesis (6.1) plays no special roll in either the motivation or the derivation of any method, and the presentation of multiple comparison inference in terms of $p$-values is not recommended. Thus, his arguments (which are correct) do not apply to the methods recommended in this book.

Another abuse occurs when one obtains multiple comparison methods by

---

* Reprinted by courtesy of Marcel Dekker Inc.

Figure 6.1 *A cat food tasting thought experiment*

applying the so-called 'rank transformation' to parametric multiple comparison methods, as recommended by Conover and Iman (1981). Special cases of methods so obtained include the methods based on joint ranking that mimic Dunnett's method and Tukey's method, which we discussed in Sections 3.3.2 and 5.3.2. While the rank-transformed methods control the error rate under the null hypothesis (6.1), under which all possible ranking are equally likely, they may not be confident directions or confident inequalities methods, even if the parametric multiple comparison methods they mimic are, as we saw in Sections 3.3.2 and 5.3.2. Thus, blanket statements that multiple comparison methods can be applied to joint ranking data with good results (Conover and Iman 1981) need to be taken with a grain of salt.

## 6.3 Conditional inference

An unfortunate common practice is to pursue multiple comparisons *only* when the null hypothesis of homogeneity $H_0 : \mu_1 = \cdots = \mu_k$ (typically based on the $F$-test) is rejected, due to the mistaken beliefs that:

1. Magical 'protection' is endowed by first performing a test of homogeneity. (For example, one protects the unprotected LSD by first performing a size-$\alpha$ test for $H_0 : \mu_1 = \cdots = \mu_k$ and proceeds only if $H_0$ is rejected.)

2. No useful result will be found if the test of homogeneity accepts.

Not only does performing a test of homogeneity first not guarantee the probability of an incorrect assertion to be less than $\alpha$, as we saw in the Yummy Cat Chow example in section 6.2, it might guarantee this probability to be (conditionally) greater than $\alpha$ if multiple comparison results are only reported when the test of homogeneity rejects. That is, the *conditional* error rate of multiple comparisons, conditional on the rejection of the homogeneity hypothesis, can be much higher than the *unconditional* error rate. For example, Olshen (1973) showed that if Scheffé's $100(1-\alpha)\%$ simultaneous confidence intervals for treatment contrasts are only given when the size-$\alpha$ $F$-test rejects $H_0 : \mu_1 = \cdots = \mu_k$, then the conditional coverage probability of these confidence intervals can be substantially less than $1 - \alpha$. In other words, among the experiments in which multiple comparisons are actually performed and subsequently reported, the proportion of experiments for which some of the assertions made are wrong can be much higher than the nominal error rate of the multiple comparison method.

*Conflicting F-test and multiple comparison inference*   As we saw in Section 5.1.5, it is possible that the $F$-test rejects but Scheffé's method finds no significant pairwise difference. This also occurs when a multiple comparison method other than Scheffé's is employed upon the rejection of the $F$-test. Conversely, it is possible that the $F$-test accepts but a multiple comparison method detects significant differences. Suppose $k = 3, n_1 = n_2 = n_3 = n$, and the multiple comparison method is Tukey's method. The acceptance region of the $F$-test

$$(\hat{\mu}_1 - \hat{\bar{\mu}})^2 + (\hat{\mu}_2 - \hat{\bar{\mu}})^2 + (\hat{\mu}_3 - \hat{\bar{\mu}})^2 \le F_{\alpha,2,\nu}\hat{\sigma}^2(2/n),$$

in terms of $\hat{\mu}_1, \hat{\mu}_2, \hat{\mu}_3$, is an infinite cylinder with a circular cross-section centered at the vector $\{(\delta, \delta, \delta) : \delta \in \Re\}$. Tukey's confidence intervals for $\mu_1 - \mu_2, \mu_2 - \mu_3, \mu_3 - \mu_1$ will all contain 0 if and only if

$$\max_{i \ne j} |\hat{\mu}_i - \hat{\mu}_j| \le |q^*|\hat{\sigma}\sqrt{2/n},$$

which is an infinite cylinder with a hexagonal cross-section, also centered at the vector $\{(\delta, \delta, \delta) : \delta \in \Re\}$. The circular cylinder and the hexagonal cylinder necessarily overlap, as each has probability content $1 - \alpha$ under $\mu_1 = \mu_2 = \mu_3$. Therefore, as indicated in Figure 6.2, which takes a (parallax) view of the overlapping cylinders parallel to the vector $\{(\delta, \delta, \delta) : \delta \in \Re\}$, if $(\hat{\mu}_1, \hat{\mu}_2, \hat{\mu}_3)$ is in a 'corner' of the hexagon outside the circle, then $H_0 : \mu_1 = \mu_2 = \mu_3$ is rejected by the $F$-test, but Tukey's confidence intervals for $\mu_1 - \mu_2, \mu_2 - \mu_3, \mu_3 - \mu_1$ all contain 0. On the other hand, if $\hat{\mu}_1, \hat{\mu}_2, \hat{\mu}_3$ is in a 'crescent' of the circle outside the hexagon, then $H_0 : \mu_1 = \mu_2 = \mu_3$ is accepted by the $F$-test, but one of Tukey's confidence intervals for $\mu_1 - \mu_2, \mu_2 - \mu_3, \mu_3 - \mu_1$ does not contain 0.

   In short, to consider multiple comparisons as to be performed only if the $F$-test for homogeneity rejects is a mistake.

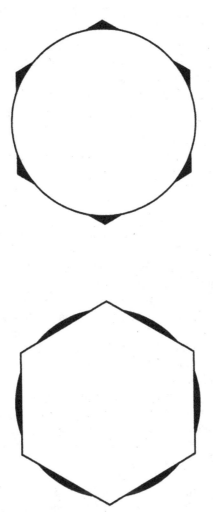

Figure 6.2 *Conflicting F-test and Tukey inference*

## 6.4 Post hoc comparisons

Except for Scheffé's method, the methods discussed in this book are for pre-planned comparisons: comparisons with a control, comparisons with the best, all pairwise comparisons. If it is desirable to allow for *post hoc* comparisons, that is, comparisons selected after the data has been collected, then Scheffé's method (which provides simultaneous confidence intervals for all contrasts) is appropriate.

A particularly insidious abuse is to report post hoc comparisons as if they were *pre-planned* comparisons. In the treatment × age example in Section

1.1.1, the subjects were divided into five age groups prior to experimentation. An example of abuse would be to conduct a completely randomized experiment and, upon failing to detect a significant overall difference between the two treatments, divide the subjects into age groups large and small until significant differences are found for some of the age groups using methods meant for pre-planned comparisons. Nowak (1995) reported that there was allegation such an abuse occured in an AIDS study. This practice is not unlike that of the so-called Texas Sharpshooter, who first shoots at the side of a barn, and then draws a bull's eye around where the bullets hit.

## 6.5 Recommendations

Several earlier studies on multiple comparison procedures compared all-pairwise comparison methods under different definitions of error rate and efficiency. Sometimes, in the same study, the authors would compare multiple comparison methods capable of different strengths of inference. For example, the oft-quoted study by Carmer and Swanson (1973) compared Scheffé's method, Tukey's method, the Newman–Keuls multiple range test, Duncan's multiple range test, Fisher's LSD methods and, after much computer simulation, recommended the protected LSD and Duncan's multiple range test. Different choices of strength of inference then naturally led to different recommendations (see Gabriel 1978).

To gain increased 'significance,' instead of choosing a low-strength all-pairwise comparisons method, leading necessarily to abuses if useful assertions are to be made, my view is that it is better to keep the level of strength useful, but concentrate on the type of inference of primary interest. (In other words, instead of going *down* column 1 of Table 2.2, go *across* row 1 of Table 2.2.)

Specifically, for confidence intervals, if all pairwise comparisons are of primary interest, then Tukey's method for balanced design and the Tukey–Kramer method for unbalanced design are recommended. If comparisons with the best are of primary interest, then constrained and unconstrained MCB methods are available. If comparisons with a control are of primary interest, then Dunnett's method is recommended.

If confident directions suffice, then for one-sided comparisons with a control, the stepdown method can be used. For all-pairwise comparisons, Bofinger's constrained confidence intervals are available for situations in which exact equalities of treatment effects are implausible. One can use the Ryan/Einot–Gabriel/Welsch methods or Peritz's method if one believes they are indeed confident directions methods. (Otherwise the recommended confidence intervals methods should be used.)

# CHAPTER 7

# Multiple comparisons in the general linear model

So far, we have only discussed multiple comparisons for experiments which consist of taking independent simple random samples under the treatments to be compared. However, in addition to measuring the response to the treatments, many real-life experiments incorporate the co-measurement of one or more quantitative variables that might impact on the response. For such experiments, the data analyst should first check whether the relative merits of the treatments depend on the values of these covariates, that is, whether the treatment effect and the covariates interact. Suppose they do not interact. Then one can meaningfully compare the treatments after adjusting for the covariates (which puts the treatments on an equal footing in the comparison). The experimenter might also attempt to increase the sensitivity of the comparison by blocking on one or more qualitative variables. Again, the data analyst should check whether the relative merits of the treatments depend on the blocks, that is, whether the treatment effect and the block effect interact. If they do not, then one can meaningfully compare the treatments after adjusting for block effects.

In experiments with fixed block effects or covariate effects or both, the appropriate model is the general linear model (GLM)

$$\mathbf{Y} = \mathbf{X}\boldsymbol{\beta} + \boldsymbol{\epsilon} \tag{7.1}$$

where $\mathbf{Y}_{N \times 1}$ is the vector of observations, $\mathbf{X}_{N \times p}$ is a known design matrix, $\boldsymbol{\beta}_{p \times 1} = (\beta_1, \ldots, \beta_p)'$ is the vector of parameters, and $\boldsymbol{\epsilon}_{N \times 1}$ is a vector of i.i.d. normally distributed errors with mean 0 and unknown variance $\sigma^2$.

For example, the one-way model

$$Y_{ih} = \mu_i + \epsilon_{ih}, \ i = 1, \ldots, k, \ h = 1, \ldots, n_i \tag{7.2}$$

is a GLM with

$$
\begin{pmatrix} Y_{11} \\ \vdots \\ Y_{1n_1} \\ \vdots \\ Y_{k1} \\ \vdots \\ Y_{kn_k} \end{pmatrix}
=
\begin{pmatrix}
1 & 0 & 0 & \cdots & 0 \\
\vdots & \vdots & \vdots & \cdots & \vdots \\
1 & 0 & 0 & \cdots & 0 \\
0 & 1 & 0 & \cdots & 0 \\
\vdots & \vdots & \vdots & \cdots & \vdots \\
0 & 1 & 0 & \cdots & 0 \\
& & \ddots & & \\
0 & 0 & \cdots & 0 & 1 \\
\vdots & \vdots & \cdots & \vdots & \vdots \\
0 & 0 & \cdots & 0 & 1
\end{pmatrix}
\begin{pmatrix} \mu_1 \\ \vdots \\ \mu_k \end{pmatrix}
+
\begin{pmatrix} \epsilon_{11} \\ \vdots \\ \epsilon_{1n_1} \\ \vdots \\ \epsilon_{k1} \\ \vdots \\ \epsilon_{kn_k} \end{pmatrix}.
$$

If a (fixed) covariate $X$ is measured along with response $Y$ and, under each treatment, the response $Y$ can be assumed to depend on $X$ linearly, then the usual analysis of covariance (ANCOVA) model

$$Y_{ih} = \theta_i + \beta_i X_{ih} + \epsilon_{ih}, \quad i = 1, \ldots, k, \quad h = 1, \ldots, n_i \tag{7.3}$$

is a GLM with

$$
\begin{pmatrix} Y_{11} \\ \vdots \\ Y_{1n_1} \\ \vdots \\ Y_{k1} \\ \vdots \\ Y_{kn_k} \end{pmatrix}
=
\begin{pmatrix}
1 & \cdots & 0 & X_{11} & \cdots & 0 \\
\vdots & \cdots & \vdots & \vdots & \cdots & \vdots \\
1 & \cdots & 0 & X_{1n_1} & \cdots & 0 \\
& \ddots & & & \ddots & \\
0 & \cdots & 1 & 0 & \cdots & X_{k1} \\
\vdots & \cdots & \vdots & \vdots & \cdots & \vdots \\
0 & \cdots & 1 & 0 & \cdots & X_{kn_k}
\end{pmatrix}
\begin{pmatrix} \theta_1 \\ \vdots \\ \theta_k \\ \beta_1 \\ \vdots \\ \beta_k \end{pmatrix}
+
\begin{pmatrix} \epsilon_{11} \\ \vdots \\ \epsilon_{1n_1} \\ \vdots \\ \epsilon_{k1} \\ \vdots \\ \epsilon_{kn_k} \end{pmatrix}.
$$

Another example of a GLM is the two-way with-interaction model

$$Y_{ihr} = \mu + \tau_i + \beta_h + (\tau\beta)_{ih} + \epsilon_{ihr}, \tag{7.4}$$
$$i = 1, \ldots, k, \quad h = 1, \ldots, b, \quad r = 1, \ldots, n_{ih}.$$

In a GLM, it is not always possible to talk about *treatment effect* $\mu_i$ unambiguously, because the (long-run) average response under the $i$th treatment may depend on the value of a covariate, or the level of a blocking factor.

For example, if

$$\beta_i \neq \beta_j$$

in the ANCOVA model

$$Y_{ih} = \theta_i + \beta_i X_h + \epsilon_{ih}; \quad i = 1, \ldots, k, \quad h = 1, \ldots, n_i,$$

then whether or not the (long-run) average response under the $i$th treatment ($= \theta_i + \beta_i X$) is larger or smaller than the (long-run) average response under the $j$th treatment ($= \theta_j + \beta_j X$) depends on the value of $X$. The parameters of interest may then be

$$\theta_1 - \theta_j + \beta_i X - \beta_j X, \text{ for all } X \in \mathcal{X}$$

where $\mathcal{X}$ is a set of discrete or continuous $X$ values.

As another example, if

$$(\tau\beta)_{ih} \neq (\tau\beta)_{jh}$$

in the two-way model (7.5), then whether the (long-run) average response at the $i$th level ($= \mu + \tau_i + \beta_h + (\tau\beta)_{ih}$) is larger or smaller than the (long-run) average response at the $j$th level ($= \mu + \tau_j + \beta_h + (\tau\beta)_{jh}$) of the first factor may depend on the level $h$ of the second factor. If both factors are controllable and the purpose of experimentation is to determine the best combination of the two factors, then the desired inference may be multiple comparison of all factor combinations. If, on the other hand, the first factor represents treatments while the second factor is a blocking factor representing age groups of patients (say), and it is important to find out which treatments are better for each age group, then the desired inference may be multiple comparisons of treatments within blocks but not comparison of treatments across the blocks.

However, in the absence of treatment–block or treatment–covariate interaction, it is meaningful to discuss treatment *contrasts* $\mu_i - \mu_j$, the difference between the (long-run) average response under $i$th treatment and the $j$th treatment. Thus, if

$$\beta_1 = \cdots = \beta_k$$

in the ANCOVA model (7.3), then the parameters

$$\theta_i - \theta_j$$

may be of interest. If

$$(\tau\beta)_{ih} = (\tau\beta)_{jh}$$

for all $h$ in the two-way model (7.5), then the parameters

$$\tau_i - \tau_j$$

may be of interest. We will denote generically by $\mu_i - \mu_j$ the multiple comparison parameters of interest in a general linear model. So $\mu_i - \mu_j$ may be $\theta_i - \theta_j$ in an ANCOVA model, or $\tau_i - \tau_j$ in a two-way model.

It should be understood that, by 'absence of interaction,' we mean that the models with and without the interaction term are practically equivalent in some suitable sense. This practical equivalence cannot be established by the acceptance of a null hypothesis of no interaction effect, which unfortunately is common practice. Fabian (1991) discussed some dire consequences this practice can lead to in multiple comparisons, in the context of a two-way model. A more sensible practice is to consider practical equivalence established if the confidence bounds on the interaction parameter(s) are sufficiently close to zero.

There are also situations where comparisons are to be made of parameters in a GLM that do not correspond to long-run average treatment effects. For example, in the ANCOVA model (7.3), $\beta_1, \ldots, \beta_k$ may correspond to degradation rates of batches of drugs, and it may be of interest to compare these rates, in which case the parameters

$$\mu_i - \mu_j = \beta_i - \beta_j$$

may be of interest.

In practice, especially when a statistical computer package is employed, the GLM model (7.1) may be parametrized to include an intercept, a $\beta_i$ for each treatment, and a $\beta_i$ for each block. Such over-parameterization results in a design matrix $\mathbf{X}$ of less than full rank, and infinitely many solutions to the normal equation

$$\mathbf{X}'\mathbf{X}\beta = \mathbf{X}'\mathbf{Y}. \tag{7.5}$$

However, if $\mu_i - \mu_j$ is *estimable*, that is, if there exists an unbiased estimate of $\mu_i - \mu_j$, then all solutions to the normal equation lead to the same estimate for $\mu_i - \mu_j$ (Searle 1971, p. 181). More specifically, let $(\mathbf{X}'\mathbf{X})^-$ denote a generalized inverse of $\mathbf{X}'\mathbf{X}$, so that

$$\hat{\beta} = \begin{pmatrix} \hat{\beta}_1 \\ \vdots \\ \hat{\beta}_p \end{pmatrix} = (\mathbf{X}'\mathbf{X})^-\mathbf{X}'\mathbf{Y}$$

is a solution to the normal equation (7.5). If

$$\mu_i - \mu_j = c\beta,$$

where $c$ is a linear combination of the rows of $\mathbf{X}$, so that $\mu_i - \mu_j$ is estimable, then $c\hat{\beta}$ is unique regardless of the choice of the generalized inverse $(\mathbf{X}'\mathbf{X})^-$.

We assume that all

$$\mu_i - \mu_j, \ i \neq j,$$

are estimable (block designs leading to models having this property are called *connected* designs in the literature; cf. John 1987). For each fixed $i$, if $\mathbf{C}_{-i}$ is the matrix such that

$$\mu_{-i} = (\mu_j - \mu_i, \ j \neq i)' = \mathbf{C}_{-i}\beta,$$

then even if the individual $\mu_i$ may not be estimable, there will be no confusion if we use the notation

$$\hat{\mu}_{-i} = (\hat{\mu}_j - \hat{\mu}_i, \ j \neq i)' = \mathbf{C}_{-i}\hat{\beta}.$$

Under the i.i.d. normal errors assumption of model (7.1) then,

$$\hat{\mu}_{-i} \sim MVN(\mathbf{C}_{-i}\beta, \sigma^2\mathbf{V}_{-i}),$$

where
$$\mathbf{V}_{-i} = \mathbf{C}_{-i}(\mathbf{X'X})^{-}\mathbf{C}'_{-i},$$

which we assume is non-singular. It will be convenient later on if we let $\mathbf{R}_{-i}$ denote the correlation matrix of $\hat{\mu}_{-i}$. Let

$$\hat{\sigma}^2 = MSE = (\mathbf{Y} - \hat{\beta}\mathbf{X})'(\mathbf{Y} - \hat{\beta}\mathbf{X})/(N - \text{rank}(\mathbf{X}))$$

denote the usual estimator of $\sigma^2$. Then $\nu\hat{\sigma}^2/\sigma^2$ has a $\chi^2$ distribution with $\nu = N - \text{rank}(\mathbf{X})$ degrees of freedom. Further, $\hat{\mu}_{-i}$ and $\hat{\sigma}^2$ are independent.

## 7.1 Models with one-way structure

Some models are one-way-like, in the sense that the joint distribution of the least squares estimators of parameters of interest in multiple comparisons has the same structure as its counterpart in a one-way model. Let $\sigma^2 v_j^i$ denote the variance of $\hat{\mu}_i - \hat{\mu}_j$, i.e., $v_j^i$ is the diagonal element of $\mathbf{V}_{-i} = \mathbf{C}_{-i}(\mathbf{X'X})^{-}\mathbf{C}'_{-i}$ corresponding to $\hat{\mu}_i - \hat{\mu}_j$. We say a model has a *one-way structure* if there exists a set of positive constants $a_1, \ldots, a_k$, such that

$$v_j^i = a_i + a_j \tag{7.6}$$

for all $i \neq j$. This important condition (7.6), which first appeared in Hayter (1989), actually completely determines the variance-covariance structure of $\hat{\mu}_i - \hat{\mu}_j, i \neq j$.

**Theorem 7.1.1** *Condition (7.6) is equivalent to the condition that there exist positive constants $a_1, \ldots, a_k$ such that the variance-covariance matrix of $\sigma^{-1}(\hat{\mu}_1 - \hat{\mu}_k, \ldots, \hat{\mu}_{k-1} - \hat{\mu}_k)$ is*

$$\begin{pmatrix} a_1 & 0 & \cdots & 0 \\ 0 & \ddots & \ddots & \vdots \\ \vdots & \ddots & \ddots & 0 \\ 0 & \cdots & 0 & a_{k-1} \end{pmatrix} + \begin{pmatrix} a_k & a_k & \cdots & a_k \\ a_k & \ddots & \ddots & \vdots \\ \vdots & \ddots & \ddots & a_k \\ a_k & \cdots & a_k & a_k \end{pmatrix}. \tag{7.7}$$

*Proof.* By the uniqueness of least squares estimators,

$$\hat{\mu}_i - \hat{\mu}_j = \hat{\mu}_i - \hat{\mu}_k - (\hat{\mu}_j - \hat{\mu}_k).$$

Therefore,

$$Var(\hat{\mu}_i - \hat{\mu}_j) = Var(\hat{\mu}_i - \hat{\mu}_k) + Var(\hat{\mu}_j - \hat{\mu}_k) - 2Cov(\hat{\mu}_i - \hat{\mu}_k, \hat{\mu}_j - \hat{\mu}_k).$$

If (7.7) is satisfied, then (7.6) is satisfied trivially for $j = k$, and for $i < j < k$ as well because

$$Var(\hat{\mu}_i - \hat{\mu}_j) = a_i + a_k + a_j + a_k - 2a_k = a_i + a_j.$$

Conversely, if (7.6) is satisfied, then trivially the diagonal elements of the variance-covariance matrix of $\sigma^{-1}(\hat{\mu}_1 - \hat{\mu}_k, \ldots, \hat{\mu}_{k-1} - \hat{\mu}_k)$ are as specified in (7.7). For $i < j < k$,

$$a_i + a_j = a_i + a_k + a_j + a_k - 2Cov(\hat{\mu}_i - \hat{\mu}_k, \hat{\mu}_j - \hat{\mu}_k),$$

so $Cov(\hat{\mu}_i - \hat{\mu}_k, \hat{\mu}_j - \hat{\mu}_k) = a_k$ and the off-diagonal elements of the variance-covariance matrix of $\sigma^{-1}(\hat{\mu}_1 - \hat{\mu}_k, \ldots, \hat{\mu}_{k-1} - \hat{\mu}_k)$ are as specified in (7.7). $\square$

If we let $\bar{Y}_i = \sum_{h=1}^{n_i} Y_{ih}/n_i$ in the one-way model (7.2), then comparing (7.7) with the variance-covariance matrix of $\sigma^{-1}(\bar{Y}_1 - \bar{Y}_k, \ldots, \bar{Y}_{k-1} - \bar{Y}_k)$ in a one-way model, which is

$$\begin{pmatrix} 1/n_1 & 0 & \cdots & 0 \\ 0 & \ddots & \ddots & \vdots \\ \vdots & \ddots & \ddots & 0 \\ 0 & \cdots & 0 & 1/n_{k-1} \end{pmatrix} + \begin{pmatrix} 1/n_k & 1/n_k & \cdots & 1/n_k \\ 1/n_k & \ddots & \ddots & \vdots \\ \vdots & \ddots & \ddots & 1/n_k \\ 1/n_k & \cdots & 1/n_k & 1/n_k \end{pmatrix}, \quad (7.8)$$

we see that in a model with a one-way structure, multiple comparisons can be executed as if one were in a one-way model, with the following substitutions

$$\bar{Y}_i - \bar{Y}_j \leftarrow \hat{\mu}_i - \hat{\mu}_j$$
$$n_i \leftarrow 1/a_i$$
$$\sum_{i=1}^{k}\sum_{a=1}^{n_i}(Y_{ia} - \bar{Y}_i)^2 / \sum_{i=1}^{k}(n_i - 1) \leftarrow \hat{\sigma}^2$$
$$\sum_{i=1}^{k}(n_i - 1) \leftarrow N - \text{rank}(\mathbf{X});$$

no new technique is needed.

You might worry that we have only considered the joint distribution of a set of treatments versus control estimators, not the joint distribution of estimators of all elementary contrasts involved in MCA, for example. The fact is, by the uniqueness of least squares estimators of estimable parameters, any set of treatments versus control estimators determines the estimators of any set of contrasts. Thus, the joint distribution of a set of treatments versus control estimators determines the joint distribution of any set of contrast estimators. As a simple illustration, we show that, in a model with a one-way structure, estimators of disjoint pairwise differences are uncorrelated.

**Corollary 7.1.1** *If a model has a one-way structure, then*

$$Cov(\hat{\mu}_i - \hat{\mu}_j, \hat{\mu}_k - \hat{\mu}_m) = 0 \text{ if } i \neq j \neq k \neq m.$$

*Proof.* By the uniqueness of least squares estimators,

$$\hat{\mu}_i - \hat{\mu}_m = \hat{\mu}_i - \hat{\mu}_j + (\hat{\mu}_j - \hat{\mu}_k) + (\hat{\mu}_k - \hat{\mu}_m).$$

Therefore,

$$
\begin{aligned}
Var(\hat\mu_i - \hat\mu_m) = \;& Var(\hat\mu_i - \hat\mu_j) + Var(\hat\mu_j - \hat\mu_k) + Var(\hat\mu_k - \hat\mu_m) + \\
& 2Cov(\hat\mu_i - \hat\mu_j, \hat\mu_j - \hat\mu_k) + 2Cov(\hat\mu_j - \hat\mu_k, \hat\mu_k - \hat\mu_m) + \\
& 2Cov(\hat\mu_i - \hat\mu_j, \hat\mu_k - \hat\mu_m)
\end{aligned}
$$

or

$$
a_i + a_m = a_i + a_j + a_j + a_k + a_k + a_m - 2a_j - 2a_k + 2Cov(\hat\mu_i - \hat\mu_j, \hat\mu_k - \hat\mu_m)
$$

The assertion follows.  $\square$

At least two possible characteristics of an experimental design are individually sufficient to ensure a one-way structure for the model: *balance* in the assignment of treatments to blocks ensures an equal covariance one-way structure for the model; and *orthogonality* of effects ensures a one-way structure of the model, although not necessarily an equal covariance one-way structure.

### 7.1.1 Variance-balanced designs

A design is *varianced-balanced* if all elementary contrasts $\mu_i - \mu_j$ are estimated with the same precision:

$$
Var(\hat\mu_i - \hat\mu_j) = 2\sigma^2/e \text{ for all } i \neq j, \tag{7.9}
$$

where $e$ is a design dependent known constant. Examples of varianced-balanced designs include balanced incomplete block designs (BIBD), Latin squares, Youden square designs (cf. John 1987, which gives additional examples), and Williams designs (cf. Chow and Liu 1992, Section 10.4), under the usual models. Hedayat and Stufken (1989) show the construction of pairwise balanced designs (for which there is an extensive literature) can be used to find varianced-balanced designs, and gives further references.

By Theorem 7.1.1, the variance-covariance matrix of $\sqrt{e}\sigma^{-1}(\hat\mu_1 - \hat\mu_k, \ldots, \hat\mu_{k-1} - \hat\mu_k)'$ is

$$
\begin{pmatrix}
1 & 0 & \cdots & 0 \\
0 & \ddots & \ddots & \vdots \\
\vdots & \ddots & \ddots & 0 \\
0 & \cdots & 0 & 1
\end{pmatrix}
+
\begin{pmatrix}
1 & 1 & \cdots & 1 \\
1 & \ddots & \ddots & \vdots \\
\vdots & \ddots & \ddots & 1 \\
1 & \cdots & 1 & 1
\end{pmatrix}. \tag{7.10}
$$

This is the same as the variance-covariance matrix of $\sqrt{n}\sigma^{-1}(\bar Y_1 - \bar Y_k, \ldots, \bar Y_{k-1} - \bar Y_k)$ in a balanced one-way model. Therefore, in a variance-balanced model, multiple comparisons can be executed as if one were in a balanced one-way model, with the following substitutions:

$$
\bar Y_i - \bar Y_j \;\leftarrow\; \hat\mu_i - \hat\mu_j
$$

$$n \leftarrow e$$

$$\sum_{i=1}^{k}\sum_{a=1}^{n}(Y_{ia} - \bar{Y}_i)^2/[k(n-1)] \leftarrow \hat{\sigma}^2$$

$$k(n-1) \leftarrow N - \text{rank}(\mathbf{X}).$$

(Hochberg 1974a showed this for all-pairwise comparisons.)

### 7.1.2 Orthogonal designs

Latin square designs are not only variance-balanced but orthogonal as well in the sense that the blocking effects are orthogonal to the treatment effect. Therefore, their estimates of treatment effect parameters are based on treatment means and no adjustment for blocking effects is necessary except in the estimation of experiment error (John 1987, p. 101 and p. 103). An example of a linear model that is not variance-balanced, but is orthogonal, is the two-way proportional cell frequencies model.

### 7.1.2.1 Two-way model with proportional cell frequencies

Consider the two-way no-interaction model

$$Y_{ihr} = \mu + \tau_i + \beta_h + \epsilon_{ihr}, \quad i = 1, \ldots, k, h = 1, \ldots, b, r = 1, \ldots, n_{ih}, \quad (7.11)$$

where

$$
\begin{aligned}
Y_{ihr} &= r\text{th observation under } i\text{th treatment in } h\text{th block}\\
\tau_i &= i\text{th treatment effect}\\
\beta_h &= h\text{th block effect}\\
\epsilon_{ihr} &= \text{random error associated with } Y_{ihr}
\end{aligned}
$$

and it is assumed that $\epsilon_{111}, \ldots, \epsilon_{kbn_{kb}}$ are i.i.d. normally distributed with mean 0 and unknown variance $\sigma^2$. We assume all cells are filled ($n_{ih} \geq 1$), so all elementary contrasts

$$\mu_i - \mu_j = \tau_i - \tau_j, \quad i \neq j$$

are estimable.

Define

$$\mathbf{N} = \begin{pmatrix} n_{11} & \cdots & n_{1b} \\ \vdots & & \vdots \\ n_{k1} & \cdots & n_{kb} \end{pmatrix},$$

$$n_{i\cdot} = \sum_{h=1}^{b} n_{ih},$$

$$n_{*+} = \begin{pmatrix} n_{1.} \\ \vdots \\ n_{k.} \end{pmatrix},$$

$$\bar{Y}_{i..} = \sum_{h=1}^{b} \sum_{r=1}^{n_{ih}} Y_{ihr}/n_{i.},$$

$$n_{.h} = \sum_{i=1}^{k} n_{ih},$$

$$n_{+*} = \begin{pmatrix} n_{.1} \\ \vdots \\ n_{.b} \end{pmatrix},$$

$$\bar{Y}_{.h.} = \sum_{i=1}^{k} \sum_{r=1}^{n_{ih}} Y_{ihr}/n_{.h},$$

$$N = \sum_{i=1}^{k} \sum_{h=1}^{b} n_{ih},$$

$$\bar{Y}_{...} = \sum_{i=1}^{k} \sum_{h=1}^{b} \sum_{r=1}^{n_{ih}} Y_{ihr}/N.$$

If we write the two-way no-interaction model as

$$Y = 1\mu + X\tau + Z\beta + \epsilon$$

then

$$\begin{align}
X'1 &= n_{*+} & (7.12) \\
X'X &= \mathrm{diag}(n_{*+}) & (7.13) \\
Z'1 &= n_{+*} & (7.14) \\
Z'Z &= \mathrm{diag}(n_{+*}) & (7.15) \\
X'Z &= N & (7.16)
\end{align}$$

where $\mathrm{diag}(v)$ is the diagonal matrix with elements of $v$ as its diagonal. This is an over-parametrized model and, for our present purpose, a convenient reparametrization is to set

$$\sum_{i=1}^{k} n_{i.}\tau_i = \sum_{h=1}^{b} n_{.h}\beta_h = 0$$

or equivalently letting

$$X^* = X - \frac{1}{N}1n'_{*+}$$

and

$$Z^* = Z - \frac{1}{N} 1 n'_{+*}$$

and rewriting the two-way no-interaction model as

$$Y = 1\mu + X^* \tau + Z^* \beta + \epsilon.$$

For the one-way model

$$Y = 1\mu + X^* \tau + \epsilon,$$

the least squares estimates for the treatment contrasts $\tau_i - \tau_k, i = 1, \ldots, k$, are the differences of the marginal means $\bar{Y}_{i..} - \bar{Y}_{k..}, i = 1, \ldots, k$. Clearly these differences of marginal sample means will remain the least squares estimates of $\tau_i - \tau_k, i = 1, \ldots, k$, if $Z^*$ is orthogonal to $X^*$ which, using (7.12)–(7.16), can be seen to occur if

$$N = \frac{1}{N} n_{*+} n'_{+*}.$$

That is, if the *proportions* of observations allocated to the $k$ treatments within a block remain the same from block to block,

$$n_{ih} = n_{i.} n_{.h} / N$$

for all $ih$, then

$$\hat{\tau}_i - \hat{\tau}_j = \bar{Y}_{i..} - \bar{Y}_{j..}.$$

Since $\bar{Y}_{i..}, i = 1, \ldots, k$, are independently distributed, in the case of comparing treatment effects in a two-way model with proportional cells frequencies, multiple comparisons can be executed as if the observations were taken under the treatments without blocking, but with the following substitutions:

$$\bar{Y}_i - \bar{Y}_j \quad \leftarrow \quad \bar{Y}_{i..} - \bar{Y}_{j..}$$

$$\sum_{i=1}^{k} \sum_{h=1}^{n_i} (Y_{ih} - \bar{Y}_i)^2 / \sum_{i=1}^{k} (n_i - 1) \quad \leftarrow \quad \hat{\sigma}^2$$

$$\sum_{i=1}^{k} (n_i - 1) \quad \leftarrow \quad N - k - b + 1$$

where

$$\hat{\sigma}^2 = MSE = \sum_{i=1}^{k} \sum_{h=1}^{b} \sum_{r=1}^{n_{ih}} (Y_{ihr} - \bar{Y}_{i..} - \bar{Y}_{.h.} + \bar{Y}_{...})^2 / (N - k - b + 1).$$

Can a model have the one-way structure if the least squares estimators for the parameters of interest are not differences of marginal sample means? The answer is 'yes.' An example of such is the one-way ANCOVA model with different slopes and intercepts.

*7.1.2.2 One-way ANCOVA model with different slopes*

Consider the ANCOVA model (7.3)

$$Y_{ih} = \theta_i + \beta_i X_{ih} + \epsilon_{ih}, \quad i = 1, \ldots, k, \quad h = 1, \ldots, n_i, \qquad (7.17)$$

where

$$\begin{array}{rcl}
Y_{ih} & = & h\text{th response under the }i\text{th treatment} \\
\theta_i & = & \text{intercept of the }i\text{th treatment} \\
\beta_i & = & \text{slope of the }i\text{th treatment} \\
X_{ih} & = & \text{covariate value at which the sample }Y_{ih}\text{ is taken} \\
\epsilon_{ih} & = & \text{random error associated with }Y_{ih}.
\end{array}$$

It is assumed that $\epsilon_{11}, \ldots, \epsilon_{kn_k}$ are i.i.d. normally distributed with mean 0 and unknown variance $\sigma^2$.

The model (7.17) is a full-rank model. If we define

$$\bar{X}_i = \sum_{h=1}^{n_i} X_{ih}/n_i,$$

$$\bar{Y}_i = \sum_{h=1}^{n_i} Y_{ih}/n_i,$$

$$S_{X_i}^2 = \sum_{h=1}^{n_i} (X_{ih} - \bar{X}_i)^2,$$

then the least squares estimates are

$$\hat{\beta}_i = \sum_{h=1}^{n_i} (Y_{ih} - \bar{Y}_i)(X_{ih} - \bar{X}_i)/S_{X_i}^2,$$

$$\hat{\theta}_i = \bar{Y}_i - \hat{\beta}_i \bar{X}_i,$$

$$\hat{\sigma}^2 = MSE = \sum_{i=1}^{k} \sum_{h=1}^{n_i} (Y_{ih} - \hat{\theta}_i - \hat{\beta}_i X_{ih})^2 / \sum_{i=1}^{k} (n_i - 2).$$

The $\hat{\beta}_i$ are independently distributed with

$$Var(\hat{\beta}_i) = \sigma^2 / S_{X_i}^2.$$

Therefore, multiple comparisons of slopes in this ANCOVA model can be executed as if one were comparing treatment means in a one-way model, with the following substitutions:

$$\bar{Y}_i \leftarrow \hat{\beta}_i$$
$$n_i \leftarrow S_{X_i}^2$$

$$\sum_{i=1}^{k}\sum_{h=1}^{n_i}(Y_{ih} - \bar{Y}_i)^2 / \sum_{i=1}^{k}(n_i - 1) \quad \leftarrow \quad \hat{\sigma}^2$$

$$\sum_{i=1}^{k}(n_i - 1) \quad \leftarrow \quad \sum_{i=1}^{k}(n_i - 2).$$

### 7.1.2.3 Example: Pooling batches in drug stability studies

Pharmaceutical products are routinely monitored for their stability over time. Stability studies often consist of a random sample of dosage units (e.g. tablets, capsules, vials) from several batches placed in a storage room and periodically assayed for their contents. The model most commonly used to analyze degradation of batches of drugs over time is (7.17), where

$Y_{ih}$ = $h$th response from the $i$th batch

$\theta_i$ = $i$th batch effect

$\beta_i$ = degradation rate of $i$th batch

$X_{ih}$ = time at which the stability sample $Y_{ih}$ is taken

$\epsilon_{ih}$ = random error associated with $Y_{ih}$.

According to the Food and Drug Administration's *Guideline for Submitting Documentation for Stability Studies of Human Drugs and Biologics* (FDA, 1987), the shelf-life is calculated as the time point at which the lower 95% confidence limit about the fitted line crosses the lowest acceptable limit for drug content, expressed as a percentage of the labeled amount of drug. Currently the Food and Drug Administration (FDA) Guideline states that if the null hypothesis

$$H_0 : \beta_i = \beta_j \text{ for all } i \neq j \tag{7.18}$$

is accepted at the 25% level, then a reduced model with all $\beta_i = \beta$ is used for data with all bathes pooled. If (7.18) is rejected at the 25% level, then the FDA Guideline suggests that subsets of batches with similar slopes may be considered together and retested as above. Ultimately, shelf-lives are computed for each subgroup of batches, and the shortest shelf-life is used for the drug product.

It is fairly clear that the intent of the FDA Guideline is batches which are practically equivalent to the worst batch can be pooled. Therefore, as pointed out in Ruberg and Hsu (1992), the current FDA Guideline is unreasonable, because a large $p$-value associated with testing (7.18) or a homogeneity hypothesis of any subset of $\beta_1, \ldots, \beta_k$ does not necessarily imply that the corresponding $\beta_i$ are approximately equal. Further, small variations within a batch, frequent sampling times, and replication of samples at a given time make it easier to reject (7.18). Thus, a penalty is paid for doing a good study. Inappropriate pooling of too many batches is a risk to the consumer, since shelf-lives will tend to be too long as data are pooled.

Consider the two data sets cited in Ruberg and Hsu (1992), with summary statistics given in Tables 7.1 and 7.2. The mean square errors $\hat{\sigma}^2$ corresponding to Tables 7.1 and 7.2 are $1.07/25$ with 25 degrees of freedom and $18.28/23$ with 23 degrees of freedom, respectively. Even though the degradation rates of the six batches given in Table 7.1 are reasonably close to each other, pooling of the batches is not allowed under the current FDA Guideline since the small $\hat{\sigma}^2$ causes the null hypothesis (7.18) to be rejected at the 25% level. On the other hand, even though the degradation rates of the six batches given in Table 7.2 are quite different, pooling of the batches is allowed under the current FDA Guideline since the large $\hat{\sigma}^2$ causes the null hypothesis (7.18) to be accepted at the 25% level.

Table 7.1 *Summary statistics for Experiment 1*

| Batch | $\hat{\theta}_i$ | $\hat{\beta}_i$ | $S_{X_i}^2$ |
|-------|------|--------|--------|
| 1 | 100.49 | −1.515 | 14.629 |
| 2 | 100.66 | −1.449 | 7.830 |
| 3 | 100.25 | −1.682 | 5.141 |
| 4 | 100.45 | −1.393 | 1.379 |
| 5 | 100.45 | −1.999 | 0.608 |
| 6 | 99.98 | −1.701 | 0.608 |

Table 7.2 *Summary statistics for Experiment 2*

| Batch | $\hat{\theta}_i$ | $\hat{\beta}_i$ | $S_{X_i}^2$ |
|-------|------|--------|--------|
| 1 | 102.48 | −0.109 | 14.629 |
| 2 | 103.63 | −0.449 | 2.963 |
| 3 | 101.24 | −0.778 | 5.141 |
| 4 | 102.21 | 0.194 | 1.260 |
| 5 | 102.21 | −2.218 | 0.608 |
| 6 | 100.07 | −1.046 | 0.608 |

Given that the desired inference is practical equivalence with the worst batch, an alternative to the FDA Guideline is to formulate the problem as one of constrained multiple comparison with the best (MCB), with the 'best' being the batch with the worst degradation rate. Ruberg and Hsu (1992) proposed that simultaneous confidence intervals for $\beta_i - \min_{j \neq i} \beta_j$

be examined, and batches with

$$\beta_i - \min_{j \neq i} \beta_j < \delta,$$

that is, batches with degradation rates practically equivalent to the worst degradation rate, be pooled. If $L$ is the desired shelf-life, and $\Delta$ is the maximum allowable difference in mean drug content between batches at time $L$, then $\delta$ can be defined as $\Delta/L$. For illustration, suppose the desired shelf-life is 3 years and the maximum allowable difference in mean drug content between batches after 3 years is 4%, then $\delta = 1.33\%/$year.

The qmcb computer program described in Appendix D, with inputs $\lambda = (\lambda_{ij}, \ j \neq i)$, $\nu$ and $\alpha$, where

$$\lambda_{ij} = \left( 1 + \frac{S_{X_i}^2}{S_{X_j}^2} \right)^{-1/2},$$

computes $d^i, i = 1, \ldots, k$. Tables 7.3 and 7.4 display MCB simultaneous confidence intervals for the difference between each degradation rate and the worst degradation rate for the two experiments.

For Experiment 1, all batches would be pooled under the Ruberg and Hsu (1992) proposal since the MCB upper confidence bounds are all less than 1.33%, and the computed shelf-life turns out to be 6.2 years, which is not that much different from the shelf-life computed under the current FDA Guideline.

For Experiment 2, none of the batches should be pooled under the Ruberg and Hsu (1992) proposal, since the MCB upper confidence bounds are all greater than 1.33%. Computing the shelf-life based on batch 6 which had the worst degradation rate, the shelf-life turns out to be 1.7 years, which is drastically shorter than the shelf-life of 15.7 years computed based on the current FDA Guideline, which allows all batches to be pooled.

Table 7.3  *95% MCB confidence intervals for Experiment 1*

| Batch | $d^i$ | $D_i^-$ | $D_i^+$ |
|-------|-------|---------|---------|
| 1 | 2.452 | −0.09 | 1.07 |
| 2 | 2.433 | −0.05 | 1.15 |
| 3 | 2.411 | −0.36 | 0.92 |
| 4 | 2.288 | −0.12 | 1.30 |
| 5 | 2.168 | −0.92 | 0.52 |
| 6 | 2.168 | −0.52 | 1.11 |

Beyond variance-balanced designs and orthogonal designs, new techniques are needed to effect multiple comparisons properly. These techniques are discussed separately in the contexts of multiple comparisons with a con-

Table 7.4 *95% MCB confidence intervals for Experiment 2*

| Batch | $d^i$ | $D_i^-$ | $D_i^+$ |
|-------|-------|---------|---------|
| 1 | 2.469 | −0.46 | 4.68 |
| 2 | 2.401 | −1.23 | 4.53 |
| 3 | 2.437 | −1.51 | 4.10 |
| 4 | 2.311 | −0.81 | 5.48 |
| 5 | 2.202 | −4.10 | 2.39 |
| 6 | 2.202 | −2.39 | 4.73 |

trol (MCC), multiple comparisons with the best (MCB), and all-pairwise comparisons (MCA).

## 7.2 Multiple comparisons with a control

Suppose treatment versus control comparisons are our primary concern, so the parameters of interests are

$$\mu_i - \mu_k, \ i = 1, \ldots, k - 1.$$

Recall that $\sigma^2 v_i^k$ denotes the variance of $\hat{\mu}_i - \hat{\mu}_k$.

In theory, one can generalize Dunnett's one-sided MCC confidence intervals for one-way designs to the general linear model in a straightforward fashion. Using the following theorem, we obtain one-sided MCC confidence intervals centered at $\hat{\mu}_1 - \hat{\mu}_k, \ldots, \hat{\mu}_{k-1} - \hat{\mu}_k$, scaled by $\hat{\sigma}\sqrt{v_1^k}, \ldots, \hat{\sigma}\sqrt{v_{k-1}^k}$.

**Theorem 7.2.1** *Suppose the constant d satisfies*

$$P\left\{\max_{1 \leq i \leq k-1} \frac{\hat{\mu}_i - \hat{\mu}_k - (\mu_i - \mu_k)}{\hat{\sigma}\sqrt{v_i^k}} < d\right\} = 1 - \alpha;$$

*then*

$$P\{\hat{\mu}_i - \hat{\mu}_k - d\hat{\sigma}\sqrt{v_i^k} < \mu_i - \mu_k \ for \ i = 1, \ldots, k - 1\} = 1 - \alpha$$

*and*

$$P\{\mu_i - \mu_k < \hat{\mu}_i - \hat{\mu}_k + d\hat{\sigma}\sqrt{v_i^k} \ for \ i = 1, \ldots, k - 1\} = 1 - \alpha.$$

*Proof.*

$$
\begin{aligned}
1 - \alpha &= P\left\{\max_{1 \leq i \leq k-1} \frac{\hat{\mu}_i - \hat{\mu}_k - (\mu_i - \mu_k)}{\hat{\sigma}\sqrt{v_i^k}} < d\right\} \\
&= P\{\hat{\mu}_i - \hat{\mu}_k - (\mu_i - \mu_k) < d\hat{\sigma}\sqrt{v_i^k} \ for \ i = 1, \ldots, k - 1\}
\end{aligned}
$$

$$= P\{\hat{\mu}_i - \hat{\mu}_k - d\hat{\sigma}\sqrt{v_i^k} < \mu_i - \mu_k \text{ for } i = 1, \ldots, k-1\}.$$

Similarly,

$$1 - \alpha = P\left\{\max_{1 \leq i \leq k-1} \frac{-(\hat{\mu}_i - \hat{\mu}_k - (\mu_i - \mu_k))}{\hat{\sigma}\sqrt{v_i^k}} < d\right\}$$

$$= P\{-(\hat{\mu}_i - \hat{\mu}_k - (\mu_i - \mu_k)) < d\hat{\sigma}\sqrt{v_i^k} \text{ for } i = 1, \ldots, k-1\}$$

$$= P\{\mu_i - \mu_k < \hat{\mu}_i - \hat{\mu}_k + d\hat{\sigma}\sqrt{v_i^k} \text{ for } i = 1, \ldots, k-1\}. \quad \square$$

Thus, in theory, one infers

$$\mu_i - \mu_k > \hat{\mu}_i - \hat{\mu}_k - d\hat{\sigma}\sqrt{v_i^k} \text{ for } i = 1, \ldots, k-1 \qquad (7.19)$$

or

$$\mu_i - \mu_k < \hat{\mu}_i - \hat{\mu}_k + d\hat{\sigma}\sqrt{v_i^k} \text{ for } i = 1, \ldots, k-1. \qquad (7.20)$$

Similarly, in theory, one can generalize Dunnett's two-sided MCC confidence intervals for one-way designs to the general linear model in a straightforward fashion. Using the following theorem, we obtain two-sided MCC confidence intervals centered at $\hat{\mu}_1 - \hat{\mu}_k, \ldots, \hat{\mu}_{k-1} - \hat{\mu}_k$, the best linear unbiased estimator of $\mu_1 - \mu_k, \ldots, \mu_{k-1} - \mu_k$, scaled by $\hat{\sigma}\sqrt{v_1^k}, \ldots, \hat{\sigma}\sqrt{v_{k-1}^k}$.

**Theorem 7.2.2** *Suppose the constant $|d|$ satisfies*

$$P\left\{\max_{1 \leq i \leq k-1} \frac{|\hat{\mu}_i - \hat{\mu}_k - (\mu_i - \mu_k)|}{\hat{\sigma}\sqrt{v_i^k}} < |d|\right\} = 1 - \alpha;$$

*then*

$$P\{\hat{\mu}_i - \hat{\mu}_k - |d|\hat{\sigma}\sqrt{v_i^k} < \mu_i - \mu_k < \hat{\mu}_i - \hat{\mu}_k + |d|\hat{\sigma}\sqrt{v_i^k}$$
$$\text{for } i = 1, \ldots, k-1\} = 1 - \alpha.$$

*Proof.*

$$1 - \alpha = P\left\{\max_{1 \leq i \leq k-1} \frac{|\hat{\mu}_i - \hat{\mu}_k - (\mu_i - \mu_k)|}{\hat{\sigma}\sqrt{v_i^k}} < |d|\right\}$$

$$= P\{-|d|\hat{\sigma}\sqrt{v_i^k} < \hat{\mu}_i - \hat{\mu}_k - (\mu_i - \mu_k) < +|d|\hat{\sigma}\sqrt{v_i^k}$$
$$\text{for } i = 1, \ldots, k-1\}$$

$$= P\{\hat{\mu}_i - \hat{\mu}_k - |d|\hat{\sigma}\sqrt{v_i^k} < \mu_i - \mu_k < \hat{\mu}_i - \hat{\mu}_k + |d|\hat{\sigma}\sqrt{v_i^k}$$
$$\text{for } i = 1, \ldots, k-1\}. \quad \square$$

Thus, in theory, one infers

$$\hat{\mu}_i - \hat{\mu}_k - |d|\hat{\sigma}\sqrt{v_i^k} < \mu_i - \mu_k < \hat{\mu}_i - \hat{\mu}_k + |d|\hat{\sigma}\sqrt{v_i^k}, \ i = 1, \ldots, k-1. \ (7.21)$$

However, to solve for $d$, the $1 - \alpha$ quantile of $\max_{1 \leq i \leq k-1} \{ \hat{\mu}_i - \hat{\mu}_k - (\mu_i - \mu_k)/\hat{\sigma} \sqrt{v_i^k} \}$ needed for one-sided MCC, the probability

$$P \left\{ \max_{1 \leq i \leq k-1} (\hat{\mu}_i - \hat{\mu}_k - (\mu_i - \mu_k)/\hat{\sigma} \sqrt{v_i^k}) < d \right\}$$

$$= \int_0^\infty \int_{-\infty}^{ds} \cdots \int_{-\infty}^{ds} \phi_{\mathbf{R}_{-k}}(z_1, \ldots, z_{k-1}) dz_1 \cdots dz_{k-1} \gamma(s) ds \quad (7.22)$$

has to be computed for candidates $d$. Here $\phi_{\mathbf{R}_{-k}}(z_1, \cdots, z_{k-1})$ is the multivariate normal density with means 0 and covariance matrix $\mathbf{R}_{-k} = \{\rho_{ij}^k\}$, the correlation matrix of $\hat{\mu}_i - \hat{\mu}_k, i = 1, \ldots, k - 1$, and $\gamma$ is the density of $\hat{\sigma}/\sigma$. Similarly, to solve for $|d|$, the $1 - \alpha$ quantile of $\max_{1 \leq i \leq k-1} \{ |\hat{\mu}_i - \hat{\mu}_k - (\mu_i - \mu_k)|/\hat{\sigma} \sqrt{v_i^k} \}$ needed for two-sided MCC, the probability

$$P \left\{ \max_{1 \leq i \leq k-1} (|\hat{\mu}_i - \hat{\mu}_k - (\mu_i - \mu_k)|/\hat{\sigma} \sqrt{v_i^k}) < |d| \right\}$$

$$= \int_0^\infty \int_{-|d|s}^{|d|s} \cdots \int_{-|d|s}^{|d|s} \phi_{\mathbf{R}_{-k}}(z_1, \ldots, z_{k-1}) dz_1 \cdots dz_{k-1} \gamma(s) ds \quad (7.23)$$

has to be computed for candidates $|d|$. If one integrates in turn over $s$ and $z_{k-1}, \ldots, z_1$, then the nesting of the integrations causes the computation to be too slow for interactive data analysis except for the smallest $k$ ($k \leq 3$ perhaps).

### 7.2.1 Deterministic approximation methods

#### 7.2.1.1 Methods based on probabilistic inequalities

For one-sided MCC, let

$$E_i = \{ (\hat{\mu}_i - \hat{\mu}_k - (\mu_i - \mu_k))/\hat{\sigma} \sqrt{v_i^k} > -d \} \quad (7.24)$$

or

$$E_i = \{ (\hat{\mu}_i - \hat{\mu}_k - (\mu_i - \mu_k))/\hat{\sigma} \sqrt{v_i^k} < d \}, \quad (7.25)$$

depending on whether upper confidence bounds or lower confidence bounds for $\mu_i - \mu_k$ are desired, respectively. The Bonferroni inequality (A.3) states

$$P(\bigcup_{i=1}^{k-1} E_i^c) \leq \sum_{i=1}^{k-1} P(E_i^c).$$

If each $E_i^c$ is such that

$$P(E_i^c) = \alpha/(k-1),$$

then

$$P(\bigcup_{i=1}^{k-1} E_i^c) \leq \alpha.$$

Thus a conservative approximation to $d$ is

$$d_{\text{Bonferroni}} = t_{\frac{\alpha}{k-1}, \nu}. \tag{7.26}$$

If the correlations among the treatment versus control estimators are all non-negative, that is, if

$$\rho_{ij}^k = Corr(\hat{\mu}_i - \hat{\mu}_k, \hat{\mu}_j - \hat{\mu}_k) \geq 0, \ 1 \leq i < j \leq k - 1, \tag{7.27}$$

then applying Slepian's inequality (Corollary A.3.1) to the one-sided MCC events in (7.24) or (7.25) gives

$$P(\bigcap_{i=1}^{k-1} E_i \mid \hat{\sigma}) \geq \prod_{i=1}^{k-1} P(E_i \mid \hat{\sigma}).$$

Now $P(E_i|\hat{\sigma}), i = 1, \ldots, k - 1$, are monotone in $\hat{\sigma}$ in the same direction, thus one can further apply Corollary A.1.1 to get

$$P(\bigcap_{i=1}^{k-1} E_i) \geq E_{\hat{\sigma}}[\prod_{i=1}^{k-1} P(E_i \mid \hat{\sigma})] \geq \prod_{i=1}^{k-1} E_{\hat{\sigma}}[P(E_i|\hat{\sigma})] = \prod_{i=1}^{k-1} P(E_i).$$

If each $E_i$ is such that

$$P(E_i) = (1 - \alpha)^{1/(k-1)},$$

then

$$P(\bigcap_{i=1}^{k-1} E_i) \geq 1 - \alpha.$$

So a conservative approximation $d_{\text{Slepian}}$ to $d$ results if one pretends $E_1, \ldots, E_{k-1}$ are independent:

$$d_{\text{Slepian}} = t_{1-(1-\alpha)^{1/(k-1)}, \nu}.$$

Note, however, in contrast to the situation of an unbalanced one-way model, (7.27) is not always satisfied in a GLM, as for example in an AN-COVA model with common slope

$$Y_{ih} = \theta_i + \beta X_{ih} + \epsilon_{ih}, \ i = 1, \ldots, k, \ h = 1, \ldots, n_i,$$

or a three-way no-interaction model

$$Y_{ihr} = \mu + \tau_i + \beta_h + \gamma_r + \epsilon_{ihr} \tag{7.28}$$

generated by row-column designs. The following example (communicated by Jane Cheng), concerns a row-column design with four treatments, three rows and three columns specified by

$$
\begin{array}{ccc}
1 & 4 & 3 \\
4 & 3 & 2 \\
1 & 2 & 4 \\
\end{array}
$$

(row 1 column 1 gets treatment 1, row 1 column 2 gets treatment 4, etc.). Then the variance-covariance matrix (cf. John 1987, p. 100) of $\sigma^{-1}(\hat{\tau}_1 - \hat{\tau}_4, \hat{\tau}_2 - \hat{\tau}_4, \hat{\tau}_3 - \hat{\tau}_4)$ under (7.28) is

$$
\begin{pmatrix}
2.503 & -0.501 & -0.501 \\
-0.501 & 1.301 & 0.701 \\
-0.501 & 0.701 & 1.301
\end{pmatrix}.
$$

Thus Slepian's inequality is not always applicable to one-sided MCC in the GLM.

For two-sided MCC, let

$$
E_i = \{|\hat{\mu}_i - \hat{\mu}_k - (\mu_i - \mu_k)|/\hat{\sigma}\sqrt{v_i^k} < |d|\}.
$$

The Bonferroni inequality (A.3) states

$$
P(\bigcup_{i=1}^{k-1} E_i^c) \le \sum_{i=1}^{k-1} P(E_i^c).
$$

If each $E_i^c$ is such that

$$
P(E_i^c) = \alpha/(k-1),
$$

then

$$
P(\bigcup_{i=1}^{k-1} E_i^c) \le \alpha.
$$

Thus, a conservative approximation to $|d|$ is

$$
|d|_{\text{Bonferroni}} = t_{\frac{\alpha}{2(k-1)}, \nu}. \tag{7.29}
$$

Šidák's inequality (Corollary A.4.1) states

$$
P(\bigcap_{i=1}^{k-1} E_i \mid \hat{\sigma}) \ge \prod_{i=1}^{k-1} P(E_i \mid \hat{\sigma}).
$$

Now $P(E_i|\hat{\sigma}), i = 1, \ldots, k-1$, are monotone in $\hat{\sigma}$ in the same direction, so one can further apply Corollary A.1.1 to give

$$
P(\bigcap_{i=1}^{k-1} E_i) \ge E_{\hat{\sigma}}[\prod_{i=1}^{k-1} P(E_i \mid \hat{\sigma})] \ge \prod_{i=1}^{k-1} E_{\hat{\sigma}}[P(E_i|\hat{\sigma})] = \prod_{i=1}^{k-1} P(E_i).
$$

If each $E_i$ is such that

$$
P(E_i) = (1-\alpha)^{1/(k-1)},
$$

then

$$
P(\bigcap_{i=1}^{k-1} E_i) \ge 1 - \alpha.
$$

Thus, a conservative approximation $|d|_{\text{Šidák}}$ to $|d|$ results if one pretends $E_1, \ldots, E_{k-1}$ are independent:

$$|d|_{\text{Šidák}} = t_{\frac{1-(1-\alpha)^{1/(k-1)}}{2}, \nu}.$$

The Hunter–Worsley inequality (A.6) states if $\{\{i, j\} \in \mathcal{T}\}$ are the edges in a spanning tree $\mathcal{T}$ of the nodes $\{1, 2, \ldots, k-1\}$, then

$$P(\bigcup_{i=1}^{k-1} E_i^c) \leq \sum_{i=1}^{k-1} P(E_i^c) - \sum_{\{i,j\} \in \mathcal{T}} P(E_i^c \cap E_j^c), \qquad (7.30)$$

which is always more accurate than the Bonferroni inequality, since the second term on the right-hand side is non-negative. Thus, conservative approximations to $d$ and $|d|$ can be obtained by finding the smallest $d = d_{HW}$ and $|d| = |d|_{HW}$ such that the right hand side of (7.30) equals $1 - \alpha$. Clearly, the best Hunter–Worsley approximations are obtained by finding the *optimal* spanning tree, the tree $\mathcal{T}$ which maximizes the second term on the right-hand side.

For one-sided MCC, since $P(E_i^c \cap E_j^c)$ is increasing in

$$\rho_{ij}^k = Corr(\hat{\mu}_i - \hat{\mu}_k, \hat{\mu}_j - \hat{\mu}_k)$$

by Slepian's inequality, the optimal spanning tree $\mathcal{T}$ can be found by applying any minimal spanning tree algorithm to the distance matrix $(\delta_{ij})$ defined by

$$\delta_{ij} = 1 - \rho_{ij}^k.$$

For two-sided MCC, since $P(E_i^c \cap E_j^c)$ is increasing in $|\rho_{ij}^k|$ (see Tong 1980, p. 26), the optimal spanning tree $\mathcal{T}$ can be found by applying any minimal spanning tree algorithm to the distance matrix $(\delta_{ij})$ defined by

$$\delta_{ij} = 1 - |\rho_{ij}^k|.$$

### 7.2.1.2 The factor analytic method

Notice that Slepian's and Šidák's inequalities approximate an intersection of events as an intersection of events associated with *independent* random variables. Therefore, one way of improving Slepian's and Šidák's approximations is to approximate dependent random variables as *conditionally* independent random variables, as follows.

Suppose there exist constants $\lambda_1, \ldots, \lambda_{k-1}$ such that

$$\rho_{ij}^k = \lambda_i \lambda_j$$

for all $1 \leq i < j < k$ or, equivalently,

$$
\mathbf{R}_{-k} = \begin{pmatrix} 1 - \lambda_1^2 & 0 & \cdots & & 0 \\ 0 & \ddots & \ddots & & \vdots \\ \vdots & & \ddots & \ddots & 0 \\ 0 & \cdots & & 0 & 1 - \lambda_{k-1}^2 \end{pmatrix} + \begin{pmatrix} \lambda_1 \\ \vdots \\ \lambda_{k-1} \end{pmatrix} \begin{pmatrix} \lambda_1 & \cdots & \lambda_{k-1} \end{pmatrix}.
$$
(7.31)

Then

$$
\sigma^{-1}((\hat{\mu}_i - \hat{\mu}_k - (\mu_i - \mu_k))/\sqrt{v_i^k}, \ i < k)'
$$
(7.32)

has the same distribution as

$$
\begin{pmatrix} \sqrt{1 - \lambda_1^2} Z_1 \\ \vdots \\ \sqrt{1 - \lambda_{k-1}^2} Z_{k-1} \end{pmatrix} + \begin{pmatrix} \lambda_1 \\ \vdots \\ \lambda_{k-1} \end{pmatrix} Z_0
$$

where $Z_1, \ldots, Z_{k-1}, Z_0$ are i.i.d. standard normal random variables. The meaning of the factorization (7.31) is that there exists a random variable $Z_0$ such that, conditional on $Z_0$, the random variables $\{\hat{\mu}_i - \hat{\mu}_k, \ i < k\}$ are independent. Therefore, by conditioning on $\hat{\sigma}$ and $Z_0$, equation (7.22) defining $d$ can be written as

$$
\int_0^\infty \int_{-\infty}^\infty \prod_{i=1}^{k-1} \Phi\left(\frac{\lambda_i z + ds}{\sqrt{1 - \lambda_i^2}}\right) d\Phi(z)\gamma(s)ds = 1 - \alpha
$$
(7.33)

and equation (7.23) defining $|d|$ can be written as

$$
\int_0^\infty \int_{-\infty}^\infty \prod_{i=1}^{k-1} \left[\Phi\left(\frac{\lambda_i z + |d|s}{\sqrt{1 - \lambda_i^2}}\right) - \Phi\left(\frac{\lambda_i z - |d|s}{\sqrt{1 - \lambda_i^2}}\right)\right] d\Phi(z)\gamma(s)ds = 1 - \alpha,
$$
(7.34)

where $\Phi$ is the standard normal distribution, and $\gamma$ again is the density of $\hat{\sigma}/\sigma$. The point is, with conditional independence, the $k$-dimensional integral equations (7.22) and (7.23) reduce to two-dimensional integral equations. Thus, with efficient numerical integration and root finding algorithms, $d$ and $|d|$ can be obtained quickly enough for interactive data analysis.

Models with the one-way structure (7.7) satisfy (7.31) with

$$
\lambda_i = (1 + a_i/a_k)^{-1/2}, \ i < k.
$$
(7.35)

However, experiments in the field often have covariates and/or missing observations, leading to $\mathbf{R}_{-k}$ that do not satisfy (7.31). (The claim in Bristol 1993 that they do is incorrect.)

One possible strategy, therefore, is to replace $\mathbf{R}_{-k}$ by an $\mathbf{R}_{-k}^{FA}$ which satisfies (7.31). For example, $|d|_{\text{Šidák}}$ and $d_{\text{Slepian}}$ are obtained by replacing $\mathbf{R}_{-k}$ by the identity matrix, which trivially satisfies (7.31). One would expect, however, if a better approximation to $\mathbf{R}_{-k}$ than the identity matrix is found, then a better approximations to $|d|$ or $d$ would result.

Iyengar (1988) proposed to replace all the correlations by the average of the correlations, but the equal correlation structure is computationally unnecessarily restrictive. Royen (1987) and Hsu (1992) independently proposed to approximate $\mathbf{R}_{-k}$ by the 'closest' $\mathbf{R}_{-k}^{FA}$ satisfying (7.31). If no constraint is placed upon $\mathbf{R}_{-k}^{FA}$ other than (7.31), then Hsu (1992) pointed out that existing factor analytic algorithms can be utilized to find the $\mathbf{R}_{-k}^{FA}$ closest to $\mathbf{R}_{-k}$. For example, the iterated principal factor method in factor analysis would find the $\mathbf{R}_{-k}^{FA}$ that minimizes $\| \mathbf{R}_{-k} - \mathbf{R}_{-k}^{FA} \|$, where $\| \cdot \|$ is the Euclidean norm defined by

$$\| (a_{ij}) \| = \sum_{i<j} a_{ij}^2.$$

The generalized least squares method would find the $\mathbf{R}_{-k}^{FA}$ that minimizes $\| \mathbf{R}_{-k}\mathbf{R}_{-k}^{FA^{-1}} - \mathbf{I} \|$. In essence, a factor analytic algorithm (which extracts one 'factor') finds the auxiliary random variable $Z_0$ which has correlations $\lambda_1, \ldots, \lambda_{k-1}$ with $\hat{\mu}_1 - \hat{\mu}_k, \ldots, \hat{\mu}_{k-1} - \hat{\mu}_k$ such that, conditional on $Z_0$, $\hat{\mu}_1 - \hat{\mu}_k, \ldots, \hat{\mu}_{k-1} - \hat{\mu}_k$ are 'as independent as possible.' Let us denote generically by $d_{FA}$ and $|d|_{FA}$ the approximations to $d$ and $|d|$ obtained by replacing $\mathbf{R}_{-k}$ with an $\mathbf{R}_{-k}^{FA}$

Note that if $\mathbf{R}_{-k}$ in fact satisfies (7.31), then the factor analytic algorithm will recover it, that is,

$$\mathbf{R}_{-k}^{FA} = \mathbf{R}_{-k}.$$

Thus, in contrast to methods based on probabilistic inequalities, the factor analytic method will automatically produce exact simultaneous MCC confidence intervals whenever possible. On the other hand, the factor analytic approximation is not known to be conservative in general.

Hsu (1992) presented some evidence that $|d|_{FA}$ outperforms the Hunter–Worsley approximation $|d|_{HW}$ and can be very close to $|d|$. In one simulation, 100 two-way designs (7.11) with $k = b = 10$ and disproportional cell frequencies were generated by starting with $n_{1,1} = \cdots = n_{10,10} = 10$ and then giving each of the original observations independently a 0.2 probability of being missing. A variance reduction technique was used to estimate the bias in the coverage probability caused by using $|d|_{FA}$ instead of $|d|$, defined as

$$bias = P_{\mathbf{R}_{-k}}\{\mu_i - \mu_k \in \hat{\mu}_i - \hat{\mu}_k \pm |d|_{FA}\hat{\sigma}\sqrt{v_i^k}, i = 1, \ldots, k-1\} - (1-\alpha),$$
$$(7.36)$$

for each of the 100 designs. Boxplots of $\widehat{bias}$ are given in Figure 7.1. The

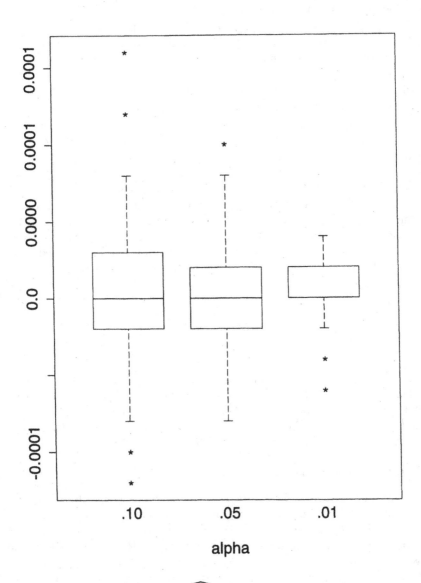

Figure 7.1 *Boxplots of* $\widehat{bias}$ *in two-way design simulation*

simulation also kept track of the upper bounds on $\alpha$ given by the Hunter–Worsley inequality when one sets $|d| = |d|_{FA}$. Figure 7.2 displays boxplots of the estimated true $\alpha$ and the upper bounds. It appears that the Hunter–Worsley bounds are rather conservative relative to the factor analytic approximations. In another simulation, 100 one-way analysis of covariance models (7.3) with $k = 10$, $n_1 = \cdots = n_{10} = 10$, and $X_1, \ldots, X_{10}$ i.i.d. standard normal random variables were randomly generated. Again, bias in the coverage probability caused by using $|d|_{FA}$ instead of $|d|$ was estimated with variance reduction. Boxplots of $\widehat{bias}$ are given in Figure 7.3. That simulation also kept track of the upper bounds on $\alpha$ given by the Hunter–Worsley inequality when one sets $|d| = |d|_{FA}$. Figure 7.4 displays boxplots of the estimated true $\alpha$ and the upper bounds, showing the Hunter–Worsley bounds are rather conservative relative to the factor analytic approximation.

However, the factor analytic approach fails when, even though the distribution of $\hat{\mu}_i - \hat{\mu}_k, 1 \leq i \leq k - 1$, is non-degenerate, an unfavorable structure of the correlation matrix $\mathbf{R}_{-k}$ causes the factor analysis algorithms to return an $\mathbf{R}_{-k}^{FA}$ which is degenerate. For example, the generalized least squares algorithm returns $(\lambda_1, \ldots, \lambda_5) = (0, 0.5, 1, 0.5, 0)$ for the correlation matrix

$$
\begin{pmatrix}
1 & 0.5 & 0 & 0 & 0 \\
0.5 & 1 & 0.5 & 0 & 0 \\
0 & 0.5 & 1 & 0.5 & 0 \\
0 & 0 & 0.5 & 1 & 0.5 \\
0 & 0 & 0 & 0.5 & 1
\end{pmatrix}.
$$

In the factor analytic approximation, all $\lambda_i$'s must be strictly less than 1.

The ADJUST = DUNNETT option, together with the CL, PDIFF = CONTROLL, PDIFF = CONTROLU and PDIFF = CONTROL options under the LSMEANS statement in PROC GLM of the SAS system, implement the factor analytic approach to one- and two-sided MCC inference in the GLM.

### 7.2.1.3 The linear programming method

For one-sided MCC, if all $\rho_{ij}^k$ are positive, then one can utilize Slepian's inequality and linear programming algorithms to approximate $\mathbf{R}_{-k}$ by an $\mathbf{R}_{-k}^{LP}$ which is guaranteed to lead to a conservative $d$.

An example of a linear model for which all $\rho_{ij}^k$ in $\mathbf{R}_{-k}$ are positive is the two-way no-interaction model (7.11). Using the notation in Section 7.1.2.1, the inverse of the variance-covariance matrix of $\sigma^{-1}(\hat{\tau}_1 - \hat{\tau}_k, \ldots, \hat{\tau}_{k-1} - \hat{\tau}_k)$ is

$$
\left[ \mathrm{diag}(n_{*+}) - \mathbf{N}[\mathrm{diag}(n_{+*})]^{-1}\mathbf{N}' \right]_{k-}
$$

(cf. Searle 1971, p. 267), where $\mathbf{M}_{k-}$ denotes the matrix $\mathbf{M}$ with its $k$th row and column deleted. Since the off-diagonal elements of this matrix are all negative, it is a so-called *M-matrix*. Consequently, all covariances of $\hat{\tau}_1 - \hat{\tau}_k, \ldots, \hat{\tau}_{k-1} - \hat{\tau}_k$ are positive (cf. Graybill 1983, Section 11.4). (In

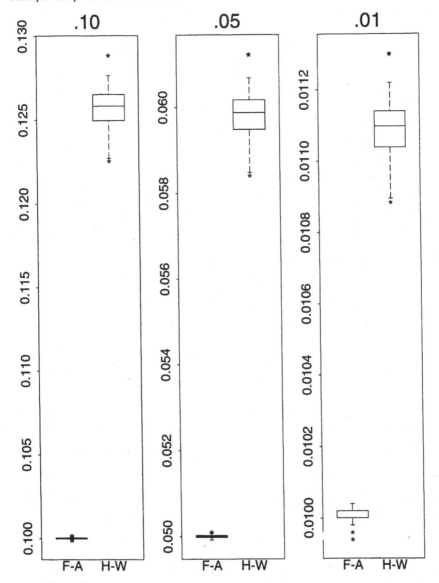

Figure 7.2 *Boxplots of estimated true α versus upper bounds in two-way design simulation*

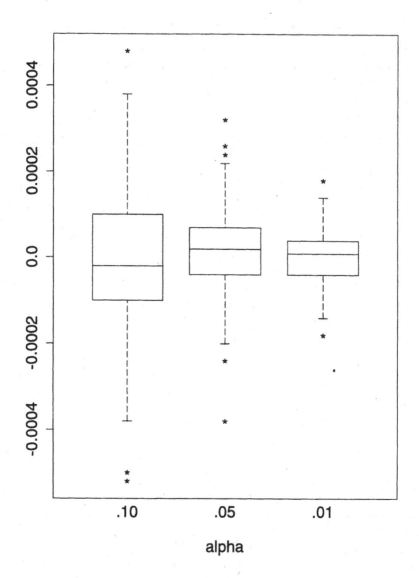

Figure 7.3 *Boxplots of* $\widehat{bias}$ *in ANCOVA simulation*

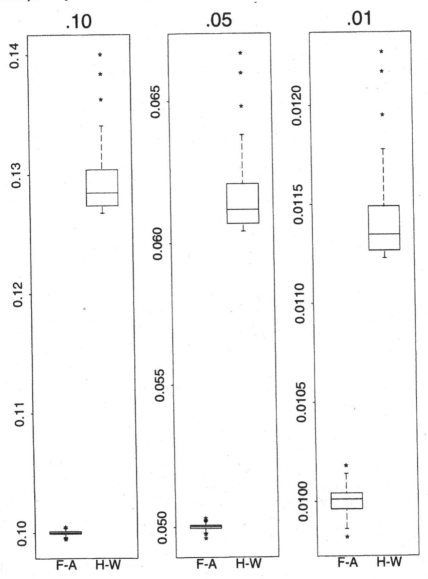

Figure 7.4 *Boxplots of estimated true α versus upper bounds in ANCOVA simulation*

fact, this shows the joint distribution of $\hat{\tau}_1 - \hat{\tau}_k, \ldots, \hat{\tau}_{k-1} - \hat{\tau}_k$ has the stronger property of multivariate total positivity of order 2 ($MTP_2$); see Karlin and Rinott 1980.)

The linear programming technique of Hsu and Nelson (1998) approximates $\mathbf{R}_{-k}$ by a correlation matrix

$$\mathbf{R}_{-k}^{LP} = \begin{pmatrix} 1 & \lambda_1\lambda_2 & \cdots & \lambda_1\lambda_{k-1} \\ & 1 & \cdots & \lambda_2\lambda_{k-1} \\ & & \ddots & \vdots \\ & & & 1 \end{pmatrix}$$

where

$$\lambda_i\lambda_j \leq \rho_{ij}^k, \forall j > i \tag{7.37}$$

and

$$0 < \lambda_i \leq 1 \text{ for } i = 1, \ldots, k-1, \tag{7.38}$$

while keeping $\lambda_i\lambda_j$ close to $\rho_{ij}^k$. One measure of closeness is the ratio $\rho_{ij}^k/\lambda_i\lambda_j$, and one overall objective is to

$$\text{minimize} \prod_{j>i} \frac{\rho_{ij}^k}{\lambda_i\lambda_j}. \tag{7.39}$$

The objective function (7.39) combined with the constraints (7.37)–(7.38) form a nonlinear program. By taking logarithms of (7.37)–(7.39) and letting $x_i = -\log \lambda_i$ we obtain the linear program

$$\text{minimize} \sum_{j>i} (x_i + x_j) \quad + \quad \text{constant} \tag{7.40}$$

$$x_i + x_j \geq -\log \rho_{ij}^k, \forall j > i$$
$$x_i \geq 0, \forall i.$$

This linear program has $k-1$ variables and $(k-1)(k-2)/2$ constraints, excluding the non-negativity constraints.

Formulation (7.40) attempts to minimize the average ratio $\rho_{ij}^k/(\lambda_i\lambda_j)$. Another reasonable objective is to minimize the largest ratio. To formulate this objective we add the constraints

$$\frac{\rho_{ij}^k}{\lambda_i\lambda_j} \leq w, \forall j > i \tag{7.41}$$

and replace the objective function (7.39) with

$$\text{minimize } w. \tag{7.42}$$

After the log transformation and the substitution $x = \log w$ we obtain the linear program

$$\text{minimize } x \tag{7.43}$$

$$x_i + x_j \geq -\log \rho_{ij}^k, \forall j > i$$

$$sx - x_i - x_j \geq \log \rho_{ij}^k, \forall j > i$$
$$x_i \geq 0, \forall i$$
$$x \geq 0.$$

This linear program has $k$ variables and $(k-1)(k-2)$ constraints, excluding the non-negativity constraints.

When $\rho_{ij}^k > 0, \forall i \neq j$, both linear programs are guaranteed to have a feasible solution (that is, a one-factor approximation exists).

## 7.2.2 Stochastic approximation methods

We discuss two approaches to approximating a quantile by simulation. The first is based on confidence intervals for the true error rate, while the second is based on confidence intervals for the true quantile.

For notational simplicity, we shall discuss the case of approximating $d$; the case of $|d|$ is analogous.

Let

$$D = \max_{1 \leq i \leq k-1} \hat{\sigma}^{-1}(\hat{\mu}_i - \hat{\mu}_k - (\mu_i - \mu_k))/\sqrt{v_i^k}$$

and let

$$D_{[1]} < \cdots < D_{[m]}$$

be the order statistics of a simulated random sample $D_1, \ldots, D_m$ of size $m$ of the random variable $D$.

### 7.2.2.1 The method of Edwards and Berry

The approach of Edwards and Berry (1987) estimates $d$ by $D_{[r]}$, assuming $r = (1 - \alpha)(m + 1)$ is an integer.

**Lemma 7.2.1** *Let $F$ denote the cumulative distribution function of $D$. Then*

$$E[F(D_{[r]})] = 1 - \alpha,$$
$$Var[F(D_{[r]})] = \frac{\alpha(1 - \alpha)}{m + 2}.$$

*Proof.* Since $F(D_{[r]})$ can be regarded as the $r$th order statistic of a random sample of size $m$ from a Uniform$(0, 1)$ distribution, it has the Beta$(r, m - r + 1)$ distribution. $\square$

Thus, if one uses the simulated $D_{[r]}$ as the critical value for MCC inference, then the expectation (long-run average) of the actual coverage probability is $1 - \alpha$. Further, for any desired $\gamma$ ($> 0$) and $\epsilon$ ($> 0$), one can use the expression for $Var[F(D_{[r]})]$ to set the simulation run size $m$ so that the true coverage probability of simulatation-based confidence intervals will be within $\gamma$ of $1 - \alpha$ in approximately $100(1 - \epsilon)\%$ of repeated generations of simulation-based critical values. The `ADJUST = SIMULATE` option, together with the `CL`, `PDIFF = CONTROLL`, `PDIFF = CONTROLU` and `PDIFF = CONTROL` options of the `LSMEANS` statement in `PROC GLM` of the SAS system, implement the Edwards–Berry approach to one- and two-sided MCC inference in the GLM, setting $\gamma = 0.005$ and $\epsilon = 0.01$ by default.

### 7.2.2.2 A variance reduction method

A different approach to approximating a quantile stochastically is to obtain a level $1 - \gamma$ upper confidence bound for the desired quantile via simulation.

A crude level $1 - \gamma$ upper confidence bound for $d$ is $D_{(m^c)}$, where $m^c$ is the smallest integer such that the Binomial$(m, \alpha)$ probability of at most $m^c - 1$ successes is at least $1 - \gamma$. This upper confidence bound is obtained by inverting the usual level-$\gamma$ test for

$$H_0 : P\{D > d_0\} = \alpha. \tag{7.44}$$

Hsu and Nelson (1994) proposed a method for obtaining a sharper upper confidence bound, by using the variance reduction technique of *control variates*. Specifically, they propose to generate a control variate $D^{cv}$ with known $(1 - \alpha)$th quantile $d_{cv}$ along with each $D$, and inverting McNemar's conditional test (cf. Lehmann 1986, p. 169) for

$$H_0 : P\{D > d_0\} = P\{D^{cv} > d_{cv}\} \ (= \alpha) \tag{7.45}$$

to get a conditional upper confidence bound for $d$. After deriving the conditional upper confidence bound, they provide a practical way of computing it.

If one can generate $D^{cv}$ to be highly correlated with $D$, then the test for (7.45) will be substantially more powerful than the test for (7.44) and the corresponding confidence bound will be more accurate. To generate $D^{cv}$ with a known quantile, recall that if $\mathbf{R}_{-k}$ has the one-factor structure (7.31), then $d$ and $|d|$ can be computed exactly. As Hsu (1992) has demonstrated that an excellent one-factor approximation to $\mathbf{R}_{-k}$ can often be found, Hsu and Nelson (1994) proposed to generate $D^{cv}$ from the same multivariate $t$ distribution as $D$, but substituting $\mathbf{R}_{-k}$ by the $R_{-k}^{FA}$ with one-factor structure that is close to it. In their study, Hsu and Nelson (1994) found the control variate approach takes approximately 50% more CPU time than crude simulation, but reduces the variance of the quantile estimate by a factor anywhere from 10 to more than 250.

### 7.2.2.3 Example: blood pressure in dogs

Consider the measurements of increases in systolic blood pressure in dogs with three experimentally induced diseases each given one of four drugs, as shown in Table 7.5 (in which a . indicates a missing value). Even though the original data in Afifi and Azen (1972) was balanced, we present it here

in an unbalanced form which has often been used for illustration purposes
(Fleiss 1986, p. 166; SAS 1989, p. 972).

Table 7.5 *Increases in systolic pressure of dogs*

| Disease | Treatment | | | |
|---|---|---|---|---|
| | 1 | 2 | 3 | 4 |
| 1 | 42 | 28 | . | 24 |
| | 44 | . | . | . |
| | 36 | 23 | 1 | 9 |
| | 13 | 34 | 29 | 22 |
| | 19 | 42 | . | -2 |
| | 22 | 13 | 19 | 15 |
| | 33 | . | . | 27 |
| 2 | . | 34 | 11 | 12 |
| | 26 | 33 | 9 | 12 |
| | . | 31 | 7 | -5 |
| | 33 | . | 1 | 16 |
| | 21 | 36 | -6 | 15 |
| 3 | 31 | 3 | 21 | 22 |
| | -3 | 26 | 1 | 7 |
| | . | 28 | . | 25 |
| | 25 | 32 | 9 | 5 |
| | 25 | 4 | 3 | 12 |
| | 24 | 16 | . | . |

Suppose $\tau_1 - \tau_4, \ldots, \tau_3 - \tau_4$ are of interest. The least squares estimates
are

$$\hat{\tau}_1 - \hat{\tau}_4 = 12.469$$
$$\hat{\tau}_2 - \hat{\tau}_4 = 12.365 \qquad (7.46)$$
$$\hat{\tau}_3 - \hat{\tau}_4 = -4.527,$$

with $\hat{\sigma}^2 = 111.309$ and $\nu = 52$. The covariance matrix of $\sigma^{-1}(\hat{\tau}_1 - \hat{\tau}_4, \ldots, \hat{\tau}_3 - \hat{\tau}_4)$ is

$$\begin{pmatrix} 0.1302 & - & - \\ 0.0633 & 0.1302 & - \\ 0.0620 & 0.0623 & 0.1462 \end{pmatrix},$$

so the correlation matrix $\mathbf{R}_{-4}$ is

$$\begin{pmatrix} 1 & - & - \\ 0.4863 & 1 & - \\ 0.4493 & 0.4515 & 1 \end{pmatrix},$$

which has the one-factor structure (7.31) with $\lambda_1 = 0.6957, \lambda_2 = 0.6990,$

$\lambda_3 = 0.6458$. Therefore, in this case, the factor analytic and LP methods produce exact MCC inferences. Table 7.6 and Table 7.7 compare the critical values $|d|$ and $d$ given by the various deterministic methods.

Table 7.6  *Critical values* $|d|$ *for two-sided MCC*

| Method | $\alpha$ | | |
|---|---|---|---|
| | 0.10 | 0.05 | 0.01 |
| Bonferroni | 2.186 | 2.474 | 3.077 |
| Šidák | 2.171 | 2.467 | 3.076 |
| Hunter–Worsley | 2.137 | 2.440 | 3.060 |
| Exact (factor analytic) | 2.119 | 2.427 | 3.054 |

Table 7.7  *Critical values* $d$ *for one-sided MCC*

| Method | $\alpha$ | | |
|---|---|---|---|
| | 0.10 | 0.05 | 0.01 |
| Bonferroni | 1.873 | 2.186 | 2.826 |
| Slepian | 1.857 | 2.179 | 2.825 |
| Hunter–Worsley | 1.800 | 2.137 | 2.804 |
| Exact (factor analytic or LP) | 1.774 | 2.119 | 2.795 |

We remark that the MEANS statement in PROC GLM of the SAS system uses contrasts of the marginal sample means to estimate treatment contrasts. For example, the confidence intervals computed using

```
PROC GLM;
  CLASS DRUG DISEASE;
  MODEL PRESSURE = DRUG DISEASE;
  MEANS DRUG/DUNNETT('4');
```

will be centered at

$$\bar{y}_1 - \bar{y}_4 = 12.567$$
$$\bar{y}_2 - \bar{y}_4 = 12.033$$
$$\bar{y}_3 - \bar{y}_4 = -4.750,$$

which are not the best linear unbiased estimates (7.46) for $\tau_1 - \tau_4, \ldots, \tau_3 - \tau_4$. Unless the design is orthogonal, multiple comparisons should be executed using the options available under the LSMEANS statement in Release 6.10.

## 7.3 Multiple comparisons with the best

Let us first suppose a larger treatment effect implies a better treatment. A natural generalization of constrained multiple comparisons with the best for the one-way model to the GLM is first to define the critical values $d^i$, $i = 1, \ldots, k$, to be the constants satisfying the equations

$$P\{\hat{\mu}_i - \hat{\mu}_j - (\mu_i - \mu_j) > -d^i \hat{\sigma} \sqrt{v_j^i} \text{ for all } j, j \neq i\} = 1 - \alpha, \qquad (7.47)$$

then define

$$D_i^+ = +(\min_{j \neq i}\{\hat{\mu}_i - \hat{\mu}_j + d^i \hat{\sigma} \sqrt{v_j^i}\})^+,$$

$$G = \{i : D_i^+ > 0\},$$

$$D_i^- = \begin{cases} 0 \text{ if } G = \{i\} \\ \min_{j \in G, j \neq i}\{\hat{\mu}_i - \hat{\mu}_j - d^j \hat{\sigma} \sqrt{v_j^i}\} \text{ otherwise;} \end{cases}$$

and finally infer

$$\mu_i - \max_{j \neq i} \mu_j \in [D_i^-, D_i^+] \text{ for } i = 1, \ldots, k.$$

**Theorem 7.3.1 (Chang and Hsu 1992)** *For all $\beta$ and $\sigma^2$,*

$$P_{\beta,\sigma^2}\{\mu_i - \max_{j \neq i} \mu_j \in [D_i^-, D_i^+] \text{ for } i = 1, \ldots, k\} \geq 1 - \alpha.$$

*Proof.* Let $(k)$ denote the unknown index such that

$$\mu_{(k)} = \max_{1 \leq i \leq k} \mu_i.$$

Ties among $\mu_i$ can be broken in any fashion without affecting the validity of the proof.

Define the event $E$ as follows:

$$E = \{\hat{\mu}_{(k)} - \mu_{(k)} > \hat{\mu}_i - \mu_i - d^{(k)} \hat{\sigma} \sqrt{v_i^{(k)}} \text{ for all } i, i \neq (k)\}.$$

By the definition of the critical value $d^{(k)}$, $P\{E\} = 1 - \alpha$.

*Derivation of upper confidence bounds*

$$\begin{aligned} E &= \{\hat{\mu}_{(k)} - \mu_{(k)} > \hat{\mu}_i - \mu_i - d^{(k)} \hat{\sigma} \sqrt{v_i^{(k)}} \text{ for all } i, i \neq (k)\} \\ &= \{\mu_{(k)} - \mu_i < \hat{\mu}_{(k)} - \hat{\mu}_i + d^{(k)} \hat{\sigma} \sqrt{v_i^{(k)}} \text{ for all } i, i \neq (k)\} \\ &\subseteq \{\mu_{(k)} - \max_{j \neq (k)} \mu_j < \hat{\mu}_{(k)} - \hat{\mu}_i + d^{(k)} \hat{\sigma} \sqrt{v_i^{(k)}} \text{ for all } i, i \neq (k)\} \\ &= \{\mu_{(k)} - \max_{j \neq (k)} \mu_j < \min_{j \neq (k)} (\hat{\mu}_{(k)} - \hat{\mu}_j + d^{(k)} \hat{\sigma} \sqrt{v_j^{(k)}})\} \\ &= \{\mu_{(k)} - \max_{j \neq (k)} \mu_j < \min_{j \neq (k)} (\hat{\mu}_{(k)} - \hat{\mu}_j + d^{(k)} \hat{\sigma} \sqrt{v_j^{(k)}}) \text{ and} \end{aligned}$$

$$\mu_i - \max_{j \neq i} \mu_j \leq 0 \text{ for all } i, i \neq (k)\}$$

$$\subseteq \ \{\mu_i - \max_{j \neq i} \mu_j \leq (\max_{j \neq i}\{\hat{\mu}_i - \hat{\mu}_j + d^i \hat{\sigma}\sqrt{v_j^i}\})^+ \text{ for all } i\}$$

$$= \ E_1.$$

Note also that $E \subseteq \{(k) \in G\}$ because $\mu_{(k)} - \max_{j \neq (k)} \mu_i \geq 0$.

*Derivation of lower confidence bounds*

$$E \ = \ \{(k) \in G \text{ and } \hat{\mu}_{(k)} - \mu_{(k)} > \hat{\mu}_i - \mu_i - d^{(k)}\hat{\sigma}\sqrt{v_i^{(k)}} \text{ for all } i, i \neq (k)\}$$

$$= \ \{(k) \in G \text{ and } \mu_i - \mu_{(k)} > \hat{\mu}_i - \hat{\mu}_{(k)} - d^{(k)}\hat{\sigma}\sqrt{v_i^{(k)}} \text{ for all } i, i \neq (k)\}$$

$$= \ \{(k) \in G \text{ and } \mu_i - \max_{j \neq i} \mu_j > \hat{\mu}_i - \hat{\mu}_{(k)} - d^{(k)}\hat{\sigma}\sqrt{v_i^{(k)}} \text{ for all } i,$$
$$i \neq (k)\}$$

$$\subseteq \ \{\mu_i - \max_{j \neq i} \mu_j > \min_{j \in G, j \neq i} (\hat{\mu}_i - \hat{\mu}_j - d^j \hat{\sigma}\sqrt{v_j^i}) \text{ for all } i,$$
$$i \neq (k) \text{ and } (k) \in G\}$$

$$= \ \{\mu_i - \max_{j \neq i} \mu_j > \min_{j \in G, j \neq i} (\hat{\mu}_i - \hat{\mu}_j - d^j \hat{\sigma}\sqrt{v_j^i}) \text{ for all } i, i \neq (k) \text{ and}$$
$$(k) \in G \text{ and } \mu_i - \max_{j \neq i} \mu_j \geq 0 \text{ for } i = (k)\}$$

$$\subseteq \ \{\mu_i - \max_{j \neq i} \mu_j \geq D_i^- \text{ for all } i\}$$

$$= \ E_2.$$

We have shown $E \subseteq E_1 \cap E_2$. Therefore,

$$1 - \alpha \ = \ P\{E\}$$
$$\leq \ P_{\boldsymbol{\beta}, \sigma^2}\{E_1 \cap E_2\}$$
$$= \ P_{\boldsymbol{\beta}, \sigma^2}\{D_i^- \leq \mu_i - \max_{j \neq i} \mu_j \leq D_i^+ \text{ for all } i\}. \ \square$$

Now suppose either a smaller treatment effect implies a better treatment, or a larger treatment effect is better but multiple comparisons with the *worst* treatment are of interest (an example of which was given in Section 7.1.2.3); then the parameters of interest become $\mu_i - \min_{j \neq i} \mu_j$, $i = 1, \ldots, k$. Using the same definitions of $d^i$ as in (7.47), but redefining

$$D_i^- = -(\max_{j \neq i}\{\hat{\mu}_i - \hat{\mu}_j - d^i \hat{\sigma}\sqrt{v_j^i}\})^-,$$

$$G = \{i : D_i^- < 0\},$$

$$D_i^+ = \begin{cases} 0 \text{ if } G = \{i\} \\ \max_{j \in G, j \neq i}\{\hat{\mu}_i - \hat{\mu}_j + d^j \hat{\sigma}\sqrt{v_j^i}\} \text{ otherwise,} \end{cases}$$

one can analogously prove that, for all $\boldsymbol{\beta}$ and $\sigma^2$,

$$P_{\boldsymbol{\beta},\sigma^2}\{\mu_i - \min_{j \neq i} \mu_j \in [D_i^-, D_i^+] \text{ for } i = 1, \ldots, k\} \geq 1 - \alpha.$$

Notice that the MCB critical value $d^i$ is the same as the one-sided MCC critical value with the $i$th treatment as the control. Therefore, the methods for computing MCC critical values discussed in the previous section apply to the computation of MCB critical values without change.

## 7.4 All-pairwise comparisons

A natural generalization of Tukey's simultaneous confidence intervals inference for the balanced one-way model to the GLM is

$$\mu_i - \mu_j \in \hat{\mu}_i - \hat{\mu}_j \pm |q^e|\hat{\sigma}\sqrt{v_j^i} \text{ for all } i \neq j, \qquad (7.48)$$

where $|q^e|$ is the critical value such that the probability of (7.48) is exactly equal to $1 - \alpha$. Recall, however, from Chapter 5 that even for the one-way model, if the sample sizes are unequal, the exact critical value $|q^e|$ for Tukey's method is difficult to compute, and the conservative Tukey–Kramer approximation is recommended instead.

One may thus consider a natural generalization of the Tukey–Kramer simultaneous confidence intervals inference from the one-way model to the GLM:

$$\mu_i - \mu_j \in \hat{\mu}_i - \hat{\mu}_j \pm |q^*|\hat{\sigma}\sqrt{v_j^i} \text{ for all } i \neq j, \qquad (7.49)$$

where $|q^*|$ is the solution of

$$k \int_0^\infty \int_{-\infty}^{+\infty} [\Phi(z + \sqrt{2}|q^*|s) - \Phi(z)]^{k-1} d\Phi(z) \, \gamma(s)ds = 1 - \alpha,$$

with $\gamma$ again being the density of $\hat{\sigma}/\sigma = \sqrt{MSE/\sigma^2}$. Recall Hayter (1984) showed that, for the one-way model, the probability of (7.49) is at least $1 - \alpha$. The question is whether the probability of (7.49) is still at least $1 - \alpha$ in a GLM (of which the one-way model is a special case).

Brown (1979) showed that the answer is 'yes' for $k = 3$. However, the geometric nature of his proof makes it not easy to extend to $k > 3$.

Hayter's (1984) analytic proof of the conservatism of the Tukey–Kramer approximation was cast in the framework of a one-way model. For his proof to be applicable, as explained in Hayter (1989), the model must have the one-way structure (7.6), which many general linear models do not possess. Nevertheless, extensive numerical computations reported in Hayter (1989) strongly indicate that the Tukey–Kramer approximation is conservative when $k = 4$ for any general linear model.

If you worry about the fact that there is no mathematical proof that the extension of the Tukey–Kramer approximation to the GLM is conservative, then you can use an approximation based on the Bonferroni inequality, Šidák's inequality, or the Hunter–Worsley inequality. Šidák's inequality will be marginally more accurate (less conservative) than the Bonferroni inequality, while the Hunter–Worsley inequality will in turn generally be more accurate than Šidák's inequality.

### 7.4.1 Methods based on probabilistic inequalities

For two-sided all-pairwise comparisons, let

$$E_{ij} = \{|\hat{\mu}_i - \hat{\mu}_j - (\mu_i - \mu_j)|/\hat{\sigma}\sqrt{v_j^i} < |q^e|\}, \tag{7.50}$$

while for Hayter's one-sided all-pairwise comparisons let

$$E_{ij} = \{((\hat{\mu}_i - \hat{\mu}_j - (\mu_i - \mu_j))/\hat{\sigma}\sqrt{v_j^i} < q^e\}. \tag{7.51}$$

The Bonferroni inequality (A.3) states

$$P(\bigcup_{i<j} E_{ij}^c) \leq \sum_{i<j} P(E_{ij}^c).$$

If each $E_{ij}^c$ is such that

$$P(E_{ij}^c) = \alpha/[k(k-1)/2],$$

then

$$P(\bigcup_{i<j} E_{ij}^c) \leq \alpha.$$

Thus, a conservative approximation to $|q^e|$ is

$$|q^*|_{\text{Bonferroni}} = t_{\frac{\alpha}{k(k-1)},\nu}, \tag{7.52}$$

while a conservative approximation to $q^e$ is

$$q^*_{\text{Bonferroni}} = t_{\frac{2\alpha}{k(k-1)},\nu}. \tag{7.53}$$

The ADJUST = BON option, with the CL and the default PDIFF = ALL options under the LSMEANS statement in PROC GLM of the SAS system, implements this approximate form of two-sided MCA in the GLM.

Šidák's inequality Theorem A.4.1 states

$$P(\bigcap_{i<j} E_{ij} \mid \hat{\sigma}) \geq \prod_{i<j} P(E_{ij} \mid \hat{\sigma}).$$

Thus, a conservative approximation to $|q^e|$, proposed by Hochberg (1974b), is

$$|q^*|_{\text{Hochberg}} = |m|_{\alpha,k(k-1)/2,\nu}.$$

The ADJUST = SMM and ADJUST = GT2 options, with the CL and the default PDIFF = ALL options under the LSMEANS statement in PROC GLM of the SAS system, implement this approximate form of two-sided MCA in the GLM.

Now $P(E_{ij}|\hat{\sigma}), i < j$, are monotone in $\hat{\sigma}$ in the same direction, so one can further apply Corollary A.1.1 to get

$$P(\bigcap_{i<j} E_{ij}) \geq E_{\hat{\sigma}}[\prod_{i<j} P(E_{ij} \mid \hat{\sigma})] \geq \prod_{i<j} E_{\hat{\sigma}} P(E_{ij}|\hat{\sigma}) = \prod_{i<j} P(E_{ij}).$$

If each $E_{ij}$ is such that

$$P(E_{ij}) = (1-\alpha)^{1/[k(k-1)/2]},$$

then

$$P(\bigcap_{i<j} E_{ij}) \geq 1 - \alpha.$$

Thus, a conservative approximation to $|q^e|$ is

$$|q^*|_{\text{Šidák}} = t_{\frac{1-(1-\alpha)^{2/[k(k-1)]}}{2},\nu},$$

and a conservative approximation to $q^e$ is

$$q^*_{\text{Slepian}} = t_{1-(1-\alpha)^{2/[k(k-1)]},\nu}.$$

The ADJUST = SIDAK option, with the CL and the default PDIFF = ALL options under the LSMEANS statement in PROC GLM of the SAS system, implements this approximate form of two-sided MCA in the GLM.

The Hunter–Worsley inequality (A.6) states if $\{\{(i,j),(i^*,j^*)\} \in \mathcal{T}\}$ represent the edges in a spanning tree $\mathcal{T}$ of the nodes $\{(i,j), i < j\}$, then

$$P(\bigcup_{i<j} E^c_{ij}) \leq \sum_{i<j} P(E^c_{ij}) - \sum_{\{(i,j),(i^*,j^*)\}\in\mathcal{T}} P(E^c_{ij} \cap E^c_{i^*j^*}), \qquad (7.54)$$

which is always more accurate than the Bonferroni inequality, since the second term on the right-hand side is non-negative. Thus, conservative approximations to $|q^e|$ and $q^e$ can be obtained by finding the smallest $|q^e| = |q|_{HW}$ and $q^e = q_{HW}$ such that the right hand side of (7.54) equals $1 - \alpha$. Again, the best Hunter–Worsley approximations are obtained by finding the *optimal* spanning tree, the tree $\mathcal{T}$ which maximizes the second term on the right-hand side.

For two-sided MCA, since $P(E^c_{ij} \cap E^c_{i^*j^*})$ is increasing in $|\rho_{ij,i^*j^*}|$ (see Tong 1980, p. 26) where

$$\rho_{ij,i^*j^*} = Corr(\hat{\mu}_i - \hat{\mu}_j, \hat{\mu}_{i^*} - \hat{\mu}_{j^*}),$$

the optimal spanning tree $\mathcal{T}$ can be found by applying any minimal spanning tree algorithm to the distance matrix $(\delta_{ij,i^*j^*})$ defined by

$$\delta_{ij,i^*j^*} = 1 - |\rho_{ij,i^*j^*}|.$$

### 7.4.2 Stochastic approximation methods

The method of Edwards and Berry continues to apply, with

$$D = \max_{1 \leq i < j \leq k} \hat{\sigma}^{-1} |\hat{\mu}_i - \hat{\mu}_j - (\mu_i - \mu_j)| / \sqrt{v_j^i}$$

for two-sided MCA and

$$D = \max_{1 \leq i < j \leq k} \hat{\sigma}^{-1} (\hat{\mu}_i - \hat{\mu}_j - (\mu_i - \mu_j)) / \sqrt{v_j^i}$$

for Hayter's one-sided MCA.

The ADJUST = SIMULATE option, with the CL and the default PDIFF = ALL options under the LSMEANS statement in PROC GLM of the SAS system, implements this approximate form of two-sided MCA in the GLM, setting $\gamma = 0.005$ and $\epsilon = 0.01$ by default.

## 7.5 Scheffé's method for all contrasts

Recall the standard result that, since $\hat{\mu}_{-k}$ is $MVN(\mu, \sigma^2 \mathbf{V}_{-k})$, the quadratic form $(\hat{\mu}_{-k} - \mu_{-k})'(\sigma^2 \mathbf{V}_{-k})^{-1}(\hat{\mu}_{-k} - \mu_{-k})$ has a $\chi^2$ distribution with $k - 1$ degrees of freedom. Therefore $(\hat{\mu}_{-k} - \mu_{-k})'((k-1)\hat{\sigma}^2 \mathbf{V}_{-k})^{-1}(\hat{\mu}_{-k} - \mu_{-k})$ has an $F$ distribution with $k - 1$ and $\nu$ degrees of freedom. Thus, if $(\mathbf{V}_{-k}^{1/2})(\mathbf{V}_{-k}^{1/2})' = \mathbf{V}_{-k}$, then

$$P\left\{ (\mathbf{V}_{-k}^{-1/2}(\hat{\mu}_{-k} - \mu_{-k}))'(\mathbf{V}_{-k}^{-1/2}(\hat{\mu}_{-k} - \mu_{-k})) \leq (k-1)\hat{\sigma}^2 F_{\alpha, k-1, \nu} \right\}$$
$$= 1 - \alpha,$$

where $F_{\alpha, k-1, \nu}$ is the upper $\alpha$ quantile of the $F$ distribution with $k - 1$ and $\nu$ degrees of freedom, so the set

$$S = \{(\mu_1, \ldots, \mu_k) : (\mathbf{V}_{-k}^{-1/2}(\mu_{-k} - \hat{\mu}_{-k}))'(\mathbf{V}_{-k}^{-1/2}(\mu_{-k} - \hat{\mu}_{-k}))$$
$$\leq (k-1)\hat{\sigma}^2 F_{\alpha, k-1, \nu}\}$$

is a $100(1 - \alpha)\%$ confidence set for $\mu_1, \ldots, \mu_k$. The set $S$ is translation invariant, in the sense that if $(\mu_1, \ldots, \mu_k) \in S$ then $(\mu_1 + \delta, \ldots, \mu_k + \delta) \in S$ as well. Thus, as $S$ is an infinite cylinder in $\Re^k$ parallel to the vector $(1, \ldots, 1)$, it is completely described by its cross-section on a plane, any plane, perpendicular to $(1, \ldots, 1)$. The projection of $S$ on such a plane forms Scheffé's $100(1 - \alpha)\%$ simultaneous confidence intervals for all contrasts of $\mu_1, \ldots, \mu_k$:

$$\sum_{i=1}^{k} c_i \mu_i \in \sum_{i=1}^{k} c_i \hat{\mu}_i \pm \sqrt{(k-1)F_{\alpha, k-1, \nu}} \hat{\sigma} (\mathbf{c}_{-k}' \mathbf{V}_{-k} \mathbf{c}_{-k})^{1/2} \text{ for all } \mathbf{c} \in \mathcal{C}$$

$$(7.55)$$

where $\mathcal{C} = \{(c_1, \ldots, c_k)' : c_1 + \cdots + c_k = 0\}$ and $\mathbf{c}_{-k} = (c_1, \ldots, c_{k-1})'$.

**Theorem 7.5.1 (Scheffé 1953)**

$$P\{\sum_{i=1}^{k} c_i\mu_i \in \sum_{i=1}^{k} c_i\hat{\mu}_i \pm \sqrt{(k-1)F_{\alpha,k-1,\nu}}\hat{\sigma}(c'_{-k}V_{-k}c_{-k})^{1/2}$$

$$\text{for all } (c_1,\ldots,c_k)' \in \mathcal{C}\}$$

$$= 1 - \alpha.$$

*Proof.* While a more mathematically detailed proof can be found in Miller (1981), for example, our humble Lemma B.1.1 in Appendix B makes possible a geometric proof.

Letting $z = (z_1,\ldots,z_{k-1})' = V_{-k}^{-1/2}\mu_{-k}$ and $\hat{z} = (\hat{z}_1,\ldots,\hat{z}_{k-1})' = V_{-k}^{-1/2}\hat{\mu}_{-k}$, we see that $z = (z_1,\ldots,z_{k-1})'$ satisfying

$$\sum_{i=1}^{k-1}(z_i - \hat{z}_i)^2 \le (k-1)F_{\alpha,k-1,\nu}\hat{\sigma}^2$$

constitutes the interior of a $(k-1)$-dimensional sphere centered at the point $(\hat{z}_1,\ldots,\hat{z}_{k-1})'$ with radius $\sqrt{(k-1)F_{\alpha,k-1,\nu}}\hat{\sigma}$. Therefore, by applying the *if* portion of Lemma B.1.1 in Appendix B to

$$c^* = (V_{-k}^{1/2})'c_{-k},$$

noting $(c^*)'z = \sum_{i=1}^{k-1} c_i(\mu_i - \mu_k) = \sum_{i=1}^{k} c_i\mu_i$ and $(c^*)'\hat{z} = \sum_{i=1}^{k-1} c_i(\hat{\mu}_i - \hat{\mu}_k) = \sum_{i=1}^{k} c_i\hat{\mu}_i$, we see that if $(\mu_1,\ldots,\mu_k) \in S$, then

$$\sum_{i=1}^{k} c_i\mu_i \in \sum_{i=1}^{k} c_i\hat{\mu}_i \pm \sqrt{(k-1)F_{\alpha,k-1,\nu}}\hat{\sigma}(c'_{-k}V_{-k}c_{-k})^{1/2}. \tag{7.56}$$

Now suppose $(\mu_1,\ldots,\mu_k) \notin S$, then

$$\|z - \hat{z}\| > \sqrt{(k-1)F_{\alpha,k-1,\nu}}\hat{\sigma}.$$

Letting $c_{-k} = (V_{-k}^{-1/2})'(z - \hat{z})$, we see that

$$\sum_{i=1}^{k} c_i(\mu_i - \hat{\mu}_i) = \sum_{i=1}^{k-1} c_i((\mu_i - \mu_k) - (\hat{\mu}_i - \hat{\mu}_k))$$

$$= (z - \hat{z})'(V_{-k}^{-1/2}(\mu_{-k} - \hat{\mu}_{-k}))$$

$$= \|z - \hat{z}\|^2$$

$$> \sqrt{(k-1)F_{\alpha,k-1,\nu}}\hat{\sigma}(c'_{-k}V_{-k}c_{-k})^{1/2},$$

that is, (7.56) is not satisfied. $\square$

*Remark.* Noting that $\sum_{i=1}^{k} c_i\mu_i = \sum_{i=1}^{k-1} c_i(\mu_i - \mu_k)$ for $c \in \mathcal{C}$, if we again let $(X'X)^-$ be any generalized inverse of $X'X$, and let $V$ be the submatrix of $(\mathbf{X}'\mathbf{X})^-$ which corresponds to $\{\mu_1,\ldots,\mu_k\} \subseteq \{\beta_1,\ldots,\beta_p\}$, then the expression $(\mathbf{c}'_{-k}\mathbf{V}_{-k}\mathbf{c}_{-k})^{1/2}$ in (7.55) can be replaced by $(\mathbf{c}'\mathbf{V}\mathbf{c})^{1/2}$.

## 7.6 Nonparametric methods

Consider the two-way no-interaction model with one observation per cell

$$Y_{ih} = \mu + \tau_i + \beta_h + \epsilon_{ih}, \quad i = 1,\ldots,k, h = 1,\ldots,n, \qquad (7.57)$$

where

$$
\begin{aligned}
Y_{ih} &= \text{observation under } i\text{th treatment in } h\text{th block} \\
\tau_i &= i\text{th treatment effect} \\
\beta_h &= h\text{th block effect} \\
\epsilon_{ih} &= \text{random error associated with } Y_{ih}
\end{aligned}
$$

and it is assumed that $\epsilon_{11},\ldots,\epsilon_{kn}$ are i.i.d. with unknown distribution $F$.

### 7.6.1 Pairwise ranking methods based on signed ranks

Let $S_{ia}^j(\delta_i)$ denote the rank of $|Y_{ia} - Y_{ja} - \delta_i|$ in

$$|Y_{i1} - Y_{j1} - \delta_i|,\ldots,|Y_{in} - Y_{jn} - \delta_i|$$

and let

$$S_i^j(\delta_i) = \sum_{a=1}^n S_{ia}^j(\delta_i)\operatorname{sign}(Y_{ia} - Y_{ja} - \delta),$$

where

$$\operatorname{sign}(x) = \begin{cases} 1 & \text{if } x > 0 \\ 0 & \text{otherwise.} \end{cases}$$

Thus, $S_i^j(0)$ is the Wilcoxon signed rank statistic for $Y_{ia} - Y_{ja}, a = 1,\ldots,n$. In the presence of a blocking effect $\beta_h$, the marginal distribution of $S_i^j(0)$ is still independent of $F$ when $H_0 : \tau_i = \tau_j$, and inference on $\tau_i - \tau_j$ with good properties can be constructed based on $S_i^j(\delta_i)$. In fact, if we let

$$D_{i[1]}^j \leq \cdots \leq D_{i[n(n+1)/2]}^j$$

denote the ordered $n(n + 1)/2$ averages

$$Y_{ia} - Y_{ja} + Y_{ib} - Y_{jb}, 1 \leq a \leq b \leq n,$$

with the additional understanding that

$$D_{i[0]}^j = -\infty$$

and

$$D_{i[n(n+1)/2+1]}^j = +\infty,$$

then using the well-known relationship between $S_i^j(\delta_i)$ and the $D_{i[a]}^j$ (cf. Lehmann 1975, p. 129), (individual) level $1 - \alpha$ confidence intervals for $\tau_i - \tau_j$ can be constructed.

Thus, one might expect nonparametric multiple comparisons with good properties can be constructed based on $S_i^j(\delta_i), 1 \leq i < j \leq k$. Unfortunately – in contrast to the situation with the pairwise Wilcoxon rank sum statistics $R_i^j(\mu_i - \mu_j), 1 \leq i < j \leq k$, in a one-way model – the joint distribution of $S_i^j(\tau_i - \tau_j), 1 \leq i < j \leq k$, in the 2-way model (7.57) depends on the unknown distribution $F$. Specifically, if

$$\lambda = P\{Y_1 < Y_2 + Y_3 - Y_4 \text{ and } Y_1 < Y_5 + Y_6 - Y_7\}$$

where $Y_1, \ldots, Y_7$ are i.i.d. random variables with distribution $F$, then under $H_0 : \tau_1 = \cdots = \tau_k$

$$\rho = \lim_{n \to \infty} Corr(S_i^k(0), S_j^k(0)) = 12\lambda - 3, \quad 1 \leq i < j < k$$

which depends on the unknown $F$, making it difficult to obtain any exact distribution-free multiple comparison method.

For multiple comparison with the best and multiple comparison with a control, one can apply Theorem 3.3.1 of Tong (1980) and construct conservative methods by setting the error rate of each individual comparison to be $1 - (1 - \alpha)^{1/k}$ (cf. Hsu 1982). As Spurrier (1991; 1995) has shown, the range of $\lambda$ is:

$$\frac{89}{315} \leq \lambda \leq \frac{7 - (1 - \sqrt{2/3})^2/4}{24},$$

so

$$0.3905 \approx \frac{41}{105} \leq \rho \leq 0.5 - \frac{(1 - \sqrt{2/3})^2}{8} \approx 0.4958.$$

Therefore, applying Slepian's inequality (Theorem A.3.1) and Šidák's inequality (Theorem A.4.1), asymptotically conservative MCB and MCC methods can be constructed using the asymptotic normality of $S_1^k(\tau_1 - \tau_k), \ldots, S_{k-1}^k(\tau_{k-1} - \tau_k)$ and setting the critical value based on the lower bound $41/105$ for $\rho$ (cf. Hsu 1982). Note that Approximation 2 suggested in Hollander and Wolfe (1973, pp. 173–174) to use an *upper* bound on $\rho$ to set the critical value results in asymptotic liberalism instead of conservatism. Finally, one can obtain asymptotically distribution-free MCB and MCC methods by estimating $\rho$ through estimating $\lambda$ from the sample, and using asymptotic critical values based on the estimated $\rho$. (cf. Hollander 1966 Approximation 1; in Hollander and Wolfe 1973, p. 173; Hsu 1982) An efficient method of estimating $\lambda$ has been given by Mann and Pirie (1982).

For all-pairwise comparisons (MCA), one suggestion has been to scale the critical value $|q^*|$ of Tukey's method for $k$ treatments and infinite error degrees of freedom ($\nu = \infty$) by the common standard deviation of $S_i^j(\tau_i - $

$\tau_j$), which would be appropriate asymptotically if $\rho = 0.5$ (cf. Hollander and Wolfe 1973, p. 171). But, given $\rho < 0.5$, Šidák's inequality (Theorem A.4.1) implies this approximation results in asymptotic liberalism.

Other pairwise ranking methods are available. For multiple comparisons with a control, Steel (1959b) proposed a nonparametric many-one sign test for the model (7.57). Spurrier (1988) generalized Steel's procedure to allow for the possibility that the control is observed more than once in each block and that not all new treatments are observed in each block. Spurrier (1992) proposed signed ranks methods which allow the use of a more general set of designs than (7.57).

### 7.6.2 Joint ranking methods based on Friedman rank sums

Under the two-way location model with one observation per cell (7.57), let $\delta = (\delta_1, \ldots, \delta_k)$ and let $R_{ih}(\delta)$ denote the rank of $Y_{ih} - \delta_i$ in the sample of $k$ observations in the $h$th block

$$Y_{1h} - \delta_1, \ldots, Y_{kh} - \delta_k.$$

Let $R_i(\delta)$ denote the rank sum of the $i$th treatment

$$R_i(\delta) = \sum_{h=1}^{n} R_{ih}(\delta).$$

By ranking within blocks, the joint distribution of $R_i(\delta), i = 1, \ldots, k$, is independent of the $\beta$'s. Suppose $|r|_J$ is the smallest integer such that

$$P_{H_0}\{|R_i - R_j| \leq |r|_J \text{ for all } i > j\} \geq 1 - \alpha \qquad (7.58)$$

under

$$H_0 : \tau_1 = \cdots = \tau_k,$$

where $R_i = R_i(0, \ldots, 0), i = 1, \ldots, k$; then since

$$(7.58) = P_{\boldsymbol{\tau}}\{|R_i(\boldsymbol{\tau}) - R_j(\boldsymbol{\tau})| \leq |r|_J \text{ for all } i > j\},$$

a $100(1 - \alpha)\%$ confidence set for $\boldsymbol{\tau} = (\tau_1, \ldots, \tau_k)$ is

$$C = \{(\delta_1, \ldots, \delta_k) : |R_i(\delta) - R_j(\delta)| \leq |r|_J \text{ for all } i > j\}.$$

This set $C$ is translation invariant in the sense that if $(\delta_1, \ldots, \delta_k) \in C$, then $(\delta_1 + \delta, \ldots, \delta_k + \delta) \in C$ as well. Therefore, by examining a cross-section of $C$ perpendicular to $(1, \ldots, 1)$, simultaneous confidence intervals for $\tau_i - \tau_j, i \neq j$, can be deduced. Unfortunately, no general technique for computing this cross-section is known. The difficulty is that, in contrast to methods based on signed ranks, changes in the value of $\delta_i$ not only induce changes in $R_i(\delta) - R_j(\delta), j \neq i$, but changes in $R_m(\delta) - R_j(\delta), m \neq j$ as well.

A popular two-sided MCA method (cf. Hollander and Wolfe 1973, p. 151;

Hettmansperger p. 197) asserts

$$\tau_i \neq \tau_j \text{ for all } i \neq j \text{ such that } |R_i - R_j| > |r|_J.$$

When $H_0$ is true, clearly

$$P_{H_0}\{\text{at least one incorrect assertion}\} \leq \alpha. \tag{7.59}$$

However, Oude Voshaar (1980) showed that at least for some $k$,

$$\lim_{n \to \infty} \sup_{F, \tau} P_{F, \tau}\{\text{at least one incorrect assertion}\} > \alpha$$

at the usual $\alpha$ levels. So this MCA method is not even a confident inequalities method. Therefore, this method and others based on Friedman-type joint ranking are not recommended for multiple comparisons.

## 7.7 Two-way mixed models

Consider the two-way model with a fixed treatment effect $\tau$ and a random blocking $B$ with no interaction:

$$Y_{ihr} = \mu + \tau_i + B_h + \epsilon_{ihr}, \tag{7.60}$$

$$i = 1, \ldots, k, h = 1, \ldots, b, r = 1, \ldots, n_{ih},$$

where

$$B_h \sim N(0, \sigma_B^2)$$
$$E_{ihr} \sim N(0, \sigma^2)$$

and all $B_h$, $\epsilon_{ihr}$ are mutually independent. If one uses the ordinary least squares estimates for treatment contrasts $\sum_i c_i \tau_i$ (with $\sum_i c_i = 0$) as in Section 7.1.2.1 where block effects are fixed, then the estimators, which are the usual intra-block estimators, are free of block effects. Also, as in the fixed effects case, the error mean square $MSE$ is independent of estimates of treatment effects (and block effects). Therefore, in this case multiple comparisons can be effected using ordinary least squares estimates (which are not necessarily the best linear unbiased estimators) in exactly the same way as in the fixed effects no-interaction case of Section 7.1.2.1.

Yang and Nelson (1991) extended the method of constrained multiple comparisons with the best to a mixed no-interaction model with a random component more complex than the one in (7.61). Their method is particularly applicable to multiple comparisons of systems by stochastic simulation, in which the random component arises from common random numbers and control variates introduced for the purpose of variance reduction.

Now consider a mixed two-way model with a fixed treatment effect $\tau$, a random blocking $B$, and a random interaction effect $\tau B$:

$$Y_{ihr} = \mu + \tau_i + B_h + (\tau B)_{ih} + \epsilon_{ihr}, \tag{7.61}$$

$$i = 1, \ldots, k, \quad h = 1, \ldots, b, \quad r = 1, \ldots, n_{ih},$$

where

$$
\begin{aligned}
B_h &\sim N(0, \sigma_B^2) \\
(\tau B)_{ih} &\sim N(0, \sigma_{\tau B}^2) \\
\epsilon_{ihr} &\sim N(0, \sigma^2)
\end{aligned}
$$

and all $B_h$, $(\tau B)_{ih}$, $\epsilon_{ihr}$ are mutually independent.

In contrast to the fixed effects with-interaction case, since $E(Y_{ihr} - Y_{jhr}) = \tau_i - \tau_j$, multiple comparisons on the treatment effects $\tau_1, \ldots, \tau_k$ are now meaningful. Using the notation in Section 7.1.2.1, we refer to

$$\sum_{i=1}^{k} c_i (\sum_{h=1}^{b} \bar{Y}_{ih.}/b) \tag{7.62}$$

as the least squares estimator of main-effect contrast $\sum_i c_i \tau_i$ with $\sum_i c_i = 0$.

For an equireplicate design ($n_{11} = \cdots = n_{kb} = n$), define

$$MSAB = n \sum_{i=1}^{k} \sum_{h=1}^{b} (\bar{Y}_{ih.} - \bar{Y}_{i..} - \bar{Y}_{.j.} + \bar{Y}_{...})^2 / [(k-1)(b-1)]. \tag{7.63}$$

Then $MSAB$ is independent of treatment contrast estimators (7.62) with an expected value proportional to the variance of those treatment contrast estimators. Therefore, it provides an appropriate variance estimate for the least squares estimator of each treatment contrast, and multiple comparisons can be executed as if the observations were taken under the treatments without blocking, but with the following substitutions:

$$\bar{Y}_i - \bar{Y}_j \quad \leftarrow \quad \bar{Y}_{i..} - \bar{Y}_{j..}$$

$$\sum_{i=1}^{k} \sum_{r=1}^{n_i} (Y_{ir} - \bar{Y}_i)^2 / \sum_{i=1}^{k} (n_i - 1) \quad \leftarrow \quad MSAB$$

$$\sum_{i=1}^{k} (n_i - 1) \quad \leftarrow \quad (k-1)(b-1).$$

See Bhargava and Srivastava (1973) and Hochberg and Tamhane (1983).

Recall from Section 7.1.2.1 that when factor $B$ is fixed, the condition of proportional frequencies ensures multiple comparison methods apply, using the error mean square $MSE$ as a variance estimate. However, as Voss and Hsu (1998) point out, when $B$ is random, for treatment contrasts to be estimated independently of the usual $MSAB$, the condition of proportional frequencies is not sufficient. In fact, they show the surprising result that even though when the design consists of identical blocks (a strong condition of proportional frequencies) the statistic $F = MSA/MSAB$ is appropriate for testing equality of the treatment effects, $MSAB$ is not an appropriate quantity with which to Studentize the treatment contrasts estimates. In short, when factor $B$ is random, only when the design has equireplications do standard methods of multiple comparisons apply using the interaction mean square $MSAB$ as a variance estimator. Nevertheless, they identify appropriate variance estimates for certain non-equireplicate designs, and construct some conservative and approximate multiple comparison methods.

## 7.8 One-way repeated measurement models

Consider the single-replication one-way repeated measurement model

$$Y_{ih} = \mu_i + \epsilon_{ih}, \quad i = 1, \ldots, k, \quad h = 1, \ldots, b, \tag{7.64}$$

where $(Y_{1h}, \ldots, Y_{kh}), h = 1, \ldots, b$, are i.i.d. multivariate normal with mean vector $(\mu_1, \ldots, \mu_k)$ and unknown variance-covariance matrix $\Sigma$. Suppose $\Sigma$ satisfies the so-called *sphericity* condition of Huynh and Feldt (1970), that is, $\Sigma$ is of the form

$$\Sigma = \begin{pmatrix} 2\gamma_1 + \sigma_R^2 & \gamma_1 + \gamma_2 & \cdots & \gamma_1 + \gamma_k \\ \gamma_2 + \gamma_1 & 2\gamma_2 + \sigma_R^2 & \cdots & \gamma_2 + \gamma_k \\ \vdots & \vdots & \ddots & \vdots \\ \gamma_k + \gamma_1 & \gamma_k + \gamma_2 & \cdots & 2\gamma_k + \sigma_R^2 \end{pmatrix}; \tag{7.65}$$

then multiple comparisons can be effected exactly as in the two-way model (7.61) with $n_{ih} = 1$ for all $ih$. To see this, observe that contrast estimators (7.62) and $MSAB$ (7.63) remain unchanged if a constant $\delta_h$ is added to each of $Y_{1h}, \ldots, Y_{kh}$. Therefore, contrast estimators (7.62) and $MSAB$ can be considered functions of $(Y_{1h} - Y_{kh}, \ldots, Y_{(k-1)h} - Y_{kh}), h = 1, \ldots, b$. The i.i.d. random vectors $(Y_{1h} - Y_{kh}, \ldots, Y_{(k-1)h} - Y_{kh}), h = 1, \ldots, b$, have mean vector $(\mu_1 - \mu_k, \ldots, \mu_{k-1} - \mu_k)$ and variance-covariance matrix

$$\begin{pmatrix} 2\sigma_R^2 & \sigma_R^2 & \cdots & \sigma_R^2 \\ \sigma_R^2 & 2\sigma_R^2 & \cdots & \sigma_R^2 \\ \vdots & \vdots & \ddots & \vdots \\ \sigma_R^2 & \sigma_R^2 & \cdots & 2\sigma_R^2 \end{pmatrix},$$

as they did under the model (7.61) (with $\sigma_R^2 = \sigma_{\tau B}^2 + \sigma^2$).

When the sphericity condition (7.65) is not known to be satisfied, one might contemplate constructing multiple comparison methods based on $\hat{\Sigma} = (\hat{\sigma}_{ij})$, the sample variance-covariance matrix, where

$$\hat{\sigma}_{ij} = \frac{\sum_{h=1}^{b}(\bar{Y}_{ih} - \bar{Y}_{i.})(\bar{Y}_{jh} - \bar{Y}_{j.})}{b - 1}$$

with

$$\bar{Y}_{i.} = \sum_{h=1}^{b} Y_{ih}/b, \quad i = 1, \ldots, k.$$

Alberton and Hochberg (1984) considered simultaneous confidence intervals of the form

$$\mu_i - \mu_j \in \bar{Y}_{i.} - \bar{Y}_{j.} \pm Q\sqrt{\hat{\sigma}_{ii} + \hat{\sigma}_{jj} - 2\hat{\sigma}_{ij}} \text{ for all } i \neq j. \qquad (7.66)$$

In general, the coverage probability of (7.66) depends on the unknown $\Sigma$. Based on simulation results for $k = 3$, Alberton and Hochberg (1984) extended the Tukey–Kramer conjecture to conjecture that the infimum of the coverage probability of (7.66) occurs when $\Sigma = I$, the identity matrix, that is, when $Y_{1h}, \ldots, Y_{jh}$ are actually independent. Lin, Seppänen and Uusipaikka (1990) proved this for $k = 3$, but as yet there is no proof of the validity of this conjecture for $k > 3$.

## 7.9 Exercises

1. Consider a one-way design with a covariate

$$Y_{ih} = \mu_i + \beta X_{ih} + e_{ih}, \qquad i = 1, \ldots, k, \quad h = 1, \ldots, n_i.$$

Give a condition in terms of $X_{ih}$ for (7.6) to be satisfied.

2. Stevenson, Lee and Stigler (1986) first reported on the mathematics achievement of Chinese, Japanese and American children. Summary mathematics test statistics for children in kindergarten (K) and grades 1 and 5 are given in Table 7.8. Formulate your own multiple comparison problem in the framework of a GLM.

Table 7.8 *Summary statistics on mathematics tests*

| Grade | | United States | Taiwan | Japan |
|---|---|---|---|---|
| K | Mean | 37.5 | 37.8 | 42.2 |
| | Standard deviation | 5.6 | 7.4 | 5.1 |
| | Sample size | 288 | 286 | 280 |
| 1 | Mean | 17.1 | 21.2 | 20.1 |
| | Standard deviation | 5.3 | 5.5 | 5.2 |
| | Sample size | 237 | 241 | 240 |
| 5 | Mean | 44.4 | 50.8 | 53.3 |
| | Standard deviation | 6.2 | 5.7 | 7.5 |
| | Sample size | 238 | 241 | 239 |

# Some useful probabilistic inequalities

## A.1 An inequality for conditionally independent random variables

A very useful inequality is referred to as Chebyshev's *other* inequality.

**Theorem A.1.1 (Chebyshev's other inequality)** *Let $T$ be a p-dimensional random variable. Suppose the functions $g_1, g_2 : \Re^p \to \Re^1$ satisfy*

$$[g_1(t_2) - g_1(t_1)][g_2(t_2) - g_2(t_1)] \geq 0$$

*for all $t_1, t_2$ in the support of the distribution of $T$, then*

$$E[g_1(T)g_2(T)] \geq E[g_1(T)]E[g_2(T)]$$

*provided the expectations exist, i.e., the covariance between $g_1(T)$ and $g_2(T)$ is non-negative.*

*Proof.* Following the hint in Exercise 8 of Tong (1980, p. 32), let $T_1, T_2$, and $T$ be i.i.d. random vectors. Then

$$
\begin{aligned}
0 &\leq E\{[g_1(T_2) - g_1(T_1)][g_2(T_2) - g_2(T_1)]\} \\
&= E\{[g_1(T_2)g_2(T_2) + g_1(T_1)g_2(T_1)] - [g_1(T_1)g_2(T_2) + g_1(T_2)g_2(T_1)]\} \\
&= 2\{E[g_1(T)g_2(T)] - E[g_1(T)]E[g_2(T)]\}. \quad \square
\end{aligned}
$$

**Corollary A.1.1 (Kimball's inequality)** *Let $U$ be a univariate random variable. If $g_1, \ldots, g_p$ are bounded, non-negative and monotone in the same direction, then*

$$E[\prod_{i=1}^{p} g_i(U)] \geq \prod_{i=1}^{p} E[g_i(U)]. \tag{A.1}$$

*Note.* As Tong (1980) remarked, (A.1) was proven independently by Kimball (1951), and so is often referred to as Kimball's inequality in the literature.

## A.2 The Bonferroni inequality

Let $E_1, \ldots, E_p$ be events. Then by Boole's formula

$$P(\bigcup_{i=1}^{p} E_i^c) = \sum_{i=1}^{p} P(E_i^c) - \sum_{i<j} P(E_i^c \cap E_j^c) + \cdots + (-1)^{p-1} P(\bigcap_{i=1}^{p} E_i^c). \quad (A.2)$$

The first-order approximation provides an upper bound

$$P(\bigcup_{i=1}^{p} E_i^c) \leq \sum_{i=1}^{p} P(E_i^c), \quad (A.3)$$

which is the familiar Bonferroni inequality. The second-order approximation provides a lower bound

$$P(\bigcup_{i=1}^{p} E_i^c) \geq \sum_{i=1}^{p} P(E_i^c) - \sum_{i<j} P(E_i^c \cap E_j^c). \quad (A.4)$$

## A.3 Slepian's inequality

**Theorem A.3.1 (Slepian's inequality)** *Let $Z_1, \ldots, Z_p$ be multivariate normal random variables with some mean vector and correlation matrix $\Sigma$. If $\Gamma$ and $\Lambda$ are two correlation matrices*

$$\Gamma = \begin{pmatrix} 1 & \gamma_{12} & \cdots & \gamma_{1p} \\ \gamma_{21} & \ddots & \ddots & \vdots \\ \vdots & \ddots & \ddots & \gamma_{(p-1)p} \\ \gamma_{p1} & \cdots & \gamma_{p(p-1)} & 1 \end{pmatrix},$$

$$\Lambda = \begin{pmatrix} 1 & \lambda_{12} & \cdots & \lambda_{1p} \\ \lambda_{21} & \ddots & \ddots & \vdots \\ \vdots & \ddots & \ddots & \lambda_{(p-1)p} \\ \lambda_{p1} & \cdots & \lambda_{p(p-1)} & 1 \end{pmatrix},$$

*such that $\gamma_{ij} \geq \lambda_{ij}$ for all $i \neq j$, then for any constants $c_1, \ldots, c_p$,*

$$P_{\Sigma = \Gamma}\{Z_i \leq c_i \text{ for } i = 1, \ldots, p\} \geq P_{\Sigma = \Lambda}\{Z_i \leq c_i \text{ for } i = 1, \ldots, p\}.$$

*Proof.* See Tong (1980, p. 10). □

**Corollary A.3.1** *Let $Z_1, \ldots, Z_p$ be multivariate normal random variables such that*

$$Cov(Z_i, Z_j) \geq 0$$

for all $i \neq j$; then for any constants $c_1, \ldots, c_p$,

$$P\{Z_i \leq c_i \text{ for } i = 1, \ldots, p\} \geq \prod_{i=1}^{p} P\{Z_i \leq c_i\}.$$

## A.4 Šidák's inequality

**Theorem A.4.1 (Šidák's inequality)** *Let $Z_1, \ldots, Z_p$ be multivariate normal random variables with means 0 and variance-covariance matrix $\Sigma(\lambda)$,*

$$\Sigma(\lambda) = \begin{pmatrix} \sigma_{11} & \lambda_1 \lambda_2 \sigma_{12} & \cdots & & \lambda_1 \lambda_p \sigma_{1p} \\ \lambda_2 \lambda_1 \sigma_{21} & \ddots & & \ddots & \vdots \\ \vdots & & \ddots & & \lambda_{(p-1)} \lambda_p \sigma_{(p-1)p} \\ \lambda_p \lambda_1 \sigma_{p1} & \cdots & \lambda_p \lambda_{p-1} \sigma_{p(p-1)} & & \sigma_{pp} \end{pmatrix},$$

*where $\lambda = (\lambda_1, \ldots, \lambda_p)$ with $\lambda_i \in [0, 1]$. Then $P_{\Sigma(\lambda)}\{|Z_i| \leq c_i \text{ for } i = 1, \ldots, p\}$ is non-decreasing in each $\lambda_i \in [0, 1]$.*

*Proof.* See Tong (1980, p. 22). □

**Corollary A.4.1** *Let $Z_1, \ldots, Z_p$ be multivariate normal random variables with means 0 and any variance-covariance matrix. Then for any constants $c_1, \ldots, c_p$,*

$$P\{|Z_i| \leq c_i \text{ for } i = 1, \ldots, p\} \geq \prod_{i=1}^{p} P\{|Z_i| \leq c_i\}.$$

## A.5 The Hunter–Worsley Inequality

Instead of Boole's formula (A.2), consider the decomposition

$$\begin{aligned} P(\bigcup_{i=1}^{p} E_i^c) &= P(E_1^c) + P(E_2^c \cap E_1) + \cdots + P(E_p^c \cap E_{p-1} \cdots \cap E_1) \\ &= P(E_1^c) + \sum_{i=2}^{p} P(E_i^c \cap E_{i-1} \cap \cdots \cap E_1). \end{aligned}$$

If $(i)$ denotes some arbitrary choice of indices in $\{1, \ldots, i-1\}$ (for $i > 1$), then clearly

$$\begin{aligned} P(\bigcup_{i=1}^{p} E_i^c) &\leq P(E_1^c) + \sum_{i=2}^{p} P(E_i^c \cap E_{(i)}) \\ &= \sum_{i=1}^{p} P(E_i^c) - \sum_{i=2}^{p} P(E_i^c \cap E_{(i)}^c). \end{aligned} \tag{A.5}$$

Note that there are exactly $p - 1$ terms in the second summation on the right-hand side, and every index $i, i = 1, \ldots, p$, appears in at least one term. Since which event is chosen to be $E_1$, which is chosen to be $E_2$, etc., is arbitrary, a more general way of stating (A.5), given by Hunter (1976) and Worsley (1982), is that if the summation of the second term in (A.4) is only taken over the $k - 1$ pairs of $i$ and $j$ that are connected in a spanning tree $\mathcal{T}$ of the vertices $\{1, \ldots, p\}$, then an upper bound is obtained. We define what constitutes a spanning tree. Our terminology follows Knuth (1973) more or less.

A *graph* is a set of *vertices* together with a set of *edges* joining certain pairs of distinct vertices. Two vertices are *adjacent* if there is an edge joining them. If $V$ and $V'$ are vertices, we say $(V_0, V_1, \ldots, V_m)$ is a *path* of length $m$ from $V$ to $V'$ if $V = V_0$, $V_i$ is adjacent to $V_{i+1}$ for $0 \leq i < m$, and $V_m = V'$. The path is *simple* if $V_0, V_1, \ldots, V_{m-1}$ are distinct and if $V_1, V_2, \ldots, V_m$ are distinct. A graph $\mathcal{G}$ is *connected* if there is a path between any two of its vertices. A *cycle* is a simple path of length at least three from a vertex to itself. A *spanning tree* $\mathcal{T}$ is a connected graph with no cycles. Basically, a spanning tree with $p$ vertices is a connected graph with exactly $p - 1$ edges (see Knuth 1973, p. 363)

**Theorem A.5.1 (The Hunter–Worsley inequality)** *If $\mathcal{T}$ is a spanning tree of the vertices $\{1, \ldots, p\}$, and $\{i, j\} \in \mathcal{T}$ denotes $i$ and $j$ are adjacent, then*

$$P(\bigcup_{i=1}^{p} E_i^c) \leq \sum_{i=1}^{p} P(E_i^c) - \sum_{\{i,j\} \in \mathcal{T}} P(E_i^c \cap E_j^c). \qquad (A.6)$$

Clearly, the sharpest bond is achieved when $\mathcal{T}$ is the spanning tree that maximizes the second term. Therefore, computer implementation of the Hunter–Worsley inequality should include a *maximal spanning tree* algorithm to maximize the second term and optimize the inequality.

# Some useful geometric lemmas

## B.1 Projecting spheres

**Lemma B.1.1** $z_1, \ldots, z_p$ *satisfy*

$$\sum_{i=1}^{p} a_i y_i - r(\sum_{i=1}^{p} a_i^2)^{1/2} \leq \sum_{i=1}^{p} a_i z_i \leq \sum_{i=1}^{p} a_i y_i + r(\sum_{i=1}^{p} a_i^2)^{1/2} \qquad \text{(B.1)}$$

*for all* $a = (a_1, \ldots, a_p)' \in \Re^p$ *if and only if*

$$\sum_{i=1}^{p} (z_i - y_i)^2 \leq r^2. \qquad \text{(B.2)}$$

*Proof.* Let $z = (z_1, \ldots, z_p)'$ satisfy (B.2), then by the Cauchy–Schwarz inequality,

$$| \sum_{i=1}^{p} a_i (z_i - y_i) | \leq (\sum_{i=1}^{p} (z_i - y_i)^2)^{1/2} (\sum_{i=1}^{p} a_i^2)^{1/2} \leq r(\sum_{i=1}^{p} a_i^2)^{1/2},$$

so (B.1) holds. To prove the converse, suppose $z$ is such that

$$\sum_{i=1}^{p} (z_i - y_i)^2 > r^2.$$

Then by taking $a = z - y$, we have

$$\sum_{i=1}^{p} a_i (z_i - y_i) = \|a\|^2 > r\|a\|,$$

where $\|a\| = (a_1^2 + \cdots + a_p^2)^{1/2}$ is the length of $a$, so $z$ does not satisfy (B.1). $\square$

This lemma is easy to understand geometrically if one notes that the equations defining $z = (z_1, \ldots, z_p)' \in \Re^p$ to be on the two hyperplanes perpendicular to the vector $a = (a_1, \ldots, a_p)'$ and tangent to the sphere of radius $r$, centered at $y = (y_1, \ldots, y_p)'$, are

$$\sum_{i=1}^{p} a_i z_i = \sum_{i=1}^{p} a_i y_i + r(\sum_{i=1}^{p} a_i^2)^{1/2}$$

and

$$\sum_{i=1}^{p} a_i z_i = \sum_{i=1}^{p} a_i y_i - r \left( \sum_{i=1}^{p} a_i^2 \right)^{1/2},$$

which can be seen as follows. The tangents go through

$$z_0^+ = y + r \left( \frac{a}{\|a\|} \right)$$

and

$$z_0^- = y - r \left( \frac{a}{\|a\|} \right).$$

See Figure B.1. Since the tangents are perpendicular to $a$, their equations

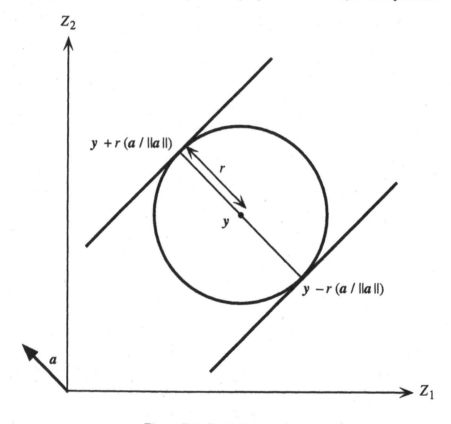

Figure B.1 *Projecting a sphere*

must be

$$a'(z - z_0^{\pm}) = 0,$$

or

$$a'z = a'z_0^{\pm},$$

or

$$a'z = a'\left[y \pm r\left(\frac{a}{||a||}\right)\right],$$

or

$$a'z = a'y \pm r\frac{a'a}{||a||},$$

or

$$\sum_{i=1}^{p} a_i z_i = \sum_{i=1}^{p} a_i y_i \pm r\left(\sum_{i=1}^{p} a_i^2\right)^{1/2}.$$

The lemma states that all points within the sphere are between any pair of such planes, and the intersection over all such planes of points between such planes is the sphere. (This geometric view was contributed by Gunnar Stefansson.)

## B.2 Projecting rectangles

### Lemma B.2.1

$$\max_{1 \le i \le k}\left\{\frac{|z_i|}{d_i}\right\} \le 1 \tag{B.3}$$

*if and only if*

$$\left|\sum_{i=1}^{k} l_i z_i\right| \le \sum_{i=1}^{k} d_i|l_i| \text{ for all } l = (l_1, \dots, l_k) \in \Re^k.$$

The set described by (B.3) is a $k$-dimensional rectangle. For each point $(z_1, \dots, z_k)$, $\sum_{i=1}^{k} l_i z_i$ is the length of its projection in the direction of $(l_1, \dots, l_k)$ times the length of $(l_1, \dots, l_k)$. Therefore, the minimum and maximum values of $\sum_{i=1}^{k} l_i z_i$ are attained at points in the rectangle (B.3) that are also on the two hyperplanes perpendicular to $(l_1, \dots, l_k)$ tangent to the rectangle. These hyperplanes touch the rectangle at its corners. A typical corner of the rectangle (B.3) has coordinates $(\pm d_1, \dots, \pm d_k)$. Therefore, the maximum (minimum) of $\sum l_i z_i$ is obtained by letting $z_i = +d_i$ or $-d_i$ so that $z_i$ has the same (opposite) sign as $l_i$.

While a more formal proof can be found in Miller (1981, p. 74), this lemma is easy to understand by taking a simple example. The maximum of value of $z_1 - 2z_2 + 3z_3$, among $z = (z_1, z_2, z_3)$ satisfying $-d \le z_i \le d$, is $6d$, attained when $z_1 = d, z_2 = -d, z_3 = d$, while the minimum value is $-6d$, attained when $z_1 = -d, z_2 = d, z_3 = -d$.

## B.3 Deriving confidence sets by pivoting tests

The connection between a family of tests for a parameter and a confidence set for that parameter is given by the following lemma. (In this section, dependency on possible nuisance parameters is suppressed notationally.)

**Lemma B.3.1 (Lehmann 1986, p. 90)** *Let* $\Theta$ *denote the parameter space and let* $\hat{\boldsymbol{\theta}}$ *be a random vector whose distribution depends on* $\boldsymbol{\theta} \in \Theta$. *If* $\{\phi_{\boldsymbol{\theta}}(\hat{\boldsymbol{\theta}}) : \boldsymbol{\theta} \in \Theta\}$ *is a family of tests such that*

$$P_{\boldsymbol{\theta}}\{\phi_{\boldsymbol{\theta}}(\hat{\boldsymbol{\theta}}) = 0\} \geq 1 - \alpha$$

*for each* $\boldsymbol{\theta} \in \Theta$, *then*

$$C(\hat{\boldsymbol{\theta}}) = \{\boldsymbol{\theta} : \phi_{\boldsymbol{\theta}}(\hat{\boldsymbol{\theta}}) = 0, \boldsymbol{\theta} \in \Theta\}$$

*is a level* $100(1 - \alpha)\%$ *confidence set for* $\boldsymbol{\theta}$.

Note that in order to obtain a confidence set via this connection, a family of tests, one for each hypothesized parameter value, is required. When the family of distributions is a location family of distributions, one can start with one or more tests for a particular hypothesized parameter value and employ equivariance to generate the family of tests. In the presence of an unknown (nuisance) scale parameter, usually this hypothesized parameter value is chosen so that a statistic whose distribution depends on neither the location parameters nor the scale parameter (i.e., a pivotal quantity) is available.

**Lemma B.3.2** *Suppose the distribution of* $\hat{\boldsymbol{\theta}} - \boldsymbol{\theta}$ *does not depend on* $\boldsymbol{\theta} = (\theta_1, \ldots, \theta_p)$. *Consider a partition* $\Theta_1, \ldots, \Theta_m$ *of the parameter space, that is,*

$$\bigcup_{i=1}^{m} \Theta_i = \Theta$$

*and*

$$\Theta_i \cap \Theta_j = \emptyset \text{ for all } i \neq j.$$

*If each* $\phi_i(\hat{\boldsymbol{\theta}}) = \phi_i(\hat{\theta}_1, \ldots, \hat{\theta}_p)$ *is a level-*$\alpha$ *test for*

$$H_0 : \theta_1 = \cdots = \theta_p = 0$$

*with acceptance region* $A_i, i = 1, \ldots, m$, *then a level* $100(1 - \alpha)\%$ *confidence region for* $\boldsymbol{\theta} = (\theta_1, \ldots, \theta_p)$ *is*

$$C(\hat{\theta}_1, \ldots, \hat{\theta}_p) = \bigcup_{i=1}^{m} \left(\{-\boldsymbol{\theta} + \hat{\boldsymbol{\theta}} : \boldsymbol{\theta} \in A_i\} \cap \Theta_i\right).$$

*Proof.* The test

$$\phi(\hat{\boldsymbol{\theta}} - \boldsymbol{\theta}^0) = \phi_i(\hat{\theta}_1 - \theta_1^0, \ldots, \hat{\theta}_p - \theta_p^0)$$

is a level-$\alpha$ test for

$$H_{\boldsymbol{\theta}^0} : \theta_1 = \theta_1^0, \ldots, \theta_p = \theta_p^0.$$

Therefore, by Lemma B.3.1, a level $100(1 - \alpha)\%$ confidence set for $\boldsymbol{\theta}$ is

$$C(\hat{\boldsymbol{\theta}}) = \{\boldsymbol{\theta}^0 : H_{\boldsymbol{\theta}^0} \text{ is accepted }\}$$

$$= \bigcup_{i=1}^{m} \{\boldsymbol{\theta}^0 : \hat{\boldsymbol{\theta}} - \boldsymbol{\theta}^0 \in A_i \text{ and } \boldsymbol{\theta}^0 \in \Theta_i\}$$

$$= \bigcup_{i=1}^{m} \left( \{-\boldsymbol{\theta} + \hat{\boldsymbol{\theta}} : \boldsymbol{\theta} \in A_i\} \cap \Theta_i \right). \quad \square$$

When the partition of the parameter space $\Theta$ is the trivial partition, an equivariant confidence set results.

**Lemma B.3.3** *Suppose the distribution of $\hat{\boldsymbol{\theta}} - \boldsymbol{\theta}$ does not depend on $\boldsymbol{\theta} = (\theta_1, \ldots, \theta_p)$. If $\phi(\hat{\boldsymbol{\theta}}) = \phi(\hat{\theta}_1, \ldots, \hat{\theta}_p)$ is a size-$\alpha$ test for*

$$H_0 : \theta_1 = \cdots = \theta_p = 0$$

*with acceptance region $A$, then a $100(1 - \alpha)\%$ confidence region for $\boldsymbol{\theta} = (\theta_1, \ldots, \theta_p)$ is*

$$C(\hat{\theta}_1, \ldots, \hat{\theta}_p) = \{-\boldsymbol{\theta} + \hat{\boldsymbol{\theta}} : \boldsymbol{\theta} \in A\}.$$

In words, $C(\hat{\theta}_1, \ldots, \hat{\theta}_p)$ is the reflection of the acceptance region $A$ through the origin, translated by the amount $\hat{\boldsymbol{\theta}}$.

# APPENDIX C

# Sample size computations

Our sample size computation discussion will be confined to the balanced one-way model (3.1)

$$Y_{ia} = \mu_i + \epsilon_{ia}, \quad i = 1, \ldots, k, \quad a = 1, \ldots, n, \qquad (C.1)$$

where $\mu_i$ is the effect of the $i$th treatment $i = 1, \ldots, k$, and $\epsilon_{11}, \ldots, \epsilon_{kn}$ are i.i.d. normal with mean 0 and variance $\sigma^2$ unknown. We use the notation

$$\hat{\mu}_i = \bar{Y}_i = \sum_{a=1}^{n} Y_{ia}/n,$$

$$\hat{\sigma}^2 = MSE = \sum_{i=1}^{k} \sum_{a=1}^{n} (Y_{ia} - \bar{Y}_i)^2 / [k(n-1)]$$

for the sample means and the pooled sample variance.

A common sample size computation technique is based on the power of the $F$-test, as described, for example, in Section 3.3 of Scheffé (1959) and Appendix A.2 of Fleiss (1986). This computation is not necessarily appropriate for designing multiple comparison experiments, because the power of the $F$-test includes the probability of making incorrect assertions. The power of the $F$-test equals the probability that Scheffé's confidence set does not cover the vector $\{(\delta, \ldots, \delta), \delta \in \Re\}$, the parameter values specified by the hypothesis of homogeneity $H_0 : \mu_1 = \cdots = \mu_k$. Therefore, this power includes the probability that Scheffé's confidence set covers neither $\{(\delta, \ldots, \delta), \delta \in \Re\}$ nor the true parameter $(\mu_1, \ldots, \mu_k)$. When the latter happens, the assertions made by Scheffé's method can be wrong (e.g., asserting $\mu_i > \mu_j$ when $\mu_i < \mu_j$, or asserting $\mu_i - \mu_j > \delta$ when $|\mu_i - \mu_j| < \delta$). Clearly, incorrect assertions may be worse than no assertion at all.

Consider sample size computation for confidence intervals multiple comparison methods such as Tukey's MCA method (Section 5.1.1), the constrained MCB method (Section 4.1.1), and Dunnett's two-sided MCC method (Section 3.2.1.1). As the sample size $n$ increases, the factor $\sqrt{2/n}$ in the width of the confidence intervals decreases. At the same time, the variability of the width decreases since the density of $\hat{\sigma}$ becomes more and more concentrated around $\sigma$. As the sample size $n$ increases, the error degrees of freedom $\nu$ increases, so the critical value decreases as well. Thus, as suggested by Tukey (1953) (see Tukey 1994, Chapter 18, pp. 156–161), one can design the experiment with a large enough sample size so that the resulting confidence intervals will be sufficiently narrow with high probability (see also Scheffé 1959, p. 65). Nevertheless, the probability that the confidence intervals will be sufficiently narrow still includes the probability that the narrow confidence intervals fail to cover the true parameter value, leading to possibly incorrect directional assertions.

Instead, Hsu (1988) suggested the following sample size computation. Given a level $1 - \alpha$ confidence intervals method, calculate the sample size so that with a prespecified probability $1 - \beta$ ($< 1 - \alpha$) the confidence intervals will cover the true parameter value and be sufficiently narrow. This probability of simultaneous *coverage* and *narrowness* can be thought of as the probability of *correct* and *useful* inference; coverage ensures correctness of assertions (see Section 2.2), while narrowness (together with coverage) ensures useful assertions will be made.

Tukey (1953, p. 220) pointed out that if narrowness of the confidence intervals is guaranteed with a probability of $1-(\beta-\alpha)$, then by the Bonferroni inequality (A.3) the probability of simultaneous coverage and narrowness is at least $1 - \beta$. However, empowered by modern computers, we might as well compute this probability exactly, instead of resorting to the Bonferroni inequality.

## C.1 Sample size computation for Tukey's MCA method

For Tukey's MCA method discussed in Section 5.1.1, if the confidence intervals for all $\mu_i - \mu_j$ cover their respective true values and have width no more than $\delta = 2\delta^*$, that is, the event

$$A = \{\mu_i - \mu_j \in \hat{\mu}_i - \hat{\mu}_j \pm |q^*|\hat{\sigma}\sqrt{2/n} \text{ for all } i \neq j \text{ and } |q^*|\hat{\sigma}\sqrt{2/n} < \delta^*\}$$

occurs, then

1. for all $\mu_i - \mu_j > \delta$, the correct and useful significant directional difference assertion $\mu_i > \mu_j$ will be made;

2. for all $\mu_i - \mu_j \approx 0$, the correct and useful practical equivalence assertion $-\delta < \mu_i - \mu_j < \delta$ will be made.

**Theorem C.1.1** *Given an error rate $\alpha$, if the sample size $n$ satisfies*

$$k \int_0^u \int_{-\infty}^{+\infty} [\Phi(z) - \Phi(z - \sqrt{2}|q^*|s)]^{k-1} d\Phi(z) \; \gamma(s)ds \geq 1 - \beta \qquad \text{(C.2)}$$

*where again $\Phi$ is the standard normal distribution function, $\gamma$ is the density of $\hat{\sigma}/\sigma$, and $u = \sqrt{n/2}(\delta^*/\sigma)/|q^*|$, then $P(A) \geq 1 - \beta$.*

*Proof.*

$$P(A)$$
$$= P\{\mu_i - \mu_j \in \hat{\mu}_i - \hat{\mu}_j \pm |q^*|\hat{\sigma}\sqrt{2/n} \text{ for all } i \neq j \text{ and } |q^*|\hat{\sigma}\sqrt{2/n} < \delta^*\}$$

$$= \sum_{i=1}^{k} P\{-|q^*|\hat{\sigma}\sqrt{2/n} < \hat{\mu}_i - \hat{\mu}_j - (\mu_i - \mu_j) < |q^*|\hat{\sigma}\sqrt{2/n}$$

$$\forall j, j \neq i, \text{ and } \hat{\mu}_i - \mu_i = \max_{j=1,\ldots,k} (\hat{\mu}_j - \mu_j) \text{ and } |q^*|\hat{\sigma}\sqrt{2/n} < \delta^*\}$$

$$= \sum_{i=1}^{k} P\{0 < \hat{\mu}_i - \hat{\mu}_j - (\mu_i - \mu_j) < |q^*|\hat{\sigma}\sqrt{2/n} \ \forall j, j \neq i,$$

$$\text{and } |q^*|\hat{\sigma}\sqrt{2/n} < \delta^*\}$$

$$= \sum_{i=1}^{k} P\{0 < \sqrt{n}(\hat{\mu}_i - \mu_i)/\sigma - \sqrt{n}(\hat{\mu}_j - \mu_j)/\sigma < \sqrt{2}|q^*|\hat{\sigma}/\sigma$$

$$\text{for all } j, j \neq i, \text{ and } |q^*|(\hat{\sigma}/\sigma)\sqrt{2/n} < \delta^*/\sigma\}$$

$$= k \cdot P\{0 < \sqrt{n}(\hat{\mu}_1 - \mu_1)/\sigma - \sqrt{n}(\hat{\mu}_j - \mu_j)/\sigma < \sqrt{2}|q^*|\hat{\sigma}/\sigma$$

$$\text{for } j = 2, \ldots, k \text{ and } |q^*|(\hat{\sigma}/\sigma)\sqrt{2/n} < \delta^*/\sigma\}$$

$$= k \int_{0}^{u} \int_{-\infty}^{+\infty} [\Phi(z) - \Phi(z - \sqrt{2}|q^*|s)]^{k-1} d\Phi(z) \ \gamma(s) ds. \quad \square$$

Given $k, \alpha, \delta^*/\sigma$ and candidate sample size $n$, the pmca computer program described in Appendix D computes $P\{A\}$.

## C.2 Sample size computation for MCB

Now assume a larger treatment effect is better. If the constrained MCB confidence intervals for all $\mu_i - \max_{j \neq i} \mu_j$ in Section 4.1.1 cover their respective true values and $d\hat{\sigma}\sqrt{2/n} < \delta^*$, that is, the event

$$B = \{-(\hat{\mu}_i - \max_{j \neq i} \hat{\mu}_j - d\hat{\sigma}\sqrt{2/n})^- \leq \mu_i - \max_{j \neq i} \mu_j \leq$$

$$+(\hat{\mu}_i - \max_{j \neq i} \hat{\mu}_j + d\hat{\sigma}\sqrt{2/n})^+ \text{ for } i = 1, \ldots, k \text{ and } d\hat{\sigma}\sqrt{2/n} < \delta^*\}$$

occurs, then

1. the upper MCB confidence bounds on $\mu_i - \max_{j \neq i} \mu_j$ will be 0 for all $\mu_i - \max_{j \neq i} \mu_j < -\delta = -2\delta^*$, that is, treatment means more than $\delta$ worse than the true best treatment mean will be correctly asserted to be no better than the true best treatment mean;

2. if $[k]$ is the random index of the sample best treatment (i.e., $\hat{\mu}_{[k]} = \max_{j=1,\ldots,k} \hat{\mu}_j$), then the lower MCB confidence bound on the difference $\mu_{[k]} - \max_{j \neq [k]} \mu_j$ will be greater than $-\delta^*$, that is, the mean of the sample best treatment will be correctly asserted to be within $\delta^*$ of the true best treatment mean.

**Theorem C.2.1** *Given an error rate $\alpha$, if the sample size $n$ satisfies*

$$\int_0^u \int_{-\infty}^{+\infty} [\Phi(z + \sqrt{2}ds)]^{k-1} d\Phi(z) \; \gamma(s)ds \geq 1 - \beta \qquad (C.3)$$

*where $u = \sqrt{n/2}(\delta^*/\sigma)/d$, then $P(B) \geq 1 - \beta$.*

*Proof.* Following the proof of Theorem 4.1.1, again denoting by $(k)$ an index such that $\mu_{(k)} = \max_{i=1,\ldots,k} \mu_i$, we see that

$$B \;\supseteq\; \{\hat{\mu}_{(k)} - \mu_{(k)} > \hat{\mu}_i - \mu_i - d\hat{\sigma}\sqrt{2/n} \text{ for all } i, i \neq (k)$$
$$\text{and } d\hat{\sigma}\sqrt{2/n} < \delta^*\} = B^*(say).$$

Thus

$$\begin{aligned}
P(B) \;\geq\;& P\{\hat{\mu}_{(k)} - \mu_{(k)} > \hat{\mu}_i - \mu_i - d\hat{\sigma}\sqrt{2/n} \text{ for all } i, i \neq (k),\\
& \text{and } d\hat{\sigma}\sqrt{2/n} < \delta^*\}\\
=\;& P\{\sqrt{n}(\hat{\mu}_{(k)} - \mu_{(k)})/\sigma > \sqrt{n}(\hat{\mu}_i - \mu_i)/\sigma - d(\hat{\sigma}/\sigma)\sqrt{2}\\
& \text{for all } i, i \neq (k), \text{ and } d(\hat{\sigma}/\sigma)\sqrt{2/n} < \delta^*/\sigma\}\\
=\;& \int_0^u \int_{-\infty}^{+\infty} [\Phi(z + \sqrt{2}ds)]^{k-1} d\Phi(z) \; \gamma(s)ds. \quad \square
\end{aligned}$$

Given $k, \alpha, \delta^*/\sigma$ and candidate sample size $n$, the pmcb computer program described in Appendix D computes $P\{B^*\}$, which is a lower bound on $P\{B\}$.

As alluded to in Section 4.1.7, when $\sigma$ is known, constrained MCB sample size computation becomes indifference zone selection sample size computation, because if we set $n$ to be the smallest integer satisfying

$$n \geq 2(d\sigma/\delta^*)^2$$

in accordance with the usual indifference zone selection formula, then the event

$$\{d\sigma\sqrt{2/n} \leq \delta^*\}$$

occurs with certainty, so (substituting $\sigma$ for $\hat{\sigma}$)

$$\begin{aligned}
1 - \alpha \;\leq\;& P\{B\}\\
=\;& P\{-(\hat{\mu}_i - \max_{j \neq i} \hat{\mu}_j - d\sigma\sqrt{2/n})^- \leq \mu_i - \max_{j \neq i} \mu_j \leq\\
& +(\hat{\mu}_i - \max_{j \neq i} \hat{\mu}_j + d\sigma\sqrt{2/n})^+ \text{ for } i = 1,\ldots,k\}\\
\leq\;& P\{-(\hat{\mu}_{[k]} - \max_{j \neq [k]} \hat{\mu}_j - d\sigma\sqrt{2/n})^- \leq \mu_{[k]} - \max_{j \neq [k]} \mu_j\}\\
\leq\;& P\{-\delta^* \leq \mu_{[k]} - \max_{j \neq [k]} \mu_j\}
\end{aligned}$$

which, as explained in Section 4.1.7, implies the usual indifference zone correct selection guarantee.

Sample size computation for the unconstrained MCB method of Edwards and Hsu (1984) discussed in Section 4.2.1.2 is essentially the same as that for Dunnett's two-sided MCC method, discussed in Section C.3.

## C.3 Sample size computation for Dunnett's MCC method

For Dunnett's two-sided MCC method discussed in Section 3.2.1.1, if the confidence intervals for all $\mu_i - \mu_k$ cover their respective true value and have width no more than $\delta = 2\delta^*$, that is, the event

$$C = \{\mu_i - \mu_k \in \hat{\mu}_i - \hat{\mu}_k \pm |d|\hat{\sigma}\sqrt{2/n} \text{ for all } i = 1, \ldots, k \text{ and } |d|\hat{\sigma}\sqrt{2/n} < \delta^*\} \tag{C.4}$$

occurs, then

1. for all $\mu_i - \mu_k > \delta$ $(< -\delta)$, the correct and useful significant directional difference assertion $\mu_i > (<) \mu_k$ will be made;

2. for all $\mu_i - \mu_k \approx 0$, the correct and useful practical equivalence assertion $-\delta < \mu_i - \mu_k < \delta$ will be made.

**Theorem C.3.1** *Given an error rate $\alpha$, if the sample size $n$ satisfies*

$$\int_0^u \int_{-\infty}^{+\infty} [\Phi(z + \sqrt{2}|d|s) - \Phi(z - \sqrt{2}|d|s)]^{k-1} d\Phi(z) \, \gamma(s)ds \geq 1 - \beta \tag{C.5}$$

*where $u = \sqrt{n/2}(\delta^*/\sigma)/|d|$, then $P(C) \geq 1 - \beta$.*

*Proof.*

$$\begin{aligned}
P(C) &= P\{\hat{\mu}_i - \mu_i - |d|\hat{\sigma}\sqrt{2/n} < \hat{\mu}_k - \mu_k < \hat{\mu}_i - \mu_i + |d|\hat{\sigma}\sqrt{2/n} \\
&\qquad \text{for } i = 1, \ldots, k - 1 \text{ and } |d|\hat{\sigma}\sqrt{2/n} < \delta^*\} \\
&= P\{\sqrt{n}(\hat{\mu}_i - \mu_i)/\sigma - |d|(\hat{\sigma}/\sigma)\sqrt{2} < \sqrt{n}(\hat{\mu}_k - \mu_k)/\sigma < \\
&\qquad \sqrt{n}(\hat{\mu}_i - \mu_i)/\sigma + |d|(\hat{\sigma}/\sigma)\sqrt{2} \text{ for } i = 1, \ldots, k - 1 \\
&\qquad \text{and } |d|(\hat{\sigma}/\sigma)\sqrt{2/n} < \delta^*/\sigma\} \\
&= \int_0^u \int_{-\infty}^{+\infty} [\Phi(z + \sqrt{2}|d|s) - \Phi(z - \sqrt{2}|d|s)]^{k-1} d\Phi(z) \, \gamma(s)ds.
\end{aligned}$$

$\square$

Given $k, \alpha, \delta^*/\sigma$ and candidate sample size $n$, the pmcc computer program described in Appendix D computes $P\{C\}$.

Sample size computation for Dunnett's one-sided method discussed in Section 3.1.1.1 is essentially the same as that for the constrained MCB method, discussed in Section C.2.

## C.4 An Example

Scheffé (1959, Section 3.3) considered designing an experiment to compare the tensile strength of eight types of alloy steel. It is expected that the standard deviation $\sigma$ of duplicate specimens from the same batch of steel will be about 3000 psi. It was considered of economic importance to detect differences in tensile strength greater than 10,000 psi ($= \delta = 2\delta^*$) with probability 0.90 or greater. Based on the power of the $F$-test computation, five specimens are required for each alloy.

Figure C.1 shows $P\{A\}, P\{B^*\}$, and $P\{C\}$ for $\delta^*/\sigma = 5000/3000$. As can be seen, if all pairwise tensile strength differences of 10,000 psi or more are to be correctly detected with probability 0.90 or greater, then a sample size of ten each is needed. One the other hand, to select correctly an alloy with strength within 5000 psi of the true best alloy with a probability 0.90 or greater, a sample of six each is adequate. If, for the sake of illustration, one of the eight alloys is a standard, and it is desired to detect correctly all other alloys at least 10,000 psi stronger or weaker than the standard, then a sample size of eight each would be sufficient.

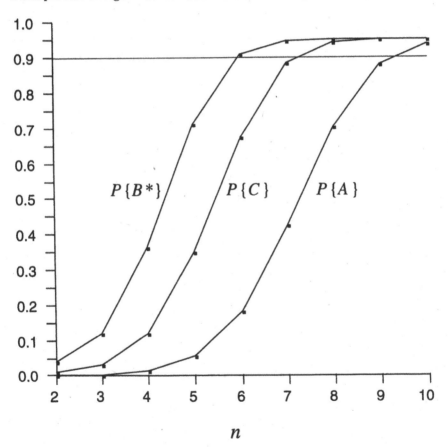

Figure C.1 *Probability of correct and useful inference for the tensile strength example ($\alpha = 0.05$)*

APPENDIX D

# Accessing computer codes

## D.1 Online access to codes

You can access codes for the computer programs mentioned in this book, those that are not otherwise available, via the **Internet**. The usual disclaimer (i.e., no warranty comes with the codes) and restriction (i.e., no profit shall be derived from the codes) apply.

Users with access to a UNIX system can download a single shell archive (shar) which contains all the codes, using either a graphics WWW (World Wide Web) browser such as Netscape or Mosaic, or by anonymous **ftp**. Users without access to a UNIX system can download the individual subroutines and driver programs using a graphics Web browser.

To download either the single **shar** file or the individual subroutines and driver programs using a graphics Web browser, open the URL (Universal Resource Locator)

http://stat.mps.ohio-state.edu

which is the home page of the Department of Statistics of the Ohio State University. Look for Jason C. Hsu's personal home page. From this homepage, follow the link to

Multiple Comparisons Resources

All clickable hypertexts associated with multiple comparisons resources, including the aforementioned link, are preceded by a miniature icon which resembles the mean-mean scatter plot on the cover of this book.

To download the **shar** file via anonymous **ftp**, **ftp** to

stat.mps.ohio-state.edu

Login with userid **ftp**, using your e-mail address as the password. Then cd to pub/jch, and get the file named mc.shar. The shar file contains a **makefile** which uses the **make** facility to automate the compilation of subroutines and driver programs, the building of libraries, and the linking/loading of object codes to obtain executables.

## D.2 Critical value computations

To compute the probabilities in (5.3), (3.7), (4.11), and (3.15) for the finite error degrees of freedom case ($\nu < \infty$), functions gha1, ghb1, and ghc1 use 48-point Gauss–Hermite quadrature for the inside integral and function glv uses 48-point Gauss–Legendre quadrature for the outside integral. In performing the outer integration, the infinite interval $(0, \infty)$ is truncated to the central $1 - 10^{-6}$ probability interval of the distribution of $\hat{\sigma}/\sigma$. For the infinite error degrees of freedom case ($\nu = \infty$), only Gauss–Hermite quadrature is involved. By plugging in the tabled critical values for the cases $\lambda_1 = \cdots = \lambda_k$ in Bechhofer and Dunnett (1988), which are reported to give probabilities well within $10^{-6}$ of $1 - \alpha$, we found the maximum absolute difference between $1 - \alpha$ and the probability computed by our Gaussian quadrature functions to be about $10^{-5}$. The subroutines qmcav, qmcbv, and qmccv then solve for $|q^*|$, $d$ and $|d|$ iteratively, as follows. Denote by $t_{\beta,\nu}$ the upper $\beta$ quantile of the $t$ distribution with $\nu$ degrees of freedom. Clearly, $t_{\alpha/2,\nu}$ is a lower bound for $|q^*|$ and $|d|$, and $t_{\alpha,\nu}$ is a lower bound for $d$. On the other hand, by the Bonferroni inequality (A.3), $t_{\alpha/[k(k-1)],\nu}$, $t_{\alpha/(k-1),\nu}$ and $t_{\alpha/[2(k-1)],\nu}$ are upper bounds for $|q^*|$, $d$ and $|d|$ respectively. Starting with an interval which brackets the critical value, the *regula falsi* method (also known as the Illinois method) is then used to solve for the critical value. Commensurate with the accuracy of the Gaussian quadrature subroutines, a candidate critical value is considered a solution if the probability evaluates to within $2 \times 10^{-5}$ of $1 - \alpha$, or if the half-width of the interval bracketing the critical value is no more than $5 \times 10^{-4}$.

At present, the codes switch to computing critical values for the infinite error degrees of freedom case when $\nu > 240$. You can change this by changing the value of the dinfnu variable in the codes.

## D.3 Sample size computations

Computing $P\{A\}$, $P\{B\}$ or $P\{C\}$ is a two-step process. The subroutine pmcav, pmcbv or pmccv first computes the critical value $|q^*|$, $d$ or $|d|$, which depends on the sample size $n$. It then plugs in the critical value to compute $P\{A\}$, $P\{B\}$ or $P\{C\}$.

# Tables of critical values

Table E.1  *Values of $|q^*|$ for MCA*
$\alpha = 0.05$

|  | | | | $k$ | | | | |
|---|---|---|---|---|---|---|---|---|
| $\nu$ | 2 | 3 | 4 | 5 | 6 | 7 | 8 | 9 |
| 3 | 3.182 | 4.179 | 4.826 | 5.305 | 5.682 | 5.995 | 6.261 | 6.489 |
| 4 | 2.776 | 3.564 | 4.071 | 4.446 | 4.743 | 4.987 | 5.195 | 5.376 |
| 5 | 2.571 | 3.254 | 3.690 | 4.012 | 4.266 | 4.476 | 4.655 | 4.810 |
| 6 | 2.448 | 3.068 | 3.462 | 3.751 | 3.980 | 4.169 | 4.329 | 4.468 |
| 7 | 2.366 | 2.945 | 3.310 | 3.578 | 3.790 | 3.964 | 4.112 | 4.241 |
| 8 | 2.306 | 2.858 | 3.203 | 3.455 | 3.654 | 3.818 | 3.957 | 4.078 |
| 9 | 2.262 | 2.792 | 3.122 | 3.363 | 3.552 | 3.708 | 3.841 | 3.956 |
| 10 | 2.228 | 2.741 | 3.059 | 3.291 | 3.473 | 3.623 | 3.751 | 3.861 |
| 11 | 2.201 | 2.701 | 3.010 | 3.234 | 3.410 | 3.556 | 3.679 | 3.785 |
| 12 | 2.179 | 2.668 | 2.969 | 3.188 | 3.359 | 3.500 | 3.620 | 3.723 |
| 13 | 2.160 | 2.641 | 2.935 | 3.149 | 3.316 | 3.454 | 3.570 | 3.671 |
| 14 | 2.145 | 2.617 | 2.907 | 3.116 | 3.280 | 3.415 | 3.529 | 3.628 |
| 15 | 2.131 | 2.598 | 2.882 | 3.088 | 3.249 | 3.381 | 3.493 | 3.590 |
| 16 | 2.120 | 2.580 | 2.861 | 3.064 | 3.222 | 3.352 | 3.462 | 3.558 |
| 17 | 2.110 | 2.565 | 2.843 | 3.043 | 3.199 | 3.327 | 3.435 | 3.529 |
| 18 | 2.101 | 2.552 | 2.826 | 3.024 | 3.178 | 3.304 | 3.411 | 3.504 |
| 19 | 2.093 | 2.541 | 2.812 | 3.007 | 3.160 | 3.285 | 3.390 | 3.482 |
| 20 | 2.086 | 2.530 | 2.799 | 2.992 | 3.143 | 3.267 | 3.371 | 3.462 |
| 21 | 2.080 | 2.521 | 2.787 | 2.979 | 3.129 | 3.251 | 3.354 | 3.444 |
| 22 | 2.074 | 2.512 | 2.777 | 2.967 | 3.115 | 3.236 | 3.339 | 3.427 |
| 23 | 2.069 | 2.505 | 2.767 | 2.956 | 3.103 | 3.223 | 3.325 | 3.413 |
| 24 | 2.064 | 2.498 | 2.759 | 2.946 | 3.092 | 3.211 | 3.312 | 3.399 |
| 26 | 2.056 | 2.485 | 2.743 | 2.929 | 3.073 | 3.190 | 3.289 | 3.375 |
| 28 | 2.048 | 2.475 | 2.730 | 2.914 | 3.056 | 3.172 | 3.270 | 3.355 |
| 30 | 2.042 | 2.465 | 2.719 | 2.901 | 3.042 | 3.157 | 3.254 | 3.338 |
| 35 | 2.030 | 2.447 | 2.697 | 2.875 | 3.013 | 3.126 | 3.221 | 3.303 |
| 40 | 2.021 | 2.434 | 2.680 | 2.856 | 2.992 | 3.103 | 3.197 | 3.277 |
| 45 | 2.014 | 2.424 | 2.667 | 2.842 | 2.976 | 3.086 | 3.178 | 3.257 |
| 50 | 2.009 | 2.416 | 2.658 | 2.830 | 2.963 | 3.072 | 3.163 | 3.241 |
| 60 | 2.000 | 2.403 | 2.643 | 2.813 | 2.944 | 3.051 | 3.140 | 3.218 |
| 70 | 1.994 | 2.395 | 2.632 | 2.800 | 2.930 | 3.036 | 3.125 | 3.201 |
| 80 | 1.990 | 2.388 | 2.624 | 2.791 | 2.920 | 3.025 | 3.113 | 3.188 |
| 100 | 1.984 | 2.379 | 2.613 | 2.778 | 2.906 | 3.009 | 3.096 | 3.171 |
| 120 | 1.980 | 2.373 | 2.605 | 2.770 | 2.897 | 2.999 | 3.085 | 3.159 |
| 160 | 1.975 | 2.366 | 2.596 | 2.759 | 2.885 | 2.986 | 3.072 | 3.145 |
| $\infty$ | 1.960 | 2.344 | 2.569 | 2.728 | 2.850 | 2.948 | 3.031 | 3.102 |

Table E.1 *Values of $|q^*|$ for MCA*
$$\alpha = 0.05$$

| | | | | | k | | | |
|---|---|---|---|---|---|---|---|---|
| $\nu$ | 10 | 11 | 12 | 13 | 14 | 15 | 16 | 17 |
| 3 | 6.691 | 6.871 | 7.033 | 7.181 | 7.316 | 7.441 | 7.556 | 7.664 |
| 4 | 5.534 | 5.676 | 5.804 | 5.921 | 6.028 | 6.127 | 6.218 | 6.303 |
| 5 | 4.947 | 5.068 | 5.179 | 5.279 | 5.371 | 5.456 | 5.535 | 5.609 |
| 6 | 4.592 | 4.702 | 4.801 | 4.891 | 4.974 | 5.051 | 5.122 | 5.189 |
| 7 | 4.354 | 4.456 | 4.548 | 4.631 | 4.708 | 4.779 | 4.845 | 4.907 |
| 8 | 4.185 | 4.280 | 4.367 | 4.446 | 4.517 | 4.584 | 4.646 | 4.704 |
| 9 | 4.058 | 4.149 | 4.231 | 4.306 | 4.374 | 4.438 | 4.496 | 4.551 |
| 10 | 3.959 | 4.046 | 4.125 | 4.197 | 4.263 | 4.324 | 4.380 | 4.433 |
| 11 | 3.880 | 3.964 | 4.040 | 4.109 | 4.173 | 4.232 | 4.286 | 4.338 |
| 12 | 3.815 | 3.896 | 3.970 | 4.038 | 4.099 | 4.157 | 4.210 | 4.259 |
| 13 | 3.761 | 3.840 | 3.912 | 3.978 | 4.038 | 4.094 | 4.145 | 4.194 |
| 14 | 3.715 | 3.793 | 3.863 | 3.927 | 3.986 | 4.040 | 4.091 | 4.138 |
| 15 | 3.676 | 3.752 | 3.821 | 3.884 | 3.941 | 3.995 | 4.044 | 4.090 |
| 16 | 3.642 | 3.717 | 3.784 | 3.846 | 3.903 | 3.955 | 4.004 | 4.049 |
| 17 | 3.612 | 3.686 | 3.752 | 3.813 | 3.869 | 3.920 | 3.968 | 4.013 |
| 18 | 3.585 | 3.658 | 3.724 | 3.784 | 3.839 | 3.890 | 3.937 | 3.981 |
| 19 | 3.562 | 3.634 | 3.699 | 3.758 | 3.812 | 3.862 | 3.909 | 3.952 |
| 20 | 3.541 | 3.612 | 3.676 | 3.735 | 3.788 | 3.838 | 3.884 | 3.927 |
| 21 | 3.522 | 3.593 | 3.656 | 3.714 | 3.767 | 3.816 | 3.861 | 3.903 |
| 22 | 3.505 | 3.575 | 3.638 | 3.695 | 3.747 | 3.796 | 3.841 | 3.883 |
| 23 | 3.490 | 3.559 | 3.621 | 3.678 | 3.730 | 3.778 | 3.822 | 3.864 |
| 24 | 3.476 | 3.544 | 3.606 | 3.662 | 3.713 | 3.761 | 3.805 | 3.846 |
| 26 | 3.451 | 3.518 | 3.579 | 3.634 | 3.685 | 3.732 | 3.775 | 3.816 |
| 28 | 3.430 | 3.496 | 3.556 | 3.611 | 3.660 | 3.707 | 3.749 | 3.789 |
| 30 | 3.411 | 3.477 | 3.536 | 3.590 | 3.639 | 3.685 | 3.727 | 3.767 |
| 35 | 3.375 | 3.439 | 3.497 | 3.550 | 3.598 | 3.642 | 3.683 | 3.722 |
| 40 | 3.348 | 3.411 | 3.468 | 3.519 | 3.567 | 3.610 | 3.651 | 3.688 |
| 45 | 3.327 | 3.389 | 3.445 | 3.496 | 3.543 | 3.586 | 3.625 | 3.663 |
| 50 | 3.310 | 3.372 | 3.427 | 3.477 | 3.523 | 3.566 | 3.605 | 3.642 |
| 60 | 3.286 | 3.346 | 3.400 | 3.450 | 3.495 | 3.536 | 3.575 | 3.611 |
| 70 | 3.268 | 3.327 | 3.381 | 3.430 | 3.474 | 3.515 | 3.553 | 3.589 |
| 80 | 3.255 | 3.314 | 3.367 | 3.415 | 3.459 | 3.500 | 3.537 | 3.573 |
| 100 | 3.236 | 3.295 | 3.347 | 3.394 | 3.438 | 3.478 | 3.515 | 3.550 |
| 120 | 3.224 | 3.282 | 3.334 | 3.381 | 3.424 | 3.463 | 3.500 | 3.534 |
| 160 | 3.209 | 3.266 | 3.317 | 3.364 | 3.406 | 3.445 | 3.482 | 3.515 |
| $\infty$ | 3.164 | 3.219 | 3.268 | 3.313 | 3.354 | 3.391 | 3.426 | 3.458 |

Table E.2 *Values of d for constrained MCB and one-sided MCC*
$$\alpha = 0.05$$

| $\nu$ | 2 | 3 | 4 | 5 | $k$ 6 | 7 | 8 | 9 |
|---|---|---|---|---|---|---|---|---|
| 3 | 2.353 | 2.938 | 3.279 | 3.519 | 3.702 | 3.850 | 3.973 | 4.079 |
| 4 | 2.132 | 2.611 | 2.885 | 3.076 | 3.222 | 3.339 | 3.437 | 3.521 |
| 5 | 2.016 | 2.441 | 2.682 | 2.849 | 2.976 | 3.078 | 3.163 | 3.236 |
| 6 | 1.943 | 2.337 | 2.558 | 2.711 | 2.827 | 2.920 | 2.998 | 3.064 |
| 7 | 1.895 | 2.268 | 2.476 | 2.619 | 2.728 | 2.815 | 2.888 | 2.950 |
| 8 | 1.860 | 2.218 | 2.417 | 2.553 | 2.657 | 2.740 | 2.809 | 2.868 |
| 9 | 1.833 | 2.180 | 2.372 | 2.504 | 2.604 | 2.684 | 2.750 | 2.807 |
| 10 | 1.812 | 2.151 | 2.338 | 2.466 | 2.562 | 2.640 | 2.704 | 2.759 |
| 11 | 1.796 | 2.127 | 2.310 | 2.435 | 2.529 | 2.605 | 2.668 | 2.721 |
| 12 | 1.782 | 2.108 | 2.287 | 2.410 | 2.502 | 2.576 | 2.638 | 2.690 |
| 13 | 1.771 | 2.092 | 2.269 | 2.389 | 2.480 | 2.552 | 2.613 | 2.664 |
| 14 | 1.761 | 2.079 | 2.253 | 2.371 | 2.461 | 2.532 | 2.592 | 2.642 |
| 15 | 1.753 | 2.067 | 2.239 | 2.356 | 2.445 | 2.515 | 2.574 | 2.624 |
| 16 | 1.746 | 2.057 | 2.227 | 2.343 | 2.430 | 2.500 | 2.558 | 2.607 |
| 17 | 1.740 | 2.048 | 2.217 | 2.332 | 2.418 | 2.487 | 2.544 | 2.593 |
| 18 | 1.734 | 2.041 | 2.208 | 2.322 | 2.407 | 2.476 | 2.532 | 2.581 |
| 19 | 1.729 | 2.034 | 2.200 | 2.313 | 2.398 | 2.465 | 2.522 | 2.569 |
| 20 | 1.725 | 2.027 | 2.192 | 2.305 | 2.389 | 2.456 | 2.512 | 2.560 |
| 21 | 1.721 | 2.022 | 2.186 | 2.297 | 2.381 | 2.448 | 2.503 | 2.551 |
| 22 | 1.717 | 2.017 | 2.180 | 2.291 | 2.374 | 2.441 | 2.496 | 2.542 |
| 23 | 1.714 | 2.012 | 2.175 | 2.285 | 2.368 | 2.434 | 2.488 | 2.535 |
| 24 | 1.711 | 2.008 | 2.170 | 2.279 | 2.362 | 2.428 | 2.482 | 2.528 |
| 26 | 1.706 | 2.001 | 2.161 | 2.270 | 2.352 | 2.417 | 2.471 | 2.517 |
| 28 | 1.701 | 1.995 | 2.154 | 2.262 | 2.343 | 2.408 | 2.461 | 2.507 |
| 30 | 1.697 | 1.989 | 2.147 | 2.255 | 2.335 | 2.400 | 2.453 | 2.498 |
| 35 | 1.690 | 1.978 | 2.135 | 2.241 | 2.320 | 2.384 | 2.436 | 2.481 |
| 40 | 1.684 | 1.971 | 2.126 | 2.230 | 2.309 | 2.372 | 2.424 | 2.468 |
| 45 | 1.679 | 1.964 | 2.118 | 2.222 | 2.301 | 2.363 | 2.414 | 2.458 |
| 50 | 1.676 | 1.959 | 2.113 | 2.216 | 2.294 | 2.356 | 2.407 | 2.450 |
| 60 | 1.671 | 1.952 | 2.104 | 2.207 | 2.284 | 2.345 | 2.396 | 2.439 |
| 70 | 1.667 | 1.947 | 2.098 | 2.200 | 2.276 | 2.337 | 2.388 | 2.430 |
| 80 | 1.664 | 1.943 | 2.094 | 2.195 | 2.271 | 2.332 | 2.382 | 2.424 |
| 100 | 1.660 | 1.938 | 2.087 | 2.188 | 2.264 | 2.324 | 2.373 | 2.416 |
| 120 | 1.658 | 1.934 | 2.083 | 2.183 | 2.259 | 2.318 | 2.368 | 2.410 |
| 160 | 1.654 | 1.930 | 2.078 | 2.178 | 2.252 | 2.312 | 2.361 | 2.403 |
| $\infty$ | 1.645 | 1.916 | 2.062 | 2.160 | 2.234 | 2.292 | 2.341 | 2.381 |

Table E.2 *Values of d for constrained MCB and one-sided MCC*
$$\alpha = 0.05$$

| | | | | | $k$ | | | | |
|---|---|---|---|---|---|---|---|---|---|
| $\nu$ | 10 | 11 | 12 | 13 | 14 | 15 | 16 | 17 |
| 3 | 4.172 | 4.255 | 4.327 | 4.394 | 4.455 | 4.512 | 4.564 | 4.613 |
| 4 | 3.594 | 3.659 | 3.718 | 3.771 | 3.819 | 3.864 | 3.905 | 3.943 |
| 5 | 3.300 | 3.356 | 3.407 | 3.453 | 3.495 | 3.533 | 3.569 | 3.602 |
| 6 | 3.122 | 3.174 | 3.220 | 3.261 | 3.300 | 3.335 | 3.367 | 3.397 |
| 7 | 3.004 | 3.052 | 3.095 | 3.134 | 3.169 | 3.202 | 3.232 | 3.260 |
| 8 | 2.919 | 2.965 | 3.006 | 3.043 | 3.076 | 3.107 | 3.136 | 3.163 |
| 9 | 2.856 | 2.900 | 2.939 | 2.975 | 3.007 | 3.037 | 3.064 | 3.090 |
| 10 | 2.807 | 2.849 | 2.887 | 2.922 | 2.953 | 2.982 | 3.008 | 3.033 |
| 11 | 2.768 | 2.809 | 2.846 | 2.879 | 2.910 | 2.938 | 2.964 | 2.988 |
| 12 | 2.736 | 2.776 | 2.812 | 2.845 | 2.874 | 2.902 | 2.927 | 2.951 |
| 13 | 2.709 | 2.748 | 2.784 | 2.816 | 2.845 | 2.872 | 2.897 | 2.920 |
| 14 | 2.686 | 2.725 | 2.760 | 2.792 | 2.820 | 2.847 | 2.871 | 2.894 |
| 15 | 2.667 | 2.705 | 2.740 | 2.771 | 2.799 | 2.825 | 2.849 | 2.871 |
| 16 | 2.650 | 2.688 | 2.722 | 2.753 | 2.781 | 2.806 | 2.830 | 2.852 |
| 17 | 2.636 | 2.673 | 2.707 | 2.737 | 2.765 | 2.790 | 2.813 | 2.835 |
| 18 | 2.623 | 2.660 | 2.693 | 2.723 | 2.750 | 2.776 | 2.799 | 2.820 |
| 19 | 2.611 | 2.648 | 2.681 | 2.711 | 2.738 | 2.763 | 2.786 | 2.807 |
| 20 | 2.601 | 2.638 | 2.670 | 2.700 | 2.726 | 2.751 | 2.774 | 2.795 |
| 21 | 2.592 | 2.628 | 2.660 | 2.690 | 2.716 | 2.741 | 2.764 | 2.785 |
| 22 | 2.583 | 2.619 | 2.652 | 2.681 | 2.707 | 2.732 | 2.754 | 2.775 |
| 23 | 2.576 | 2.612 | 2.644 | 2.672 | 2.699 | 2.723 | 2.745 | 2.766 |
| 24 | 2.569 | 2.604 | 2.636 | 2.665 | 2.691 | 2.715 | 2.738 | 2.758 |
| 26 | 2.557 | 2.592 | 2.624 | 2.652 | 2.678 | 2.702 | 2.724 | 2.744 |
| 28 | 2.546 | 2.581 | 2.613 | 2.641 | 2.667 | 2.690 | 2.712 | 2.732 |
| 30 | 2.537 | 2.572 | 2.603 | 2.631 | 2.657 | 2.680 | 2.702 | 2.722 |
| 35 | 2.520 | 2.554 | 2.584 | 2.612 | 2.637 | 2.660 | 2.682 | 2.702 |
| 40 | 2.506 | 2.540 | 2.571 | 2.598 | 2.623 | 2.646 | 2.667 | 2.686 |
| 45 | 2.496 | 2.530 | 2.560 | 2.587 | 2.612 | 2.634 | 2.655 | 2.675 |
| 50 | 2.488 | 2.522 | 2.551 | 2.578 | 2.603 | 2.625 | 2.646 | 2.665 |
| 60 | 2.476 | 2.509 | 2.539 | 2.565 | 2.590 | 2.612 | 2.632 | 2.651 |
| 70 | 2.467 | 2.500 | 2.530 | 2.556 | 2.580 | 2.602 | 2.623 | 2.642 |
| 80 | 2.461 | 2.494 | 2.523 | 2.549 | 2.573 | 2.595 | 2.615 | 2.634 |
| 100 | 2.452 | 2.485 | 2.514 | 2.540 | 2.563 | 2.585 | 2.605 | 2.624 |
| 120 | 2.446 | 2.479 | 2.507 | 2.533 | 2.557 | 2.579 | 2.599 | 2.617 |
| 160 | 2.439 | 2.471 | 2.500 | 2.525 | 2.549 | 2.570 | 2.590 | 2.609 |
| $\infty$ | 2.417 | 2.448 | 2.476 | 2.502 | 2.525 | 2.546 | 2.565 | 2.584 |

Table E.3  *Values of |d| for unconstrained MCB and two-sided MCC*
$$\alpha = 0.05$$

| $\nu$ | 2 | 3 | 4 | 5 | 6 | 7 | 8 | 9 |
|---|---|---|---|---|---|---|---|---|
| 3 | 3.182 | 3.867 | 4.263 | 4.538 | 4.748 | 4.917 | 5.057 | 5.177 |
| 4 | 2.776 | 3.310 | 3.618 | 3.832 | 3.995 | 4.126 | 4.235 | 4.328 |
| 5 | 2.571 | 3.031 | 3.293 | 3.476 | 3.615 | 3.727 | 3.821 | 3.901 |
| 6 | 2.448 | 2.863 | 3.099 | 3.264 | 3.389 | 3.489 | 3.573 | 3.645 |
| 7 | 2.366 | 2.752 | 2.971 | 3.123 | 3.238 | 3.331 | 3.409 | 3.475 |
| 8 | 2.306 | 2.673 | 2.880 | 3.023 | 3.132 | 3.219 | 3.292 | 3.354 |
| 9 | 2.262 | 2.614 | 2.812 | 2.948 | 3.052 | 3.135 | 3.205 | 3.264 |
| 10 | 2.228 | 2.569 | 2.759 | 2.891 | 2.990 | 3.071 | 3.137 | 3.194 |
| 11 | 2.201 | 2.532 | 2.717 | 2.845 | 2.941 | 3.019 | 3.084 | 3.139 |
| 12 | 2.179 | 2.503 | 2.683 | 2.807 | 2.901 | 2.977 | 3.040 | 3.094 |
| 13 | 2.160 | 2.478 | 2.655 | 2.776 | 2.868 | 2.942 | 3.004 | 3.056 |
| 14 | 2.145 | 2.457 | 2.631 | 2.750 | 2.840 | 2.913 | 2.973 | 3.024 |
| 15 | 2.131 | 2.439 | 2.610 | 2.728 | 2.816 | 2.887 | 2.947 | 2.997 |
| 16 | 2.120 | 2.424 | 2.593 | 2.708 | 2.795 | 2.866 | 2.924 | 2.974 |
| 17 | 2.110 | 2.411 | 2.577 | 2.691 | 2.777 | 2.847 | 2.904 | 2.953 |
| 18 | 2.101 | 2.399 | 2.563 | 2.676 | 2.762 | 2.830 | 2.887 | 2.935 |
| 19 | 2.093 | 2.388 | 2.551 | 2.663 | 2.747 | 2.815 | 2.871 | 2.919 |
| 20 | 2.086 | 2.379 | 2.541 | 2.651 | 2.735 | 2.802 | 2.857 | 2.905 |
| 21 | 2.080 | 2.370 | 2.531 | 2.641 | 2.723 | 2.790 | 2.845 | 2.892 |
| 22 | 2.074 | 2.363 | 2.522 | 2.631 | 2.713 | 2.779 | 2.834 | 2.881 |
| 23 | 2.069 | 2.356 | 2.514 | 2.622 | 2.704 | 2.769 | 2.824 | 2.870 |
| 24 | 2.064 | 2.350 | 2.507 | 2.614 | 2.695 | 2.760 | 2.814 | 2.860 |
| 26 | 2.056 | 2.338 | 2.494 | 2.600 | 2.681 | 2.745 | 2.798 | 2.844 |
| 28 | 2.048 | 2.329 | 2.483 | 2.589 | 2.668 | 2.731 | 2.784 | 2.829 |
| 30 | 2.042 | 2.321 | 2.474 | 2.578 | 2.657 | 2.720 | 2.772 | 2.817 |
| 35 | 2.030 | 2.305 | 2.455 | 2.558 | 2.635 | 2.697 | 2.749 | 2.792 |
| 40 | 2.021 | 2.293 | 2.442 | 2.543 | 2.619 | 2.680 | 2.731 | 2.774 |
| 45 | 2.014 | 2.284 | 2.431 | 2.531 | 2.607 | 2.667 | 2.718 | 2.760 |
| 50 | 2.009 | 2.276 | 2.423 | 2.522 | 2.597 | 2.657 | 2.707 | 2.749 |
| 60 | 2.000 | 2.265 | 2.410 | 2.509 | 2.583 | 2.642 | 2.691 | 2.733 |
| 70 | 1.994 | 2.258 | 2.401 | 2.499 | 2.572 | 2.631 | 2.679 | 2.721 |
| 80 | 1.990 | 2.252 | 2.395 | 2.492 | 2.565 | 2.623 | 2.671 | 2.712 |
| 100 | 1.984 | 2.244 | 2.385 | 2.482 | 2.554 | 2.611 | 2.659 | 2.700 |
| 120 | 1.980 | 2.239 | 2.379 | 2.475 | 2.547 | 2.604 | 2.651 | 2.692 |
| 160 | 1.975 | 2.232 | 2.372 | 2.467 | 2.538 | 2.595 | 2.642 | 2.682 |
| $\infty$ | 1.960 | 2.212 | 2.349 | 2.442 | 2.512 | 2.567 | 2.613 | 2.652 |

Table E.3  *Values of |d| for unconstrained MCB and two-sided MCC*
$\alpha = 0.05$

| $\nu$ | 10 | 11 | 12 | 13 | 14 | 15 | 16 | 17 |
|---|---|---|---|---|---|---|---|---|
| 3 | 5.281 | 5.373 | 5.455 | 5.530 | 5.598 | 5.659 | 5.718 | 5.772 |
| 4 | 4.410 | 4.482 | 4.547 | 4.605 | 4.658 | 4.707 | 4.753 | 4.795 |
| 5 | 3.970 | 4.032 | 4.088 | 4.138 | 4.183 | 4.226 | 4.264 | 4.301 |
| 6 | 3.707 | 3.763 | 3.812 | 3.858 | 3.899 | 3.937 | 3.971 | 4.004 |
| 7 | 3.533 | 3.584 | 3.630 | 3.671 | 3.709 | 3.744 | 3.777 | 3.807 |
| 8 | 3.409 | 3.457 | 3.500 | 3.539 | 3.575 | 3.608 | 3.638 | 3.667 |
| 9 | 3.316 | 3.362 | 3.403 | 3.440 | 3.474 | 3.506 | 3.535 | 3.562 |
| 10 | 3.244 | 3.288 | 3.328 | 3.364 | 3.397 | 3.427 | 3.454 | 3.480 |
| 11 | 3.187 | 3.230 | 3.268 | 3.303 | 3.334 | 3.364 | 3.391 | 3.416 |
| 12 | 3.141 | 3.182 | 3.219 | 3.253 | 3.284 | 3.312 | 3.338 | 3.363 |
| 13 | 3.102 | 3.142 | 3.179 | 3.212 | 3.242 | 3.269 | 3.295 | 3.319 |
| 14 | 3.069 | 3.109 | 3.145 | 3.177 | 3.206 | 3.233 | 3.258 | 3.282 |
| 15 | 3.041 | 3.080 | 3.115 | 3.147 | 3.176 | 3.203 | 3.227 | 3.250 |
| 16 | 3.017 | 3.056 | 3.090 | 3.121 | 3.150 | 3.176 | 3.200 | 3.223 |
| 17 | 2.996 | 3.034 | 3.068 | 3.099 | 3.127 | 3.153 | 3.176 | 3.199 |
| 18 | 2.978 | 3.015 | 3.048 | 3.079 | 3.107 | 3.132 | 3.156 | 3.178 |
| 19 | 2.961 | 2.998 | 3.031 | 3.061 | 3.089 | 3.114 | 3.137 | 3.159 |
| 20 | 2.946 | 2.983 | 3.016 | 3.046 | 3.073 | 3.098 | 3.121 | 3.142 |
| 21 | 2.933 | 2.969 | 3.002 | 3.031 | 3.058 | 3.083 | 3.106 | 3.127 |
| 22 | 2.921 | 2.957 | 2.989 | 3.019 | 3.045 | 3.070 | 3.092 | 3.113 |
| 23 | 2.911 | 2.946 | 2.978 | 3.007 | 3.034 | 3.058 | 3.080 | 3.101 |
| 24 | 2.901 | 2.936 | 2.968 | 2.997 | 3.023 | 3.047 | 3.069 | 3.090 |
| 26 | 2.883 | 2.918 | 2.950 | 2.978 | 3.004 | 3.028 | 3.050 | 3.070 |
| 28 | 2.868 | 2.903 | 2.934 | 2.962 | 2.988 | 3.011 | 3.033 | 3.053 |
| 30 | 2.856 | 2.890 | 2.921 | 2.949 | 2.974 | 2.997 | 3.019 | 3.039 |
| 35 | 2.831 | 2.864 | 2.894 | 2.922 | 2.947 | 2.969 | 2.991 | 3.010 |
| 40 | 2.812 | 2.845 | 2.875 | 2.902 | 2.926 | 2.949 | 2.970 | 2.989 |
| 45 | 2.798 | 2.830 | 2.860 | 2.886 | 2.911 | 2.933 | 2.954 | 2.973 |
| 50 | 2.786 | 2.819 | 2.848 | 2.874 | 2.898 | 2.920 | 2.941 | 2.960 |
| 60 | 2.769 | 2.801 | 2.830 | 2.856 | 2.880 | 2.901 | 2.922 | 2.940 |
| 70 | 2.757 | 2.789 | 2.817 | 2.843 | 2.867 | 2.888 | 2.908 | 2.927 |
| 80 | 2.748 | 2.780 | 2.808 | 2.833 | 2.857 | 2.878 | 2.898 | 2.916 |
| 100 | 2.736 | 2.767 | 2.795 | 2.820 | 2.843 | 2.864 | 2.884 | 2.902 |
| 120 | 2.727 | 2.758 | 2.786 | 2.811 | 2.834 | 2.855 | 2.875 | 2.893 |
| 160 | 2.717 | 2.748 | 2.775 | 2.800 | 2.823 | 2.844 | 2.863 | 2.881 |
| $\infty$ | 2.686 | 2.716 | 2.743 | 2.767 | 2.790 | 2.810 | 2.829 | 2.846 |

Table E.4  *Values of* $|q|^*$ *for MCA*
$\alpha = 0.01$

|   |   |   |   |   | $k$ |   |   |   |   |
| $\nu$ | 2 | 3 | 4 | 5 | 6 | 7 | 8 | 9 |
|---|---|---|---|---|---|---|---|---|
| 3 | 5.841 | 7.514 | 8.607 | 9.425 | 10.071 | 10.608 | 11.068 | 11.456 |
| 4 | 4.604 | 5.744 | 6.491 | 7.043 | 7.488 | 7.851 | 8.164 | 8.437 |
| 5 | 4.032 | 4.934 | 5.521 | 5.956 | 6.306 | 6.594 | 6.838 | 7.053 |
| 6 | 3.707 | 4.477 | 4.975 | 5.344 | 5.639 | 5.883 | 6.091 | 6.273 |
| 7 | 3.500 | 4.186 | 4.627 | 4.954 | 5.215 | 5.431 | 5.614 | 5.775 |
| 8 | 3.356 | 3.986 | 4.388 | 4.686 | 4.922 | 5.118 | 5.285 | 5.431 |
| 9 | 3.251 | 3.839 | 4.213 | 4.490 | 4.709 | 4.890 | 5.046 | 5.180 |
| 10 | 3.171 | 3.727 | 4.080 | 4.340 | 4.546 | 4.716 | 4.863 | 4.989 |
| 11 | 3.109 | 3.639 | 3.975 | 4.223 | 4.418 | 4.580 | 4.718 | 4.838 |
| 12 | 3.055 | 3.569 | 3.891 | 4.128 | 4.315 | 4.469 | 4.602 | 4.718 |
| 13 | 3.012 | 3.510 | 3.821 | 4.050 | 4.230 | 4.379 | 4.506 | 4.617 |
| 14 | 2.977 | 3.461 | 3.763 | 3.985 | 4.159 | 4.303 | 4.426 | 4.533 |
| 15 | 2.947 | 3.420 | 3.714 | 3.929 | 4.099 | 4.238 | 4.358 | 4.462 |
| 16 | 2.921 | 3.384 | 3.672 | 3.882 | 4.047 | 4.183 | 4.299 | 4.401 |
| 17 | 2.898 | 3.353 | 3.635 | 3.840 | 4.002 | 4.135 | 4.248 | 4.347 |
| 18 | 2.878 | 3.326 | 3.603 | 3.804 | 3.962 | 4.093 | 4.204 | 4.300 |
| 19 | 2.861 | 3.302 | 3.574 | 3.772 | 3.927 | 4.055 | 4.164 | 4.259 |
| 20 | 2.845 | 3.281 | 3.549 | 3.743 | 3.896 | 4.022 | 4.129 | 4.222 |
| 21 | 2.831 | 3.262 | 3.526 | 3.718 | 3.868 | 3.992 | 4.098 | 4.189 |
| 22 | 2.819 | 3.244 | 3.505 | 3.695 | 3.843 | 3.965 | 4.069 | 4.160 |
| 23 | 2.807 | 3.229 | 3.487 | 3.674 | 3.821 | 3.941 | 4.044 | 4.133 |
| 24 | 2.797 | 3.215 | 3.470 | 3.655 | 3.800 | 3.919 | 4.020 | 4.108 |
| 26 | 2.779 | 3.190 | 3.440 | 3.622 | 3.764 | 3.880 | 3.979 | 4.065 |
| 28 | 2.763 | 3.169 | 3.415 | 3.594 | 3.733 | 3.848 | 3.945 | 4.029 |
| 30 | 2.750 | 3.150 | 3.394 | 3.570 | 3.707 | 3.819 | 3.915 | 3.998 |
| 35 | 2.724 | 3.115 | 3.352 | 3.522 | 3.655 | 3.764 | 3.856 | 3.936 |
| 40 | 2.704 | 3.088 | 3.320 | 3.487 | 3.617 | 3.723 | 3.813 | 3.891 |
| 45 | 2.690 | 3.068 | 3.296 | 3.460 | 3.588 | 3.692 | 3.780 | 3.856 |
| 50 | 2.678 | 3.052 | 3.277 | 3.439 | 3.564 | 3.667 | 3.753 | 3.828 |
| 60 | 2.660 | 3.028 | 3.249 | 3.407 | 3.530 | 3.630 | 3.714 | 3.787 |
| 70 | 2.648 | 3.012 | 3.229 | 3.385 | 3.505 | 3.604 | 3.687 | 3.759 |
| 80 | 2.639 | 2.999 | 3.214 | 3.368 | 3.487 | 3.584 | 3.666 | 3.737 |
| 100 | 2.626 | 2.982 | 3.194 | 3.345 | 3.462 | 3.557 | 3.638 | 3.707 |
| 120 | 2.617 | 2.970 | 3.180 | 3.330 | 3.446 | 3.540 | 3.619 | 3.687 |
| 160 | 2.607 | 2.956 | 3.163 | 3.311 | 3.425 | 3.518 | 3.596 | 3.663 |
| $\infty$ | 2.576 | 2.913 | 3.113 | 3.255 | 3.364 | 3.452 | 3.527 | 3.590 |

Table E.4  *Values of* $|q|^*$ *for MCA*
$\alpha = 0.01$

| $\nu$ | 10 | 11 | 12 | 13 | 14 | 15 | 16 | 17 |
|---|---|---|---|---|---|---|---|---|
| | | | | | *k* | | | |
| 3 | 11.804 | 12.115 | 12.400 | 12.650 | 12.885 | 13.107 | 13.299 | 13.485 |
| 4 | 8.677 | 8.890 | 9.082 | 9.257 | 9.419 | 9.570 | 9.703 | 9.836 |
| 5 | 7.242 | 7.412 | 7.564 | 7.701 | 7.835 | 7.952 | 8.063 | 8.168 |
| 6 | 6.434 | 6.577 | 6.708 | 6.828 | 6.937 | 7.037 | 7.132 | 7.220 |
| 7 | 5.917 | 6.045 | 6.160 | 6.267 | 6.363 | 6.452 | 6.536 | 6.614 |
| 8 | 5.561 | 5.677 | 5.782 | 5.879 | 5.966 | 6.049 | 6.124 | 6.195 |
| 9 | 5.300 | 5.407 | 5.505 | 5.595 | 5.675 | 5.752 | 5.821 | 5.887 |
| 10 | 5.101 | 5.202 | 5.293 | 5.378 | 5.453 | 5.525 | 5.592 | 5.652 |
| 11 | 4.945 | 5.040 | 5.127 | 5.207 | 5.279 | 5.346 | 5.409 | 5.467 |
| 12 | 4.818 | 4.910 | 4.992 | 5.069 | 5.137 | 5.202 | 5.262 | 5.319 |
| 13 | 4.714 | 4.802 | 4.881 | 4.955 | 5.021 | 5.083 | 5.141 | 5.195 |
| 14 | 4.627 | 4.712 | 4.789 | 4.859 | 4.923 | 4.983 | 5.039 | 5.091 |
| 15 | 4.554 | 4.635 | 4.711 | 4.778 | 4.841 | 4.898 | 4.952 | 5.003 |
| 16 | 4.490 | 4.570 | 4.643 | 4.708 | 4.769 | 4.825 | 4.878 | 4.927 |
| 17 | 4.435 | 4.513 | 4.583 | 4.648 | 4.707 | 4.762 | 4.813 | 4.861 |
| 18 | 4.386 | 4.462 | 4.531 | 4.594 | 4.652 | 4.706 | 4.756 | 4.803 |
| 19 | 4.343 | 4.418 | 4.486 | 4.547 | 4.603 | 4.656 | 4.705 | 4.751 |
| 20 | 4.305 | 4.378 | 4.445 | 4.505 | 4.560 | 4.612 | 4.661 | 4.706 |
| 21 | 4.270 | 4.343 | 4.408 | 4.468 | 4.521 | 4.573 | 4.620 | 4.665 |
| 22 | 4.239 | 4.311 | 4.375 | 4.434 | 4.487 | 4.537 | 4.584 | 4.628 |
| 23 | 4.211 | 4.282 | 4.345 | 4.403 | 4.455 | 4.505 | 4.551 | 4.594 |
| 24 | 4.186 | 4.255 | 4.318 | 4.375 | 4.427 | 4.476 | 4.521 | 4.564 |
| 26 | 4.141 | 4.209 | 4.270 | 4.326 | 4.377 | 4.425 | 4.469 | 4.510 |
| 28 | 4.103 | 4.170 | 4.230 | 4.284 | 4.334 | 4.381 | 4.424 | 4.465 |
| 30 | 4.071 | 4.136 | 4.195 | 4.249 | 4.298 | 4.343 | 4.386 | 4.426 |
| 35 | 4.007 | 4.070 | 4.127 | 4.178 | 4.226 | 4.270 | 4.310 | 4.349 |
| 40 | 3.959 | 4.021 | 4.076 | 4.126 | 4.172 | 4.215 | 4.255 | 4.292 |
| 45 | 3.923 | 3.983 | 4.037 | 4.086 | 4.131 | 4.173 | 4.212 | 4.248 |
| 50 | 3.894 | 3.953 | 4.006 | 4.055 | 4.099 | 4.140 | 4.178 | 4.213 |
| 60 | 3.852 | 3.909 | 3.961 | 4.008 | 4.051 | 4.091 | 4.127 | 4.162 |
| 70 | 3.822 | 3.878 | 3.928 | 3.974 | 4.017 | 4.056 | 4.092 | 4.126 |
| 80 | 3.799 | 3.855 | 3.904 | 3.950 | 3.991 | 4.030 | 4.065 | 4.099 |
| 100 | 3.768 | 3.822 | 3.871 | 3.915 | 3.956 | 3.994 | 4.028 | 4.061 |
| 120 | 3.747 | 3.801 | 3.849 | 3.893 | 3.933 | 3.970 | 4.004 | 4.036 |
| 160 | 3.722 | 3.774 | 3.822 | 3.865 | 3.904 | 3.940 | 3.974 | 4.005 |
| $\infty$ | 3.646 | 3.697 | 3.741 | 3.781 | 3.818 | 3.853 | 3.884 | 3.914 |

Table E.5 *Values of d for constrained MCB and one-sided MCC*
$$\alpha = 0.01$$

| $\nu$ | 2 | 3 | 4 | 5 | 6 | 7 | 8 | 9 |
|---|---|---|---|---|---|---|---|---|
| 3 | 4.541 | 5.485 | 6.045 | 6.441 | 6.749 | 6.994 | 7.202 | 7.381 |
| 4 | 3.747 | 4.409 | 4.798 | 5.070 | 5.281 | 5.453 | 5.594 | 5.716 |
| 5 | 3.365 | 3.901 | 4.212 | 4.430 | 4.598 | 4.735 | 4.849 | 4.945 |
| 6 | 3.143 | 3.608 | 3.877 | 4.064 | 4.208 | 4.325 | 4.423 | 4.506 |
| 7 | 2.998 | 3.419 | 3.661 | 3.829 | 3.958 | 4.063 | 4.150 | 4.226 |
| 8 | 2.898 | 3.287 | 3.510 | 3.665 | 3.784 | 3.880 | 3.961 | 4.030 |
| 9 | 2.823 | 3.190 | 3.400 | 3.544 | 3.657 | 3.747 | 3.822 | 3.887 |
| 10 | 2.764 | 3.115 | 3.315 | 3.452 | 3.559 | 3.645 | 3.716 | 3.778 |
| 11 | 2.718 | 3.057 | 3.248 | 3.381 | 3.482 | 3.564 | 3.633 | 3.691 |
| 12 | 2.681 | 3.009 | 3.194 | 3.322 | 3.420 | 3.499 | 3.565 | 3.622 |
| 13 | 2.650 | 2.969 | 3.149 | 3.274 | 3.369 | 3.445 | 3.509 | 3.564 |
| 14 | 2.624 | 2.936 | 3.112 | 3.233 | 3.326 | 3.400 | 3.463 | 3.516 |
| 15 | 2.602 | 2.908 | 3.080 | 3.199 | 3.289 | 3.362 | 3.423 | 3.475 |
| 16 | 2.583 | 2.884 | 3.053 | 3.169 | 3.258 | 3.329 | 3.389 | 3.440 |
| 17 | 2.567 | 2.863 | 3.029 | 3.143 | 3.230 | 3.300 | 3.359 | 3.409 |
| 18 | 2.552 | 2.844 | 3.008 | 3.121 | 3.206 | 3.275 | 3.333 | 3.382 |
| 19 | 2.539 | 2.828 | 2.989 | 3.101 | 3.185 | 3.253 | 3.310 | 3.358 |
| 20 | 2.528 | 2.813 | 2.973 | 3.083 | 3.166 | 3.233 | 3.289 | 3.337 |
| 21 | 2.518 | 2.800 | 2.958 | 3.067 | 3.149 | 3.215 | 3.271 | 3.318 |
| 22 | 2.508 | 2.788 | 2.944 | 3.052 | 3.134 | 3.199 | 3.254 | 3.301 |
| 23 | 2.500 | 2.778 | 2.932 | 3.039 | 3.120 | 3.185 | 3.239 | 3.285 |
| 24 | 2.492 | 2.768 | 2.921 | 3.027 | 3.107 | 3.172 | 3.225 | 3.271 |
| 26 | 2.479 | 2.751 | 2.902 | 3.006 | 3.085 | 3.148 | 3.201 | 3.246 |
| 28 | 2.467 | 2.736 | 2.885 | 2.988 | 3.066 | 3.129 | 3.181 | 3.225 |
| 30 | 2.457 | 2.723 | 2.871 | 2.973 | 3.050 | 3.112 | 3.163 | 3.207 |
| 35 | 2.438 | 2.699 | 2.843 | 2.943 | 3.018 | 3.078 | 3.129 | 3.172 |
| 40 | 2.423 | 2.680 | 2.823 | 2.920 | 2.994 | 3.054 | 3.103 | 3.145 |
| 45 | 2.412 | 2.666 | 2.807 | 2.903 | 2.976 | 3.035 | 3.083 | 3.125 |
| 50 | 2.403 | 2.655 | 2.794 | 2.890 | 2.962 | 3.020 | 3.068 | 3.109 |
| 60 | 2.390 | 2.639 | 2.776 | 2.869 | 2.941 | 2.997 | 3.045 | 3.085 |
| 70 | 2.381 | 2.627 | 2.762 | 2.855 | 2.926 | 2.982 | 3.028 | 3.068 |
| 80 | 2.374 | 2.618 | 2.753 | 2.845 | 2.915 | 2.970 | 3.016 | 3.056 |
| 100 | 2.364 | 2.606 | 2.739 | 2.830 | 2.899 | 2.954 | 2.999 | 3.038 |
| 120 | 2.358 | 2.598 | 2.730 | 2.820 | 2.889 | 2.943 | 2.988 | 3.027 |
| 160 | 2.350 | 2.588 | 2.719 | 2.808 | 2.876 | 2.929 | 2.974 | 3.012 |
| $\infty$ | 2.326 | 2.558 | 2.685 | 2.772 | 2.837 | 2.889 | 2.933 | 2.970 |

Table E.5 *Values of d for constrained MCB and one-sided MCC*
$$\alpha = 0.01$$

| | | | | $k$ | | | | |
|---|---|---|---|---|---|---|---|---|
| $\nu$ | 10 | 11 | 12 | 13 | 14 | 15 | 16 | 17 |
| 3 | 7.537 | 7.676 | 7.801 | 7.914 | 8.017 | 8.115 | 8.201 | 8.284 |
| 4 | 5.824 | 5.919 | 6.005 | 6.083 | 6.154 | 6.220 | 6.283 | 6.337 |
| 5 | 5.030 | 5.106 | 5.175 | 5.237 | 5.293 | 5.345 | 5.394 | 5.440 |
| 6 | 4.579 | 4.644 | 4.703 | 4.756 | 4.805 | 4.849 | 4.891 | 4.931 |
| 7 | 4.290 | 4.348 | 4.401 | 4.448 | 4.492 | 4.532 | 4.569 | 4.605 |
| 8 | 4.091 | 4.143 | 4.191 | 4.235 | 4.275 | 4.312 | 4.346 | 4.379 |
| 9 | 3.944 | 3.994 | 4.038 | 4.079 | 4.116 | 4.150 | 4.059 | 4.087 |
| 10 | 3.831 | 3.879 | 3.922 | 3.961 | 3.995 | 4.027 | 4.059 | 4.087 |
| 11 | 3.743 | 3.788 | 3.829 | 3.866 | 3.900 | 3.931 | 3.960 | 3.987 |
| 12 | 3.671 | 3.715 | 3.755 | 3.790 | 3.823 | 3.853 | 3.881 | 3.906 |
| 13 | 3.612 | 3.655 | 3.693 | 3.728 | 3.759 | 3.788 | 3.815 | 3.840 |
| 14 | 3.563 | 3.604 | 3.641 | 3.675 | 3.706 | 3.734 | 3.760 | 3.784 |
| 15 | 3.521 | 3.561 | 3.597 | 3.630 | 3.660 | 3.688 | 3.713 | 3.737 |
| 16 | 3.484 | 3.524 | 3.560 | 3.592 | 3.621 | 3.648 | 3.673 | 3.696 |
| 17 | 3.453 | 3.492 | 3.527 | 3.558 | 3.587 | 3.614 | 3.638 | 3.661 |
| 18 | 3.425 | 3.464 | 3.498 | 3.529 | 3.557 | 3.583 | 3.607 | 3.630 |
| 19 | 3.401 | 3.438 | 3.472 | 3.503 | 3.531 | 3.557 | 3.580 | 3.602 |
| 20 | 3.379 | 3.416 | 3.450 | 3.480 | 3.507 | 3.533 | 3.556 | 3.578 |
| 21 | 3.360 | 3.396 | 3.429 | 3.459 | 3.486 | 3.511 | 3.534 | 3.556 |
| 22 | 3.342 | 3.378 | 3.411 | 3.440 | 3.467 | 3.492 | 3.515 | 3.536 |
| 23 | 3.326 | 3.362 | 3.394 | 3.423 | 3.450 | 3.475 | 3.497 | 3.518 |
| 24 | 3.311 | 3.347 | 3.379 | 3.408 | 3.434 | 3.459 | 3.481 | 3.502 |
| 26 | 3.286 | 3.321 | 3.352 | 3.381 | 3.407 | 3.431 | 3.453 | 3.473 |
| 28 | 3.264 | 3.299 | 3.330 | 3.358 | 3.384 | 3.407 | 3.429 | 3.449 |
| 30 | 3.246 | 3.280 | 3.311 | 3.338 | 3.364 | 3.387 | 3.408 | 3.428 |
| 35 | 3.209 | 3.243 | 3.272 | 3.299 | 3.324 | 3.347 | 3.368 | 3.387 |
| 40 | 3.182 | 3.215 | 3.244 | 3.271 | 3.295 | 3.317 | 3.338 | 3.357 |
| 45 | 3.161 | 3.194 | 3.222 | 3.249 | 3.272 | 3.294 | 3.315 | 3.333 |
| 50 | 3.145 | 3.177 | 3.205 | 3.231 | 3.255 | 3.276 | 3.296 | 3.315 |
| 60 | 3.121 | 3.152 | 3.180 | 3.205 | 3.228 | 3.249 | 3.269 | 3.287 |
| 70 | 3.103 | 3.134 | 3.162 | 3.187 | 3.210 | 3.231 | 3.250 | 3.268 |
| 80 | 3.090 | 3.121 | 3.148 | 3.173 | 3.196 | 3.216 | 3.236 | 3.254 |
| 100 | 3.072 | 3.103 | 3.130 | 3.154 | 3.176 | 3.197 | 3.216 | 3.233 |
| 120 | 3.061 | 3.090 | 3.117 | 3.141 | 3.164 | 3.184 | 3.203 | 3.220 |
| 160 | 3.046 | 3.075 | 3.102 | 3.126 | 3.148 | 3.168 | 3.186 | 3.204 |
| $\infty$ | 3.002 | 3.031 | 3.056 | 3.079 | 3.101 | 3.120 | 3.138 | 3.155 |

Table E.6  *Values of |d| for unconstrained MCB and two-sided MCC*
$\alpha = 0.01$

| $\nu$ | 2 | 3 | 4 | 5 | 6 | 7 | 8 | 9 |
|---|---|---|---|---|---|---|---|---|
| 3 | 5.841 | 6.977 | 7.643 | 8.106 | 8.465 | 8.748 | 8.987 | 9.192 |
| 4 | 4.604 | 5.365 | 5.811 | 6.122 | 6.363 | 6.558 | 6.717 | 6.855 |
| 5 | 4.032 | 4.628 | 4.976 | 5.219 | 5.407 | 5.559 | 5.684 | 5.793 |
| 6 | 3.707 | 4.213 | 4.507 | 4.712 | 4.870 | 4.998 | 5.106 | 5.196 |
| 7 | 3.500 | 3.949 | 4.208 | 4.390 | 4.529 | 4.643 | 4.737 | 4.819 |
| 8 | 3.356 | 3.766 | 4.003 | 4.169 | 4.295 | 4.398 | 4.484 | 4.558 |
| 9 | 3.251 | 3.633 | 3.853 | 4.007 | 4.124 | 4.220 | 4.300 | 4.368 |
| 10 | 3.171 | 3.532 | 3.739 | 3.884 | 3.995 | 4.084 | 4.159 | 4.224 |
| 11 | 3.109 | 3.452 | 3.650 | 3.787 | 3.893 | 3.978 | 4.049 | 4.111 |
| 12 | 3.055 | 3.388 | 3.578 | 3.710 | 3.811 | 3.893 | 3.961 | 4.020 |
| 13 | 3.012 | 3.335 | 3.519 | 3.646 | 3.744 | 3.822 | 3.888 | 3.945 |
| 14 | 2.977 | 3.291 | 3.469 | 3.592 | 3.687 | 3.764 | 3.828 | 3.882 |
| 15 | 2.947 | 3.253 | 3.427 | 3.547 | 3.640 | 3.714 | 3.776 | 3.829 |
| 16 | 2.921 | 3.221 | 3.391 | 3.508 | 3.598 | 3.671 | 3.732 | 3.784 |
| 17 | 2.898 | 3.193 | 3.359 | 3.474 | 3.563 | 3.634 | 3.693 | 3.744 |
| 18 | 2.878 | 3.168 | 3.332 | 3.444 | 3.532 | 3.601 | 3.660 | 3.710 |
| 19 | 2.861 | 3.146 | 3.307 | 3.418 | 3.504 | 3.573 | 3.630 | 3.679 |
| 20 | 2.845 | 3.127 | 3.286 | 3.395 | 3.480 | 3.547 | 3.603 | 3.652 |
| 21 | 2.831 | 3.110 | 3.266 | 3.374 | 3.458 | 3.524 | 3.580 | 3.628 |
| 22 | 2.819 | 3.094 | 3.249 | 3.355 | 3.438 | 3.504 | 3.558 | 3.606 |
| 23 | 2.807 | 3.080 | 3.233 | 3.338 | 3.420 | 3.485 | 3.539 | 3.586 |
| 24 | 2.797 | 3.067 | 3.219 | 3.323 | 3.404 | 3.468 | 3.522 | 3.568 |
| 26 | 2.779 | 3.044 | 3.194 | 3.296 | 3.375 | 3.438 | 3.491 | 3.536 |
| 28 | 2.763 | 3.025 | 3.172 | 3.273 | 3.351 | 3.413 | 3.465 | 3.509 |
| 30 | 2.750 | 3.009 | 3.154 | 3.253 | 3.330 | 3.391 | 3.443 | 3.486 |
| 35 | 2.724 | 2.977 | 3.118 | 3.215 | 3.289 | 3.349 | 3.399 | 3.441 |
| 40 | 2.704 | 2.953 | 3.091 | 3.186 | 3.259 | 3.318 | 3.366 | 3.408 |
| 45 | 2.690 | 2.934 | 3.071 | 3.165 | 3.236 | 3.294 | 3.341 | 3.382 |
| 50 | 2.678 | 2.920 | 3.055 | 3.147 | 3.218 | 3.275 | 3.322 | 3.362 |
| 60 | 2.660 | 2.898 | 3.031 | 3.122 | 3.191 | 3.246 | 3.293 | 3.332 |
| 70 | 2.648 | 2.883 | 3.014 | 3.104 | 3.172 | 3.226 | 3.272 | 3.311 |
| 80 | 2.639 | 2.872 | 3.001 | 3.090 | 3.157 | 3.212 | 3.257 | 3.295 |
| 100 | 2.626 | 2.856 | 2.984 | 3.071 | 3.138 | 3.191 | 3.235 | 3.273 |
| 120 | 2.617 | 2.846 | 2.972 | 3.059 | 3.125 | 3.177 | 3.221 | 3.259 |
| 160 | 2.607 | 2.833 | 2.958 | 3.043 | 3.108 | 3.161 | 3.204 | 3.241 |
| $\infty$ | 2.576 | 2.794 | 2.915 | 2.998 | 3.060 | 3.111 | 3.152 | 3.188 |

Table E.6  *Values of |d| for unconstrained MCB and two-sided MCC*
$\alpha = 0.01$

| $\nu$ | | | | | $k$ | | | |
|---|---|---|---|---|---|---|---|---|
| | 10 | 11 | 12 | 13 | 14 | 15 | 16 | 17 |
| 3 | 9.370 | 9.527 | 9.669 | 9.796 | 9.912 | 10.019 | 10.121 | 10.211 |
| 4 | 6.976 | 7.084 | 7.180 | 7.267 | 7.347 | 7.420 | 7.490 | 7.551 |
| 5 | 5.888 | 5.972 | 6.048 | 6.116 | 6.179 | 6.237 | 6.292 | 6.341 |
| 6 | 5.277 | 5.348 | 5.412 | 5.470 | 5.523 | 5.572 | 5.619 | 5.660 |
| 7 | 4.889 | 4.952 | 5.008 | 5.060 | 5.107 | 5.152 | 5.191 | 5.228 |
| 8 | 4.623 | 4.681 | 4.731 | 4.777 | 4.820 | 4.861 | 4.897 | 4.931 |
| 9 | 4.428 | 4.482 | 4.530 | 4.573 | 4.613 | 4.649 | 4.683 | 4.714 |
| 10 | 4.280 | 4.331 | 4.376 | 4.416 | 4.454 | 4.488 | 4.520 | 4.549 |
| 11 | 4.164 | 4.212 | 4.255 | 4.294 | 4.329 | 4.362 | 4.392 | 4.420 |
| 12 | 4.071 | 4.117 | 4.158 | 4.195 | 4.229 | 4.260 | 4.289 | 4.316 |
| 13 | 3.995 | 4.039 | 4.078 | 4.114 | 4.147 | 4.177 | 4.205 | 4.230 |
| 14 | 3.930 | 3.973 | 4.011 | 4.046 | 4.078 | 4.107 | 4.134 | 4.159 |
| 15 | 3.876 | 3.918 | 3.955 | 3.989 | 4.019 | 4.048 | 4.074 | 4.098 |
| 16 | 3.829 | 3.870 | 3.906 | 3.939 | 3.969 | 3.997 | 4.023 | 4.046 |
| 17 | 3.789 | 3.829 | 3.864 | 3.896 | 3.926 | 3.953 | 3.978 | 4.001 |
| 18 | 3.753 | 3.792 | 3.827 | 3.859 | 3.888 | 3.914 | 3.939 | 3.962 |
| 19 | 3.722 | 3.760 | 3.795 | 3.826 | 3.854 | 3.880 | 3.904 | 3.927 |
| 20 | 3.694 | 3.732 | 3.766 | 3.796 | 3.824 | 3.850 | 3.874 | 3.896 |
| 21 | 3.669 | 3.706 | 3.740 | 3.770 | 3.797 | 3.823 | 3.846 | 3.868 |
| 22 | 3.647 | 3.683 | 3.716 | 3.746 | 3.773 | 3.798 | 3.821 | 3.843 |
| 23 | 3.626 | 3.663 | 3.695 | 3.724 | 3.751 | 3.776 | 3.799 | 3.820 |
| 24 | 3.608 | 3.644 | 3.676 | 3.705 | 3.731 | 3.756 | 3.778 | 3.800 |
| 26 | 3.576 | 3.611 | 3.642 | 3.671 | 3.697 | 3.721 | 3.743 | 3.763 |
| 28 | 3.548 | 3.583 | 3.614 | 3.642 | 3.667 | 3.691 | 3.713 | 3.733 |
| 30 | 3.525 | 3.559 | 3.589 | 3.617 | 3.642 | 3.665 | 3.687 | 3.707 |
| 35 | 3.478 | 3.511 | 3.541 | 3.568 | 3.592 | 3.615 | 3.636 | 3.655 |
| 40 | 3.444 | 3.477 | 3.505 | 3.532 | 3.556 | 3.578 | 3.598 | 3.617 |
| 45 | 3.418 | 3.450 | 3.478 | 3.504 | 3.528 | 3.549 | 3.569 | 3.588 |
| 50 | 3.397 | 3.429 | 3.457 | 3.482 | 3.505 | 3.527 | 3.546 | 3.565 |
| 60 | 3.367 | 3.397 | 3.425 | 3.450 | 3.472 | 3.493 | 3.513 | 3.531 |
| 70 | 3.345 | 3.375 | 3.402 | 3.427 | 3.449 | 3.470 | 3.489 | 3.506 |
| 80 | 3.329 | 3.359 | 3.385 | 3.410 | 3.432 | 3.452 | 3.471 | 3.488 |
| 100 | 3.306 | 3.336 | 3.362 | 3.386 | 3.408 | 3.428 | 3.446 | 3.464 |
| 120 | 3.292 | 3.321 | 3.347 | 3.370 | 3.392 | 3.412 | 3.430 | 3.447 |
| 160 | 3.273 | 3.302 | 3.328 | 3.351 | 3.372 | 3.392 | 3.410 | 3.427 |
| $\infty$ | 3.219 | 3.247 | 3.271 | 3.293 | 3.314 | 3.333 | 3.350 | 3.366 |

# Bibliography

Afifi, A. A. and Azen, S. P. (1972). *Statistical Analysis: A Computer Oriented Approach.* Academic Press, New York.

Alberton, Y. and Hochberg, Y. (1984). Approximations for the distribution of a maximal pairwise *t* in some repeated measure designs. *Communications in Statistics - Theory and Methods*, A13(22):2847–2854.

Alexander, W. P. (1993). Testing the means of independent normal random variables. *Computational Statistics and Data Analysis*, 16:1–10.

Andrews, H. P., Snee, R. D. and Sarner, M. H. (1980). Graphical display of means. *American Statistician*, 34:195–199.

Atkinson, A. T. (1987). *Plots, Transformations and Regression.* Oxford University Press, New York.

Barlow, R., Bartholomew, D., Bremner, J. and Brunk, H. (1972). *Statistical Inference Under Order Restrictions.* Wiley, New York.

Bechhofer, R. E. (1954). A single-sample multiple decision procedure for ranking means of normal populations with known variances. *Annals of Mathematical Statistics*, 25:16–39.

Bechhofer, R. E. and Dunnett, C. W. (1988). *Percentage Points of Multivariate Student t Distributions*, Volume 11 of *Selected Tables in Mathematical Statistics*. American Mathematical Society, Providence, RI.

Bechhofer, R. E., Santner, T. J. and Goldsman, D. M. (1995). *Design and Analysis of Experiments for Statistical Selection, Screening and Multiple Comparisons.* John Wiley & Sons, New York.

Beecher, H. K. (1955). The powerful placebo. *Journal of the American Medical Association*, 159:1602–1606.

Begun, J. M. and Gabriel, R. (1981). Closure of the Newman–Keuls multiple comparisons procedure. *Journal of the American Statistical Association*, 76(374):241–245.

Belsley, D. A., Kuh, E. and Welsch, R. E. (1980). *Regression Diagnostics: Identifying Influential Data and Sources of Collinearity.* Wiley, New York.

Benjamini, Y. and Hochberg, Y. (1995). Controlling the false discovery rate: A practical and powerful approach to multiple testing. *Journal of the Royal Statistical Society* **B**, 57(2):289–300.

Berger, J. O. (1985). *Statistical Decision Theory and Bayesian Analysis.* Springer-Verlag, New York, second edition.

Berger, R. L. (1982). Multiparameter hypothesis testing and acceptance sampling. *Technometrics,* 24:295–300.

Berger, R. L. and Hsu, J. C. (1995). Bioequivalence trials, intersection-union tests, and equivalence confidence sets. *In preparation.*

Bhargava, R. P. and Srivastava, M. S. (1973). On Tukey's confidence intervals for the contrasts in the means of the intraclass correlation model. *Journal of the Royal Statistical Society Series B,* 35:147–152.

Bofinger, E. (1985). Multiple comparisons and Type III errors. *Journal of the American Statistical Association,* 80:433–438.

Bofinger, E. (1987). Stepdown procedures for comparison with a control. *Australian Journal of Statistics,* 29:348–364.

Bofinger, E. (1988). Least significant spacing for 'one versus the rest' normal populations. *Communications in Statistics – Theory and Methods,* A17:1697–1716.

Braun, H. I. and Tukey, J. W. (1983). Multiple comparisons through orderly partitions: The maximum subrange procedure. In Wainer, H. and Messick, S. (eds), *Principals of Modern Psychological Measurement: A Festschrift for Frederic M. Lord,* Chapter 4, pages 55–64. Lawrence Erlbaum Associates, Hillsdale, NJ.

Bristol, D. R. (1993). *p*-values for two-sided comparisons with a control. *Communications in Statistics – Simulation and Computation,* B22:153–158.

Brown, L. D. (1979). A proof that the Tukey–Kramer multiple comparison procedure for differences between means is level-$\alpha$ for 3, 4, or 5 treatments. Technical report, Statistics Center, Cornell University, Ithaca, NY.

Campbell, P. F. and McCabe, G. P. (1984). Predicting the success of freshmen in a computer science major. *Communications of the ACM,* 27:1108–1113.

Carmer, S. G. and Swanson, M. R. (1973). Evaluation of ten pairwise multiple comparison procedures by monte carlo methods. *Journal of the American Statistical Association,* 68:66–74.

Chang, J. Y. and Hsu, J. C. (1992). Optimal designs for multiple comparisons with the best. *Journal of Statistical Planning and Inference,* 30:45–62.

Chow, S.-C. and Liu, J.-P. (1992). *Design and Analysis of Bioavailability and Bioequivalence Studies.* Marcel Dekker, New York.

Cleveland, W. S. (1985). *The Elements of Graphing Data.* Wadsworth Advanced Books and Software, Monterey, CA.

Conover, W. J. and Iman, R. L. (1981). Rank transformations as a bridge between parametric and nonparametric statistics. *American Statistician*, 35:124–129.

Consumer Union (1992). Udder insanity. *Consumer Reports*, pages 330–332.

Cook, R. D. and Weisberg, S. (1982). *Residuals and Influence in Regression*. Chapman & Hall, New York.

Critchlow, D. E. and Fligner, M. A. (1991). Nonparametric multiple comparisons in the one-way analysis of variance. *Communications in Statistics – Theory and Methods*, A20:127–139.

Dalal, S. R. and Mallows, C. L. (1992). Buying with exact confidence. *The Annals of Applied Probability*, 2:752–765.

Dumouchel, W. (1988). A Bayesian model and a graphical elicitation procedure for multiple comparisons. In Bernardo, J. M., DeGroot, M. H., Lindley, D. V. and Smith, A. F. M. (eds), *Bayesian Statistics 3*, pages 127–145. Oxford University Press, New York and Oxford.

Duncan, D. B. (1955). Multiple range and multiple *F* tests. *Biometrics*, 11:1–42.

Dunn, O. J. (1964). Multiple comparisons using rank sums. *Technometrics*, pages 241–252.

Dunnett, C. W. (1955). A multiple comparison procedure for comparing several treatments with a control. *Journal of the American Statistical Association*, 50:1096–1121.

Dunnett, C. W. (1980). Pairwise multiple comparisons in the homogeneous variance, unequal sample size case. *Journal of the American Statistical Association*, 75:789–795.

Dunnett, C. W. and Tamhane, A. C. (1991). Step-down multiple tests for comparing treatments with a control in unbalanced one-way layouts. *Statistics in Medicine*, 10:939–947.

Dunnett, C. W. and Tamhane, A. C. (1992). A step-up multiple test procedure. *Journal of the American Statistical Association*, 87:162–170.

Dwass, M. (1960). Some *k*-sample rank-oder tests. In I. Olkin, et al (eds), *Contributions to Probability and Statistics*. Stanford University Press, Stanford, CA.

EC-GCP (1993a). *Biostatistical methodology in clinical trials in applications for marketing authorization for medical products*. CPMP Working Party on Efficacy of Medical Products, Commission of the European Communities, Brussels, Draft Guideline edition.

EC-GCP (1993b). *Good Clinical Practice for Trials on Medical Products in the European Community.* CPMP Working Party on Efficacy of Medical Products, Commission of the European Communities, Brussels, Final Guideline edition.

Edwards, D. G. (1987). Extended-Paulson sequential selection. *Annals of Statistics*, 15:449–455.

Edwards, D. G. and Berry, J. J. (1987). The efficiency of simulation-based multiple comparisons. *Biometrics*, 43:913–928.

Edwards, D. G. and Hsu, J. C. (1983). Multiple comparisons with the best treatment. *Journal of the American Statistical Association*, 78:965–971.

Edwards, D. G. and Hsu, J. C. (1984). Corrections to multiple comparisons with the best treatment. *Journal of the American Statistical Association*, 79:965.

Einot, I. and Gabriel, K. R. (1975). A study of the powers of several methods of multiple comparisons. *Journal of the American Statistical Association*, 70:574–583.

Fabian, V. (1962). On multiple decision methods for ranking population means. *Annals of Mathematical Statistics*, 33:248–254.

Fabian, V. (1991). On the problem of interactions in the analysis of variance. *Journal of the American Statistical Association*, 86:362–367.

FDA (1987). *Guideline for Submitting Documentation for Stability Studies of Human Drugs and Biologics.* Center for Drugs and Biologics, Food and Drug Administration, Rockville, MD.

FDA (1992). Bioavailability and bioequivalence requirements. In *U. S. Code of Federal Regulations*, Volume 21, Chapter 320. U. S. Government Printing Office, Washington, DC.

Finner, H. (1987). Abgeschlossene multiple Spannweitentests (closed multiple range tests) (in German). In Bauer, P., Hommel, G. and Sonnemann, E. (eds), *Multiple Hypothesenprüfung (Multiple Hypotheses Testing)*, pages 10–32. Springer-Verlag, Berlin.

Finner, H. (1990a). On the modified S-method and directional errors. *Communications in Statistics – Theory and Methods*, A19:41–53.

Finner, H. (1990b). Some new inequalities for the range distribution, with application to the determination of optimum significance levels of multiple range tests. *Journal of the American Statistical Association*, 85:191–194.

Finner, H., Hayter, A. J. and Roters, M. (1993). On the joint distribution function of order statistics with reference to step-up multiple test procedures. *Forschungsbericht Nr. 93-19, Mathematik/Informatik, University Trier.*

Fisher, R. A. (1935). *The Design of Experiments*. Oliver and Boyd, Edinburgh and London.

Fleiss, J. L. (1986). *The Design and Analysis of Clinical Experiments*. John Wiley, New York.

Fligner, M. A. (1984). A note on two-sided distribution-free treatment versus control multiple comparisons. *Journal of the American Statistical Association*, 79:208–211.

Gabriel, K. R. (1978). A simple method of multiple comparisons of means. *Journal of the American Statistical Association*, 73:724–729.

Gardner, M. J. and Altman, D. G. (1986). Confidence intervals rather than P values: estimation rather than hypothesis testing. *British Medical Journal*, 292:746–750.

Gardner, M. J., Machin, D. and Campbell, M. J. (1986). Use of check lists in assessing the statistical content of medical studies. *British Medical Journal*, 292:810–812.

Gibbons, J. D., Olkin, I. and Sobel, M. (1977). *Selecting and Ordering Populations: A New Statistical Methodology*. Wiley, New York.

Gossett, S. W. S. (1927). Errors of routine analysis. *Biometrika*, 19:151–164.

Graybill, F. A. (1983). *Matrices with Applications in Statistics*. Wadsworth, Belmont, CA, second edition.

Gupta, S. S. (1956). On a decision rule for a problem in ranking means. Mimeo Series 150, Institute of Statistics, University of North Carolina, Chapel Hill, NC.

Gupta, S. S. (1965). On some multiple decision (selection and ranking) rules. *Technometrics*, 7:225–245.

Gupta, S. S. and Panchapakesan, S. (1979). *Multiple Decision Procedures – Theory and Methodology of Selecting and Ranking Populations*. John Wiley, New York.

Halperin, M., S.W., G., Cornfield, J. and Zalokar, J. (1955). Tables of percentage points for the Studentized maximum absolute deviate in normal samples. *Journal of the American Statistical Association*, 50:185–195.

Harter, H. L. (1960). Critical values for Duncan's new multiple range test. *Biometrics*, 16:671–685.

Hayter, A. J. (1984). A proof of the conjecture that the Tukey–Kramer multiple comparisons procedure is conservative. *Annals of Statistics*, 12:61–75.

Hayter, A. J. (1986). The maximum familywise error rate of Fisher's least significant difference test. *Journal of the American Statistical Association*, 81:1000–1004.

Hayter, A. J. (1989). Pairwise comparisons of generally correlated means. *Journal of the American Statistical Association*, 84:208–213.

Hayter, A. J. (1990). A one-sided Studentized range test for testing against a simple order alternative. *Journal of the American Statistical Association*, 85:778–785.

Hayter, A. J. (1992). Multiple comparisons of three ordered normal means for unbalanced models. *Computational Statistics and Data Analysis*, 13:153–162.

Hayter, A. J. and Hsu, J. C. (1994). On the relationship between step-wise decision procedures and confidence sets. *Journal of the American Statistical Association*, 89:128–136.

Hayter, A. J. and Liu, W. (1995). Exact calculations for the one-sided Studentized range test for testing against a simple ordered alternatives. *Submitted for publication.*

Hayter, A. J. and Stone, G. (1991). Distribution free multiple comparisons for monotonically ordered treatment effects. *Australian Journal of Statistics*, 33(3):335–346.

Hedayat, A. and Stufken, J. (1989). A relation between pairwise balanced and variance balanced block designs. *Journal of the American Statistical Association*, 84:753–755.

Hettmansperger, T. P. (1984). *Statistical Inference Based on Ranks*. Wiley, New York.

Hochberg, Y. (1974a). The distribution of the range in generalized balanced models. *American Statistician*, 28:137–138.

Hochberg, Y. (1974b). Some generalizations of the $T$-method in simultaneous inference. *Journal of Multivariate Analysis*, 4:224–234.

Hochberg, Y. (1988). A sharper Bonferroni procedure for multiple tests of significance. *Biometrika*, 75:800–802.

Hochberg, Y. and Tamhane, A. C. (1983). Multiple comparisons in a mixed model. *American Statistician*, 37:305–307.

Hochberg, Y. and Tamhane, A. C. (1987). *Multiple Comparison Procedures*. John Wiley, New York.

Hochberg, Y., Weiss, G. and Hart, S. (1982). On graphical procedures for multiple comparisons. *Journal of the American Statistical Association*, 77:767–772.

Hollander, M. (1966). An asymptotically distribution-free multiple comparisons procedure – treatments vs. control. *Annals of Mathematical Statistics*, 37:735–738.

Hollander, M. and Wolfe, D. A. (1973). *Nonparametric Statistical Methods*. John Wiley, New York.

Holm, S. (1979a). A simple sequentially rejective multiple test procedure. *Scandanavian Journal of Statistics*, 6:65–70.

Holm, S. (1979b). A stagewise directional test procedure based on $t$ statistics. Unpublished report.

Hommel, G. (1988). A stagewise rejective multiple test procedure based on a modified Bonferroni test. *Biometrika*, 75:383–386.

Hommel, G. (1989). A comparison of two modified Bonferroni procedures. *Biometrika*, 76:624–625.

Horowitz, J. M. and Thompson, D. (1993). Udder insanity! *Time*, 141:52–53.

Hsu, J. C. (1981). Simultaneous confidence intervals for all distances from the 'best'. *Annals of Statistics*, 9:1026–1034.

Hsu, J. C. (1982). Simultaneous inference with respect to the best treatment in block designs. *Journal of the American Statistical Association*, 77:461–467.

Hsu, J. C. (1984a). Constrained two-sided simultaneous confidence intervals for multiple comparisons with the best. *Annals of Statistics*, 12:1136–1144.

Hsu, J. C. (1984b). Ranking and selection and multiple comparisons with the best. In Santner, T. J. and Tamhane, A. C. (eds), *Design of Experiments: Ranking and Selection*. Marcel Dekker, New York.

Hsu, J. C. (1985). A note on multiple comparisons with the best. In *45th Session of the International Statistical Institute, Book 2*, pages 445–446.

Hsu, J. C. (1988). Sample size computation for designing multiple comparison experiments. *Computational Statistics and Data Analysis*, 7:79–91.

Hsu, J. C. (1992). The factor analytic approach to simultaneous inference in the general linear model. *Journal of Graphical and Computational Statistics*, 1:151–168.

Hsu, J. C. and Edwards, D. G. (1983). Sequential multiple comparisons with the best. *Journal of the American Statistical Association*, 78:958–964.

Hsu, J. C., Hwang, J. G., Liu, H. and Ruberg, S. J. (1994). Confidence intervals associated with bioequivalence trials. *Biometrika*, 81:103–114.

Hsu, J. C. and Nelson, B. L. (1994). Multiple comparisons in the general linear model. Working Paper Series 1993-003, Department of Industrial and Systems Engineering, The Ohio State University.

Hsu, J. C. and Peruggia, M. (1994). Graphical representation of Tukey's multiple comparison method. *Journal of Computational and Graphical Statistics*, 3:143–161.

Hunter, D. (1976). An upper bound for the probability of a union. *Journal of Applied Probability*, 13:597–603.

Huynh, H. and Feldt, L. S. (1970). Conditions under which mean square ratios in repeated measurements designs have exact $f$-distributions. *Journal of the American Statistical Association*, 65:1582–1589.

Iyengar, S. (1988). Evaluation of normal probabilities of symmetric regions. *SIAM Journal on Scientific Computing*, 8:418–423.

James, P. D. (1992). Letter to the editor. *Statistical Computing and Statistical Graphics Newsletter*, 3(2):2.

John, J. A. (1987). *Cyclic Designs*. Chapman & Hall, London.

Juskevich, J. C. and Guyer, C. G. (1990). Bovine growth hormone: Human food safety evaluation. *Science*, 249:875–884.

Karlin, S. and Rinott, Y. (1980). Classes of ordering of measures and related correlation inequalities. I. Multivariate totally positive distribution. *Journal of Multivariate Analysis*, 10:467–498.

Keuls, M. (1952). The use of the 'Studentized range' in connection with an analysis of variance. *Euphytica*, 1:112–122.

Kim, W.-C. (1988). On detecting the best treatment. *Journal of the Korean Statistical Society*, 17:82–92.

Kimball, A. W. (1951). On dependent tests of significance in analysis of variance. *Annals of Mathematical Statistics*, 22:600–602.

Knuth, D. E. (1973). *The Art of Computer Programming*, Volume 1: *Fundamental Algorithms*. Addison-Wesley, Reading, MA, second edition.

Koziol, J. A. and Reid, N. (1977). On the asymptotic equivalence of two ranking methods for $k$-sample linear rank statistics. *Annals of Statistics*, 5:1099–1106.

Kramer, C. Y. (1956). Extension of multiple range tests to group means with unequal numbers of replications. *Biometrics*, 12:309–310.

Lam, K. (1986). A new procedure for selecting good populations. *Biometrika*, 73:201–206.

Langman, M. J. S. (1986). Towards estimation and confidence intervals. *British Medical Journal*, 292:716.

Lehmann, E. L. (1975). *Nonparametrics: Statistical Methods Based on Ranks*. Holden-Day, San Francisco.

Lehmann, E. L. (1986). *Testing Statistical Hypotheses*. John Wiley, New York, second edition.

Lehmann, E. L. and Shaffer, J. P. (1977). On a fundamental theorem in multiple comparisons. *Journal of the American Statistical Association*, 72:576–578.

Lin, F.-J., Seppänen, E. and Uusipaikka, E. (1990). On pairwise comparisons of the components of the mean vector in multivariate normal distribution. *Communications in Statistics - Theory and Methods*, A19:395–412.

Mann, B. L. and Pirie, W. R. (1982). Tighter bounds and simplified estimation for moments of some rank statistics. *Communications in Statistics - Theory and Methods*, A11:1107–1117.

Marcus, R. (1978). Further results on simultaneous confidence bounds in normal models with restricted alternatives. *Communications in Statistics - Theory and Methods*, A7(6):573–590.

Marcus, R., Peritz, E. and Gabriel, K. R. (1976). On closed testing procedures with special reference to ordered analysis of variance. *Biometrika*, 63:655–660.

Matejcik, F. J. and Nelson, B. L. (1995). Two-stage multiple comparisons with the best for computer simulation. *Operations Research*, 43:633–640.

McGill, R., Tukey, J. W. and Larsen, W. A. (1978). Variations of box plots. *American Statistician*, 32(1):12–16.

Mead, R. and Pike, D. J. (1975). A review of response surface methodology from a biometric viewpoint. *Biometrics*, 31:803–851.

Miller, R. G. (1981). *Simultaneous Statistical Inference*. Springer-Verlag, Heidelberg and Berlin, second edition.

Naik, U. D. (1975). Some selection rules for comparing $p$ processes with a standard. *Communications in Statistics - Theory and Methods*, A4:519–535.

Nelson, P. R. (1982). Multivariate normal and $t$ distributions with $\rho_{ij} = \alpha_i \alpha_k$. *Communications in Statistics - Simulation and Computation*, 11:239–248.

Nelson, P. R. (1993). Additional uses for the analysis of means and extended tables of critical values. *Technometrics*, 35:61–71.

Newman, D. (1939). The distribution of the range in samples from a normal population, expressed in terms of an independent estimate of standard deviation. *Biometrika*, 35:16–31.

Nowak, R. (1994). Problems in clinical trials go far beyond misconduct. *Science*, 264:1538–1541.

Olshen, R. A. (1973). The conditional level of the $F$-test. *Journal of the American Statistical Association*, 68:692–698.

Oude Voshaar, J. H. (1980). $(k-1)$-mean significance levels of nonparametric multiple comparisons procedures. *Annals of Statistics*, 8:75–86.

Peritz, E. (1970). A note on multiple comparisons. Unpublished manuscript, Hebrew University.

Pratt, J. W. (1961). Length of confidence intervals. *Journal of the American Statistical Association*, 56:541–567.

Public Broadcasting System (1995). Currents of Fear. In *Frontline*. PBS Video, Alexandria, VA.

Ramsey, P. H. (1978). Power difference between pairwise multiple comparisons. *Journal of the American Statistical Association*, 73:479–485.

Richmond, J. (1982). A general method for constructing simultaneous confidence intervals. *Journal of the American Statistical Association*, 77:455–460.

Rizvi, M. H. and Woodworth, G. G. (1970). On selection procedures based on ranks: Counterexamples concerning the least favorable configurations. *Annals of Mathematical Statistics*, 41:1942–1951.

Robertson, T., Wright, F. T. and Dykstra, R. L. (1988). *Order Restricted Statistical Inference*. John Wiley and Sons, New York.

Rom, D. M. (1990). A sequentially rejective test procedure based on a modified Bonferroni inequality. *Biometrika*, 77:663–665.

Rothman, K. (1990). No adjustments are needed for multiple comparisons. *Epidemiology*, 1:43–46.

Royen, T. (1987). An approximation for multivariate normal probabilities of rectangular regions. *Mathematische Operationsforschung und Statistik, Series Statistics*, 18:389–400.

Royen, T. (1989). Generalized maximum range tests for pairwise comparisons of several populations. *Biometrical Journal*, 31(8):905–929.

Ruberg, S. J. and Hsu, J. C. (1992). Multiple comparison procedures for pooling batches in stability studies. *Technometrics*, 34:465–472.

Ryan, T. A. (1960). Significance tests for multiple comparisons of proportions, variances, and other statistics. *Psychological Bulletin*, 57:318–328.

Sall, J. (1992a). Graphical comparison of means. *Statistical Computing and Statistical Graphics Newsletter*, 3(1):27–32.

Sall, J. (1992b). Letter to the editor. *Statistical Computing and Statistical Graphics Newsletter*, 3(2):2.

Sampson, A. R. (1980). Representations of simultaneous pairwise comparisons. In Krishnaiah, P. R. (ed.), *Handbook of Statistics*, Volume 1, pages 623–629. North-Holland.

SAS (1989). *SAS/STAT User's Guide, Release 6, Fourth Edition, Volume 2*. SAS Institute, Inc., Cary, NC.

Scheffé, H. (1953). A method for judging all contrasts in the analysis of variance. *Biometrika*, 40:87–104.

Scheffé, H. (1959). *Analysis of Variance*. John Wiley, New York.

Schoenfeld, D. A. (1986). Confidence bounds for normal means under order restrictions, with application to dose-response curves, toxicology experiments, and low-dose extrapolation. *Journal of the American Statistical Association*, 81:186–195.

Schuirmann, D. J. (1987). A comparison of the two one-sided tests procedure and the power approach for assessing the equivalence of average bioavailability. *Journal of Pharmacokinetics and Biopharmaceutics*, 15(6):657–680.

Searle, S. R. (1971). *Linear Models*. John Wiley, New York, NY.

Shaffer, J. P. (1980). Control of directional errors with stagewise multiple test procedures. *Annals of Statistics*, 8:1342–1348.

Shaffer, J. P. (1995). Multiple hypothesis testing: A review. *Annual Review of Psychology*, 46.

Simes, R. J. (1986). An improved Bonferroni procedure for multiple tests of significance. *Biometrika*, 73:751–754.

Soong, W. C. and Hsu, J. C. (1995). Using complex integration to compute multivariate normal probability integrals with negative product correlation structure. Technical report, Department of Statistics, The Ohio State University, Columbus, OH.

Spjøtvoll, E. and Stoline, M. R. (1973). An extension of the *t*-method of multiple comparisons to include the cases with unequal sample sizes. *Journal of the American Statistical Association*, 68:975–978.

SPSS (1988). *SPSS-X User's Guide*. SPSS Inc., Chicago, third edition.

Spurrier, J. D. (1988). Generalizations of Steel's treatments-versus control multivariate sign test. *Journal of the American Statistical Association*, 83:471–476.

Spurrier, J. D. (1991). Improved bounds for moments of some rank statistics. *Communications in Statistics – Theory and Methods*, A20:2603–2608.

Spurrier, J. D. (1992). Distribution-free and asymptotically distribution-free comparisons with a control in blocked experiments. In Hoppe, F. M. (ed.), *Multiple Comparisons, Selection, and Applications in Biometry: A Festschrift in Honor of Charles W. Dunnett*, Chapter 12, pages 97–119. Marcel Dekker, New York.

Spurrier, J. D. (1995). More bounds for moments of some rank statistics. *Communications in Statistics – Theory and Methods*, 23:2679–2682.

Spurrier, J. D. and Isham, S. P. (1985). Exact simultaneous confidence intervals of pairwise comparisons of three normal means. *Journal of the American Statistical Association*, 80:438–442.

Steel, R. G. D. (1959a). A multiple comparison rank sum test: Treatment versus control. *Biometrics*, 15:560–572.

Steel, R. G. D. (1959b). A multiple comparison sign test: treatments versus control. *Journal of the American Statistical Association*, 54:767–775.

Steel, R. G. D. (1960). A rank sum test for comparing all pairs of treatments. *Technometrics*, 2:197–207.

Steel, R. G. D. and Torrie, J. H. (1980). *Principles and Procedures of Statistics: A Biometrical Approach*. McGraw-Hill, New York, second edition.

Stefansson, G., Kim, W. and Hsu, J. C. (1988). On confidence sets in multiple comparisons. In Gupta, S. S. and Berger, J. O. (eds), *Statistical Decision Theory and Related Topics IV*, Volume 2, pages 89–104. Springer-Verlag, New York.

Stevenson, H. W., Chen, C. and Lee, S.-Y. (1993). Mathematics achievement of chinese, japanese, and american children: Ten years later. *Science*, 259:53–58.

Stevenson, H. W., S-Y. and Stigler, J. W. (1986). Mathematics achievement of Chinese, Japanese, and American children. *Science*, 231:693–699.

Tamhane, A. C. (1995). Multiple comparison procedures. In Ghosh, S. and Rao, C. R. (eds), *Handbook of Statistics: Design and Analysis of Experiments*, Volume 13. North-Holland, Amsterdam.

Tierney, L. (1990). *LISP-STAT: An Object-Oriented Environment for Statistical Computing and Dynamic Graphics*. Wiley, New York.

Tong, Y. L. (1980). *Probability Inequalities in Multivariate Distributions*. Academic Press, New York.

Tukey, J. W. (1953). The Problem of Multiple Comparisons. Dittoed manuscript of 396 pages, Department of Statistics, Princeton University.

Tukey, J. W. (1991). The philosophy of multiple comparisons. *Statistical Science*, 6:100–116.

Tukey, J. W. (1992). Where should multiple comparisons go next? In Hoppe, F. M. (ed.), *Multiple Comparisons, Selection, and Applications in Biometry: A Festschrift in Honor of Charles W. Dunnett*, Chapter 12, pages 187–208. Marcel Dekker, New York.

Tukey, J. W. (1993). Graphical comparisons of several linked aspects: Alternatives and suggested principles. *Journal of Computational and Graphical Statistics*, 2(1):1–33.

Tukey, J. W. (1994). The problem of multiple comparisons. In Braun, H. I. (ed.), *The Collected Works of John W. Tukey*, Volume VIII, Chapter 1, pages 1–300. Chapman & Hall, New York and London.

Uusipaikka, E. (1985). Exact simultaneous confidence intervals for multiple comparisons of three or four mean values. *Journal of the American Statistical Association*, 80:196–201.

Voss, D. T. and Hsu, J. C. (1995). Multiple comparisons for an unbalanced $a \times b$ design under mixed models with interaction. Technical Report 560, Department of Statistics, The Ohio State University, 1958 Neil Avenue, Columbus, OH 43210-1247.

Welsch, R. E. (1977). Stepwise multiple comparison procedures. *Journal of the American Statistical Association*, 72:566–575.

Westfall, P. H. and Young, S. S. (1993). *Resampling-Based Multiple Testing: Examples and Methods for P-Value Adjustment*. John Wiley & Sons, Inc., New York.

Westlake, W. J. (1976). Symmetric confidence intervals for bioequivalence trials. *Biometrics*, 32:741–744.

Westlake, W. J. (1981). Response to T.B.L. Kirkwood: Bioequivalence testing – a need to rethink. *Biometrics*, 37:589–594.

White, J. R. and Froeb, H. F. (1980). Small-airways dysfunction in non-smokers chronically exposed to tobacco smoke. *New England Journal of Medicine*, 302:720–723.

Williams, D. A. (1971). A test for differences between treatment means when several dose are compared with a zero dose. *Biometrics*, 27:103–117.

Williams, D. A. (1977). Some inference procedures for monotonically ordered normal means. *Biometrika*, 64:9–14.

Wilson, M. C. and Shade, R. E. (1967). Relative attractiveness of various luminescent colors to the cereal leaf beetle and the meadow spittlebug. *Journal of Economic Entomology*, 60:578–580.

Worsley, K. J. (1982). An improved Bonferroni inequality and applications. *Biometrika*, 69:297–302.

Yang, W.-N. and Nelson, B. L. (1991). Using common random numbers and control variates in multiple-comparison procedures. *Operations Research*, 39:583–591.

Yu, P. and Lam, K. (1991). Tightness of some confidence and predictive intervals related to selection. *Communications in Statistics – Theory and Methods*, A20(4):1401–1408.

Yuan, M. and Nelson, B. L. (1993). Multiple comparisons with the best for steady-state simulation. *ACM Transactions on Modeling and Computer Simulation*, 3:66–79.

# Author index

# Subject index